T0192445

Future Networks, Services and Management

Mehmet Toy
Editor

Future Networks, Services and Management

Underlay and Overlay, Edge, Applications, Slicing, Cloud, Space, AI/ML, and Quantum Computing

 Springer

Editor
Mehmet Toy
Associate Fellow, Verizon Communications, Inc.
Allendale, NJ, USA

ISBN 978-3-030-81963-7   ISBN 978-3-030-81961-3   (eBook)
https://doi.org/10.1007/978-3-030-81961-3

This Springer imprint is published by the registered company Springer Nature Switzerland AG
The registered company address is: Gewerbestrasse 11, 6330 Cham, Switzerland

Foreword

Over the 30 years in the networking and telecommunications industry, I have been fortunate enough to see the conceptualization, deployment, maturation, and sunset of dozens of technologies. I have also witnessed the shift in perception of networking in the business from "an unfortunate cost," to "the growth engine," and finally "the fabric that brings us together." I still remember the days when having a 1.5 Mbps DSL line to a site that served 1000 customers was frowned upon as "overkill."

Technology has shifted to a point where Network has become life blood of a functioning society. It has become a finely woven, ever-growing, ever-evolving fabric that pulls people and cultures together. Networks have allowed us to share ideas, collaborate on the future, and speak a common language.

As I think of the networks that will power our world for the next 20 years, a few themes emerge. Speed, perhaps unimaginable, is going to be required to power our future, and it is not only that speed on *wired* networks that will enable us. Wireless will continue to play a key role in meeting the connectivity needs of society. In fact, 5G and 6G cellular networks are going to completely redefine *instant*. As ultra-low latency wireless networks pave new and prioritized highways to the internet, they will combine with technologies like Multi-access Edge Computing (MEC), and the way that we consume information will be forever changed. A new network and data paradigm powered by Mobile Edge Compute, Artificial Intelligence, Computer Vision, and Machine Learning (AI/ML) will provide rich information before we even realize we need it. Through Augmented and Virtual Reality (AR and VR), we will reimagine our world in ways we can only begin to imagine. Smart cities will not only run efficiently and adapt in real time to changes in traffic and weather but our agriculture and manufacturing industries will be completely transformed by these networks. Data lakes from all of this generated information could be so large that we do not even yet have the capacity to create containers large enough to hold it.

All of this change, and this new unimagined world, has but one thing at the foundation. That foundation is the *network*. The network that connects us, not only as dots on a map, but as people in a society. I hope you will join me in imagining what the *Future Networks, Services and Management* has in store for us.

Chief Product Officer and Senior VP, Verizon Business Group   Aamir Hussain
Basking Ridge, NJ, USA
June 2021

Preface

From 2018 to 2020, I led the ITU-T Network2030 Architecture Framework specification development as the Network2030 Architecture Group Chair. After publishing the ITU specification, my colleagues suggested me to write a book on networks of 2030 and beyond. I decided not to write the book alone and invited some of the coauthors of the Network2030 Architecture Specification and more experts from industry and academia to write *Future Networks, Services and Management* together. This book with 29 coauthors provides a comprehensive view of current and future networks and services.

With 26% yearly growth in IP traffic driven mostly by cloud applications, 5G, and IoT devices, the communications industry has been experiencing dramatic changes in recent years. Substantial growth in wireline and wireless transmission rates is coupled with tremendous growth in computing power. Availabilities of a chip that can process up to 2.4 Tbps, a router line card of 12 Tbps, and Ethernet switches at 12.8 Tbps capacity with 300 Gbps port speed have been claimed. Massive parallel data transmission on 179 wavelength channels at a data rate of more than 50 Tbps has been demonstrated. Optical computing performs calculations at the speed of light (switching speed in the order of 10^{-15} s). 5G wireless is expected to support maximum peak download capacity of 20 Gbps while 6G wireless is expected to support 100 Gbps.

The growth in computing power triggered the growth in virtualization that transforms physical infrastructures into dedicated virtual resources and leads to relaxing physical hardware and operating on servers and partitions called virtual machines. Applications stored in data centers of Communication Service Providers or hyperscalers can be deployed anywhere on the network. Hyperscalers and communication service providers are teamed up to support new applications and services.

In parallel to the growth in cloud applications and services, applications and services with strict performance requirements are growing. This growth triggered Multi-access Edge Computing (MEC) applications and services where MEC devices are located either at customer premises or at nearby Communication Service Provider Points of Presence (POPs).

As new applications and services are being developed, hiding complexity from users by simplifying user interfaces and automation of service order and delivery is the primary goal of Service Providers. Intent-based Networking, Network Slicing, and Artificial Intelligence (AI)/Machine Learning (ML) techniques are employed for these purposes.

In this decade and beyond, we expect more bandwidth hungry applications with strict performance requirements such as holographic and robotic applications to be developed; and applications that are in their infancy such as those for self-driven cars are to be matured. In order to support future applications and associated services, we expect underlay and overlay networks to go through dramatic changes again. 6G and Quantum Computing are among the technologies on the horizon to make these changes possible.

Future networks will be heterogonous as the current networks. The main difference would be that the future networks will include hyperscalers and space networks and end-to-end automation. Perhaps the transaction-based usage billing for services is highly likely.

In this book, we describe future applications and services, current and future underlay and overlay networks, an overall service architecture, an overall management architecture, and various techniques in building future networks and services such as routing, security, quality of service (QoS), burst switching, 5G and 6G, MEC, AI/ML, Application Programming Interfaces (APIs), Network Slicing, and Quantum Computing.

I would like to thank the coauthors who joined me in writing this book, Mr. Aamir Hussain for writing the Foreword, and Mr. Paritosh Bajpay, Mr. Viraj Parekh, and Mr. Anoop Agrawal for their support in writing this book. I also would like to thank our editor Ms. Mary James, our project manager Ms. Cynthya Pushparaj, and our technical support Mr. Brian Halm for their help with the publication.

The ITU-T Network2030 Architecture Framework has been the main source for this book. I would like to thank Dr. Richard Li of Futurewei who initiated the ITU-T FG-NET2030 in ITU-T; and ITU-T SG13 Chairman, Dr. Leo Lehmann, and ITU-T SG13 Secretary, Dr. Tatiana Kurakova, who supported the ITU-T FG-NET2030 initiative and encouraged me to write this book.

Writing of this book required me working at nights and weekends for several months. I would like to thank my wife, Fusun, for her understanding and support.

Allendale, NJ, USA   Mehmet Toy

Contents

Chapter 1
Introduction

Mehmet Toy

1.1 Introduction

Enhancements in computing and growth in bandwidth, along with virtualization, have been driving force for the new applications and services in recent years. AI/ML technology is still in its infancy. We expect that to be one of the driving forces of future applications and services. As we move toward the year 2030, quantum computing should become a dominant factor for changes in devices, applications, networking, and services.

This book makes an attempt to describe future applications, requirements, and architectures for networks and services, and technologies that are driving changes in this decade and beyond in networking and services such as virtualization, AI/ML, and quantum computing.

Chapter 2 describes future applications such as holographic-type communications (HTC) based on 3D interactions, tactile Internet for remote operations (TIRO) which involves the real-time control of remote infrastructures, space-terrestrial integrated network, AI-enabled applications, industrial IoT and IoT advanced applications, and scientific research and big data applications, and their requirements.

Chapter 3 describes various future services such as in-time/on-time services that are performance sensitive, qualitative communication services that avoid retransmissions of less relevant portions of the payload in order to meet requirements on latency, holographic communication, haptic communication, and high-speed data delivery services.

Chapter 4 divides networks and services into underlay and overlay networks and services and provides examples. As examples to underlay networks, current and future architectures of optical and wireless networks including 5G and 6G wireless

M. Toy (✉)
Verizon Communications, Inc., Allendale, NJ, USA
e-mail: mehmet.toy@verizon.com

M. Toy (ed.), *Future Networks, Services and Management*,
https://doi.org/10.1007/978-3-030-81961-3_1

1

networks are described. As examples to overlay networks, virtualized networks, SD-WAN, and Carrier Ethernet services are described. An architecture for future services consisting of connectivity and application components with standard interfaces is proposed.

Chapter 5 describes edge applications and services, architecture, edge federation, edge-to-cloud collaboration, and key edge technologies in details.

Chapter 6 explains data center network (DCN) architectures and operational principles. The chapter describes the state-of-the-art DCN topologies, examines various operation and optimization solutions in DCNs, and provides an outlook for future DCNs and their applications.

Chapter 7 discusses general cloud networking constructs, core public cloud infrastructures, distributed cloud infrastructures of AWS Virtual Private Cloud, Microsoft's Azure (Vnet), and Google; and expected public cloud architecture and capabilities by year 2030 and beyond.

Chapter 8 provides an overview of near-to-midterm space-networking toward 2030, discusses the current state-of-art for space-terrestrial network integration, and highlights specific use cases and technical challenges. The chapter focuses on the low earth orbit (LEO) satellite system, describes different strategies for network integration, and management implications of the integrated assets and resources.

Chapter 9 focuses on network slicing and describes network slicing characteristics such as scalability, dynamicity, arbitration mechanisms to allow an efficient usage of resources, and network slicing management with examples. The chapter explains resource optimization, SLA management, programmability, and multi-domain orchestration.

Chapter 10 focuses on routing and addressing. A key set of principles and requirements for routing and addressing for future Internet are discussed in details. Special considerations are given to routing security and resilience. Novel routing protocols, Routing in Fat Trees (RIFT), Link-State Vector Routing (LSVR), and Scalability, Control, and Isolation on Next-Generation Networks (SCION) are introduced.

Chapter 11 discusses the high-speed forwarding plane aspects of QoS. The chapter describes congestion management, integrated and differentiated services, and identifies the gaps for future networks and services. Furthermore, the chapter describes QoS for the edge, Time-Sensitive Networking, Deterministic Networking, home access, IoT for industrial applications, 5G, and beyond 5G. Programmable networks and algorithms are also described.

Chapter 12 focuses on burst forwarding which is an application aware technology that optimizes both network utilization and data transmit latency where a burst is a consecutive of packets that consists an application data processing unit. This chapter describes use cases and issues with packet switching first. After that, the burst forwarding network architecture along with data packing, data forwarding mechanism flow control mechanism, network throughput, host data processing performance, data transmission latency, and router buffer requirements are described.

Chapter 13 addresses the security, anonymity, privacy, and trust of a next-generation Internet. The chapter identifies most critical security goals and requirements to pursue these goals. Finally, the chapter proposes possible pathways for achieving security and trust under these requirements.

Chapter 14 focuses on Intent-based networking. The chapter discusses policy-based management, service management, and lays out various functions of an Intent-based system, and how they can be combined into an Intent-based networking reference architecture.

Chapter 15 explores how different types of artificial intelligence can be used to enhance and improve network and service management. The chapter progressively builds an example of how AI can be incorporated into a cognitive architecture, currently being prototyped, to improve its network and service management capabilities.

Chapter 16 is devoted to quantum computing that will revolutionize networks and services in year 2030 and beyond. The chapter describes the quantum technology including superposition, entanglement, teleportation, and super-dense coding. After that, the chapter describes quantum networking landscape, current field trials, and related standard work.

1.2   Future Applications and Requirements

Chapter 2 describes future applications and requirements. The chapter introduces the following representative use cases and their analysis:

- Holographic-type communications (HTC) which is expected to be the digital teleportation of 3D images from one or multiple sources to one or more destination nodes in an interactive manner, paving the path toward a future of fully immersive 3D interactions.
- Tactile Internet for remote operations (TIRO) which involves the real-time control of remote infrastructures. Immersive video streaming applications such as telemedical services will enable real-time and immersive interaction between a human operator and remote machinery.
- Space-terrestrial integrated network and new infrastructure requirements that leverage interconnected low Earth orbit (LEO) satellites to build a parallel Internet network that can peer with its terrestrial counterpart as the legacy infrastructure.
- AI-enabled applications where AI extends along the cloud-to-things continuum, embedded in IoT devices, implemented at the network edge and in the remote cloud.
- Industrial IoT and IoT advanced applications that are fundamentally different from the IT networks in terms of performance and reliability requirements. The networks need to deliver superior performance and mandate a real-time, secure, reliable connectivity at large scale.
- Scientific research and big data applications such as astronomical telescopes and accelerators.

- Digital twins which is normally defined as a real-time representation of a physical entity in a digital world. Digital twins (DTs) add value on top of traditional analytical approaches by offering the ability to improve situational awareness and enable better responses for physical asset optimization and predictive maintenance. In the future, facilitated by vastly deployed DTs, the digital and physical worlds have the potential to be fully intertwined, contributing to the creation of a new norm, namely, a DT-enabled cyber-physical world.

Some of the key requirements of these applications are:

- High bandwidth
- Low latency
- Security and reliability
- Privacy
- Ultrahigh bandwidth
- Strict synchronization
- Differentiated prioritization levels:
- Reliable transmission
- Edge computing and storage
- Mobility
- Elasticity
- Energy efficiency
- Virtualization
- Joint network, intelligence and computing orchestration
- AI-aware addressing
- Network protocol programmability

The application requirements are described in details.

1.3 Future Services

Chapter 3 focuses on the definition of new network services in order to support emerging applications and vertical industries in the year 2030 and beyond.

The services available today are expected to co-exist with new high-precision services as illustrated in Fig. 1.1. Essentially, each network service of today or network 2030 serves a purpose and addresses application delivery requirements.

The scope of network services covers a wide variety of network functions, ranging from basic connectivity to quality of service (QoS), path control, security, telemetry, resiliency, redundancy, and performance monitoring. The delivery of services requires equal part efforts at operations and management, control, and user planes.

In this chapter, we start by briefly defining a set of fundamental concepts that apply to any network services, followed by state of the art. Furthermore, this chapter dwells on formalizing network 2030 services in greater detail.

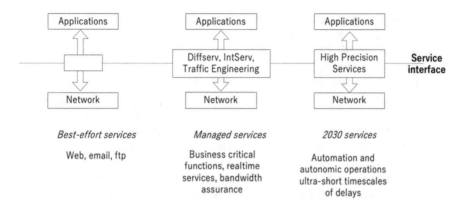

Fig. 1.1 Evolution of network services

The services are divided into foundational and compound services. A foundational network service requires dedicated support from some or all network system nodes to be able to deliver the service between two or more end users (application system nodes). An example of a foundational network service is the IP packet routing and forwarding. On the other hand, a compound (or composite) network service is composed by one or more foundational services, consisting of at least one next-generation foundational service together with a number of pre-existing foundational network services.

As part of the foundational services, the chapter describes in-time/on-time services that are performance sensitive, the coordinated network services providing coordination among multiple flows with interdependencies and qualitative communication services that avoid retransmissions of less relevant portions of the payload in order to meet requirements on latency if the application is tolerant to quality degradation (such as interactive applications). After that, the compound network services related to holographic communication, haptic communication, and high-speed data delivery are described.

1.4   Overall Network and Services Architecture

The current Internet derives mainly from the 1980s and soon after. Among the key objectives were best effort connectivity and simplicity along with the ability to survive some level of link and node failures. Private networks have been used for applications requiring more assured security and privacy, and service quality better than best effort. Transmission rates were in kbps. Nowadays, the rates are in Mbps and Gbps ranges.

New wireline and wireless technologies are pushing the transmission rates from Mbps and Gbps to Tbps. Advances in space technologies are expected to make the space communications a viable alternative to wireline communications. In parallel to the advances in communications technologies, the number of connected devices and traffic are expected to grow to 100 billion devices and 175 ZB (i.e., 175×10^{21}) by 2025, respectively. In addition, applications requiring large bandwidth and strict performance are growing rapidly.

As a result, the future networks will consist of many types of integrated networks and no longer be a vehicle only for best effort connectivity, but a programmable infrastructure of connectivity and applications supporting vital and high precision services that require low latency, appropriate security, and extremely high reliability for communications between most of the locations in the world.

The intelligence is no longer only in the end devices but rather distributed among end devices, data centers, cloud, space, edge, and core devices in the network. As a result, the complexity is increased. To help deal with this increase in complexity, the automation of operational processes for inter- and intra-networking is being worked in the industry. On-demand modifications of network elements and applications are becoming a common trend. The level of intelligence in each component is increased with the proliferation of artificial intelligence (AI)/machine learning (ML) techniques, and advances in memory and computing technologies. By 2030, it is expected that self-managed networks will be available, with substantial user controls and tremendous growth in the services supported by autonomous edge devices.

This chapter divides networks and services into underlay and overlay networks and services. As examples to underlay networks, current and future architectures of optical and wireless networks including 5G and 6G wireless networks are described. As examples to overlay networks, virtualized networks, SD-WAN, and Carrier Ethernet Services are described.

The chapter proposes an architecture for future services consisting of connectivity and application components with standard interfaces. An example of future services architecture is illustrated in Fig. 1.2. For services over public networks, the user will buy services from multiple operators and stich them together to establish an end-to-end services, therefore acts as a service provider. By 2030, we expect service establishment within and among operators to be automated to accomplish this.

The chapter also proposes a service management architecture with standard interfaces between a user and a service provider, and between operators such as connectivity operators, cloud operators, space operators, etc.

Finally, the chapter describes application programming interfaces (APIs) for the standard management interfaces.

1.5 Access and Edge Network Architecture and Management

With the advent of 5G, coupled with new computing technologies, the network edge is evolving a new paradigm where network and compute/storage combined to offer advanced services such as ultralow latency, enhanced mobile broadband, better control for users of their privacy and data, and more energy-efficient computing. User

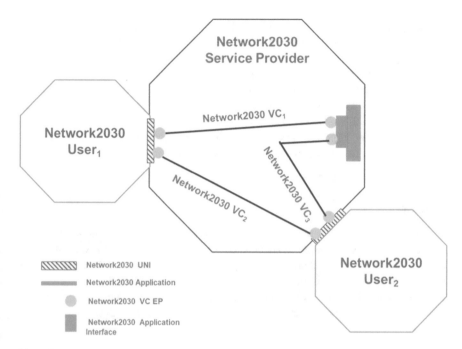

Fig. 1.2 Example Network2030 Service between Network2030 Users

applications spanning multiple vertical domains stand to benefit, from connected vehicles to intelligent fleet management, from multiplayer mobile gaming to AR/VR real-time rendering, and from industrial IoT to manufacturing. With enhancements in computing and memory technologies, it becomes possible to support these applications by devices located at the edge of future networks and/or customer premises.

These advances are being driven by the following trends:

- Densification of the edge through placing micro data center capabilities
- Innovation in future use cases (e.g., industrial automation, security and proactive monitoring, robotic surgery)
- Economics of network by optimizing backhaul and transport capacity through localization of content (e.g., augmented reality/virtual reality (AR/VR) content, HD, ultra HD media content)
- Economics of network through multi-access edge computing (MEC) federation, collaboration, and infrastructure sharing

Existing access and edge network operation is already capable of localized traffic steering. The trends above further extend such concepts in network engineering, with more innovation in technology and service domains expected.

A rapid increase in MEC deployment, localization of user plan, and data plane processing near ultradense access networks will require innovative approaches to designing future networks. These approaches need to be service oriented, adaptive to change in operating conditions including environment, secure, and capable of supporting multiple technologies at access and edge layers. Future networks need to be structured to provide easy integration with networks of multi-domains and collaboration between operators and users.

The term "network edge" refers to communication and computing infrastructure in locations such as service provider (SP) points of presence (PoPs), central offices, cell towers, stadiums, and first responder sites.

The access and edge network components are grouped as depicted in Fig. 1.3. Furthermore, future edge network devices are classified as:

- Human-operated devices
- Machine-operated devices
- Sensors

These devices must work intelligently in association with mobile or fixed-line networks and may also need to implement peer-to-peer communication. The properties of these devices and characteristics that form their role in the network access layer become important in considering future network innovation.

This chapter describes edge applications and services, architecture, edge federation, edge-to-cloud collaboration, and key edge technologies in details.

1.6 Data Center Network Architecture, Operation, and Optimization

The explosive growth of workloads driven by data-intensive applications, e.g., web search, social networks, and e-commerce, has led mankind into the era of big data. According to the IDC report, the volume of data is doubling every 2 years and thus will reach a staggering 175 ZB by 2025. Data centers have emerged as an irreplaceable and crucial infrastructure to power this ever-growing trend and became a key component of the communications infrastructure.

As the foundation of cloud computing, data centers can provide powerful parallel computing and distributed storage capabilities to manage, manipulate, and analyze massive amounts of data. Data center network (DCN) is designed to interconnect a large number of computing and storage nodes. In comparison with traditional

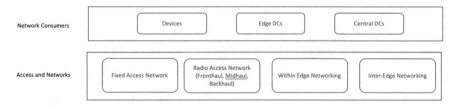

Fig. 1.3 Access and edge network components

networks, e.g., local area networks and wide area networks, the design of DCN has its unique challenges and requirements:

- **Hyperscale**: Currently, over 500 hyperscale data centers are distributed across the globe. A hyperscale data center can host over a million servers spreading across hundreds of thousands of racks. Data centers at such a large scale put forward severe challenges on system design in terms of interconnectivity, flexibility, robustness, efficiency, and overheads.
- **Huge Energy Consumption**: In 2018, global data centers consumed about 205 TWh of electricity, or 1% of global electricity consumed in that year. It has been predicted that the electricity usage of data centers will increase about 15-fold by 2030. The huge energy consumption prompts data centers to improve the energy efficiency of the hardware and system cooling. Typically, service providers operate their facilities at maximum capacity to handle the possible bursty service requests. As a result, data centers can waste 90% or more of the total consumed electricity.
- **Complex Traffic Characteristics**: Modern data centers have been applied to a wide variety of scenarios, e.g., email, video content distribution, and social networking. Furthermore, data centers are also employed to run large-scale data-intensive tasks, e.g., indexing Web pages and big data analytics. The traffic of these diversified services and applications show complex characteristics such as high fluctuation with the long-tail distribution and short flows. Furthermore, data centers suffer from fragmentation with intensive short flows. It is a challenge to handle traffic optimization tasks in hyperscale data centers.
- **Tight Service Level Agreement**: The service level agreement (SLA) plays the most crucial part in a data center lease. It has been increasingly common to include mission-critical data center services in SLAs such as power availability, interconnectivity, security, response time, and delivery service levels. Considering the inevitable network failures, congestion, or even human errors, constant monitoring, agile failure recovery, and congestion control schemes are necessary to provide tight SLAs.

To solve these significant technical challenges above, DCNs have been widely investigated in terms of network topology, routing, load balancing, green networking, optical networking, and network virtualization.

This chapter presents DCN architectures and operational principles. We start with a discussion on the state-of-the-art DCN topologies, highlighting their advantages and disadvantages in terms of network architecture and scalability. Then, we examine various operation and optimization solutions in DCNs. Thereafter, we discuss the outlook of future DCNs and their applications. The main goal of this book chapter is to highlight the salient features of existing solutions which can be utilized as guidelines in constructing future DCN architectures and operational principles.

1.7 Public Cloud Architecture

Cloud technology is one of the greatest recent innovation enablers, not just accessible to large enterprises which have traditionally had the funds to build large data centers. Cloud technology is available to anyone who has a little bit of know-how, and a credit card, with virtual server prices charged by the hour and as low as $0.0065 per hour (the T2.nano instance from AWS), and offering virtual machines that have networking speeds from the megabits per second up to 100s of gigabits per second.

Core public cloud infrastructure is depicted in Fig. 1.4 showing the breakdown of where cloud provider's responsibilities for cloud infrastructure start and end, and where the user of a cloud platforms responsibilities start and end.

In order to be closer to users, cloud operators are in the process of building distributed cloud. There are a few versions of distributed cloud available:

- AWS Outpost which is deployed as a rack or multiple racks at a user's location (i.e., usually a colocation facility or private data center)
- AWS Local Zone which is similar to AWS Outposts and built upon the same technology. It is an availability zone that sits outside the geographic area that a region is deployed.
- AWS Wavelength which uses AWS Outposts technology and deployed similar to an AWS Local Zone, however, is connected natively to a provider's mobile network. AWS Wavelength is perfect for folks that want to deploy mobile applications at the edge, with low-latency connectivity to providers that offer 5G and mobile connectivity such as Verizon.
- Microsoft Azure Stack which is similar to AWS Outposts, taking Azure on-premises on hardware that is deployed closer to a user's geographic area. The

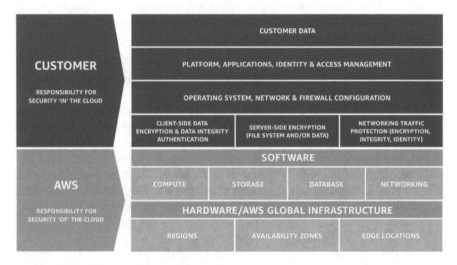

Fig. 1.4 Public cloud infrastructure

Azure Stack relies on users purchasing hardware from a specific set of vendors that are qualified to run Azure Stack.

- Google Anthos which is a Kubernetes forward stack that can run on many different types of hardware vendors and take Google Cloud Platform (GCP) capabilities on-premises. It is similar to Azure Stack, whereby users can deploy on non-Google hardware, and essentially pay by the hour for the software stack itself.

This chapter discusses general cloud networking constructs, the platform such as AWS Virtual Private Cloud and Microsoft's Azure (Vnet), and expected public cloud architecture and capabilities by year 2030 and beyond.

1.8   Integrated Space and Terrestrial Networking Toward 2030

Exponential increases in Internet speed have facilitated an entirely new set of applications and industry verticals underpinned by evolving fixed network infrastructure. The costs of deploying new fixed fiber networks are a limiting factor. As 5G and Internet infrastructure build-out continues, we must now look up both figuratively and physically, for our next networking opportunity. In the future, space communication will play a significant role in providing ubiquitous Internet communications in terms of both access and backhaul services.

Legacy satellites, probes, and space-based objects like the International Space Station (ISS) rely mostly on radiotechnology for communication. Using radio, it would take approximately 2.5 s to send data to the Moon and back to Earth, and between 5 and 20 min depending on planet alignment. In 2014, the ISS tested OPALS (Optical Payload for Lasercomm Science) system developed by NASA; this achieved a data rate of 50 Mbps. By 2015, gigabit laser-based communication was performed by the European Space Agency (ESA) and called the European Data Relay System (EDRS). The ESA system is still operational and extensively used.

In 2020, we observed a slew of next-generation meshed satellite constellations—OneWeb, SpaceX (Starlink), Viasat-4, and TeleSat—with Amazon (project Kuiper) and Facebook also developing space-based communication projects. These new space networks will be capable of providing global gigabyte Internet via Earth-to-space lasers instead of radio and, instead of bouncing signals between Earth and space and back to Earth, the signal can be transmitted in space using space-based laser communication. These new satellite constellations are positioned in a Low Earth Orbit (LEO) Earth orbit approximately ≤2000 km altitude. They number from thousands to tens of thousands, in a grid-like pattern, and will provide continuous Internet coverage. The constellation will orbit the Earth on the order of 100 min, traveling at roughly 27,000 km/h.

These new satellite constellations will form a mesh network infrastructure in space that will connect to existing network infrastructure on the ground and provide lower latency. The potential for lower latency for long-distance connectivity stems from building "nearly shortest" paths (after incurring the overhead for the uplinks and downlinks) instead of circuitous terrestrial fiber routes.

These new networks will provide connections of 100s Mbps to residential users, and multiple Gbps to enterprise users, across vast rural areas and provide competitive low-latency bandwidth in metro areas, thus significantly offloading Internet traffic from traditional terrestrial infrastructures. Current and near-future space-based networks include Telesat with 120 satellites and 40 grounds stations, OneWeb with 720 satellites and 70 ground stations, and SpaceX with planned 42,000 satellites and 120 ground stations. Also, several additional Satellite Internet projects are proposed for operational deployment by 2025.

Future space networks will also need to cooperate with the existing terrestrial network infrastructure, exploiting heterogeneous devices, systems, and networks. Thus, providing much more effective services than traditional Earth-based infrastructure, and greater reach and coverage than proprietary and isolated space-based networks.

It is envisaged that future integrated space and terrestrial networks (ISTNs) will be comprised of the following key components:

Satellite: A Low Earth Orbit (LEO) satellite has a lower physical orbit compared to legacy satellite systems, potentially bringing a short-latency benefit at the expense of constellation complexity. Medium Earth Orbit (MEO) and Geostationary Orbit (GEO) satellites can provide more physical stability, but they come with a relatively longer transmission delay than LEO systems. The current satellite systems mostly provide relay function; however, in the future, satellite systems may build up a mesh-like network to provide routing and forwarding function. LEO satellites should be organized as a routing system and work as routers covering data-plane and control-plane functions.

Ground Station and Terminal: Ground stations and terminals are physical terrestrial devices that act as gateway or interfaces between terrestrial and space networks through radio communications. The networking mechanisms and protocols used in space networks are different from those in the traditional IP framework in the terrestrial infrastructures. Hence, ground stations and terminals have been responsible for protocol translations and creation/maintenance of tunnels for data packets to traverse different network environments. It is also worth mentioning that, while ground stations use dedicated gateways between the space network and the terrestrial infrastructures today, it is envisaged that in the future network/user devices will be able to communicate direct with satellites, allowing Internet traffic to be exchanged between user devices without necessarily always going through ground stations.

- Controller (SDN architecture-based): The satellite network system may also employ a hierarchical architecture. Some of the satellites play the role not only of a router but also a controller.
- Mobile Edge Computing (MEC) Server: MEC has been a terminology used mainly in the context of 5G where local computing and storage capabilities can be embedded at the mobile network edge to provide low-latency data/computing

services to locally attached end users. It is envisaged that in emerging space and terrestrial networks, LEO satellites can also interconnect MEC servers in the satellite constellation once equipped with computing and data storage capabilities.

This chapter provides an overview of near-to-midterm space networking toward 2030, discusses the current state of art for space-terrestrial network integration and highlights specific use cases and technical challenges. A fundamental challenge will be the future seamless integration of space networks with the current terrestrial Internet infrastructure to maximize the benefits for Earth-based and space-based infrastructure. To limit the discussion's scope, we mainly focus on the Low Earth Orbit (LEO) satellite system, which can provide low end-to-end latency compared to its GEO (Geostationary Earth Orbit) counterpart. The shared vision in this scenario is that multiple (up to tens of thousands) LEO satellites can be interconnected to form a network infrastructure in space that will be further integrated with the ground's network infrastructures. On the other hand, the critical challenge, in this case, is the frequent handover between the two networks caused by the constellation behaviors at the LEO satellite side, which is considered to be the most notable feature, that incurs a wide range of technical challenges in the context of space-terrestrial network integration. The rest of this chapter aims to describe different strategies for such network integration and the specific technical issues that need to be addressed. We discuss the management implications of these integrated assets and resources, and potential technologies and capabilities that may be applied or extended.

1.9 Network Slicing and Management

Network slicing is a paradigm through which different virtual resource elements of a common shared infrastructure consisting of both connectivity and compute resources allocated to a specific customer who perceived the resulting slice as a fully dedicated, self-contained network for it. The resources are virtualized through a process of abstraction of lower-level elements, providing a great flexibility and independence of the specific element being used along the customer service lifetime, which permits exercise actions such as scalability, reliability, protection, relocation, etc.

Network slicing, despite not being a new concept, acts as a foundational concept and systems to current 5G/future networks and service delivery, with the goal of providing dedicated private networks tailored to the needs of different verticals based on the specific requirements of a diversity of new services such as high-definition (HD) video, virtual reality (VR), V2X applications, and high-precision services.

The possibility of dynamically instantiating slices through automation enables the provision of slices in an on-demand fashion, dealing to the concept of slice-as-a-service (SlaaS). The SlaaS approach is a versatile tool for trading tailored network capabilities with external third parties such as vertical customers, opening up new opportunities for service providers (SPs). The network is then transformed into a production system merging both business and operation domains.

A critical point on the overall provision of a slice is to allow control of the allocated abstract resources to the customer (e.g., the possibility of programming them). Without such control, the slice is simply made available but cannot be reconfigured by the customer, leading to a kind of static network. On the contrary, if control capabilities are enabled for the customer, the network can be then flexibly managed, for example, be reconfiguring forwarding paths adapting to changing conditions of traffic within the slice.

Different types of slices can be considered from the service provider perspective:

- *Internal slices* that are dedicated to SP's internal services and the service provider retain the total control and management capabilities.
- *External slices* that are offered to vertical customers which perceived them as dedicated networks and may run on top of shared infrastructure:
  - Slices managed by an SP, where the SP performs the control and management of the slice and the vertical customer simply runs the service on top of the capabilities and resources offered by the SP.
  - Slices managed by the vertical customer, where the customer actually has control of the resources and functions allocated. The level of control could be limited to a set of operations and/or configuration actions, but in any case, the vertical has the possibility of governing the slice behavior to some extent.

The referred control capabilities in the latter case should be enabled with care, since different actions from distinct customers could collide.

Different vertical customers with similar service needs can be accommodated in the same slice (e.g., customers requiring a generic enhanced Mobile Broadband (eMBB) service) if that slice is properly dimensioned, while in the case of external slices managed by the vertical customer, the slices should be essentially dedicated per customer.

Figure 1.5 graphically represents this distinction, showing the different responsibilities in each case.

Standards Development Organizations (SDOs) and some other industrial associations have been looking at the network slice concept from different angles and perspectives. For example, ITU-T Slicing (2011) defined slicing as a logically isolated network partitions (LINP) composed of units of programmable resources such as network, computation, and storage. More recently, ITU-T IMT2010/SG13 (2018/2019) describe the concept of network slicing and use cases of when a single user equipment (UE) simultaneously attaches to multiple network slices in the IMT-2020 network. In IETF (2017), network slicing is defined in as managed partitions of physical and/or virtual network and computation resources, network

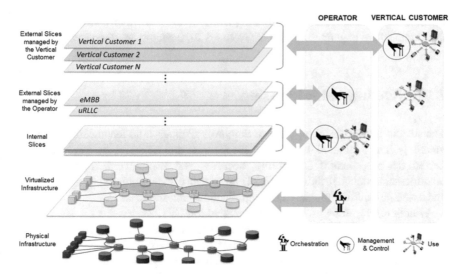

Fig. 1.5 Types of network slices according to management and control levels of responsibility

physical/virtual and service functions that can act as an independent instance of a connectivity network and/or as a network cloud.

3GPP defined network slicing architecture for 5G with charging management while ETSI E2E defined a next-generation network slicing (NGNS) framework as a generalized architecture that would allow different network SPs to coordinate and concurrently operate different services as active NS. ETSI Zero-Touch Network and Service Management Industry Specification Group (ZSM ISG) are specifically devoted to the standardization of automation technology for network slice management.

With the new control and user plane separation in 5G, particularly with the 5G Core Network (CN) Service-Based Architecture (SBA), a much finer granularity of slicing is allowed. The functions in the network become logical functions that may be instantiated in physical locations as service requirements and capabilities demand. This is further enhanced by network function virtualization (NFV) that permits the logical functions to be instantiated on a virtualization abstraction layer hardware supported on COTS hardware.

This chapter describes architectural elements of network slicing services with different mixes of low latency, ultra-reliability, massive connectivity, and enhanced mobile broadband delivered simultaneously in the same network.

The chapter describes network slicing characteristics such as scalability, dynamicity, arbitration mechanisms to allow an efficient usage of resources, and network slicing management with examples. SDN and NFV techniques can be used by a SP to orchestrate slices with full control and visibility of the nodes, topology, functions, and capabilities (such as bandwidth or compute power) to make decisions. The resource optimization, SLA management, and programmable control of the slices and their multi-domain orchestration that allows virtualized network functions to be instantiated in computing facilities available in different administrative domains are explained. Views for Network2030 slicing are discussed.

1.10   Routing and Addressing

The current Internet is facing a set of unique challenges, both technically and commercially. The exponential growth of the Internet and emerging demands for connected devices, increased mobility, security, and resilience is being met by incremental updates. Routing protocols have been critical networking technologies and essential building blocks of new applications and services.

Widely used routing protocols are depicted in Fig. 1.6.

Distance vector protocols determine best routes by distance and share information on a periodic update. Common distance vector protocols are Enhanced Interior Gateway Routing Protocol (EIGRP) and Routing Information Protocol (RIP).

With link-state protocols, each router shares state information of its directly connected links with all routers within the domain, which allows each router to make its own decision for the best path. Common link-state protocols are Open Shortest Path First (OSPF) and Intermediate System to Intermediate System (IS-IS).

A path vector protocol, such as Border Gateway Protocol (BGP), routes between autonomous systems. Each autonomous system is a set of routers under an administrative control. One of the advantages of path vector protocols is that each destination network has a path dynamically added to it. Therefore, the loop detection can be used to when it sees its own path.

Routing protocols can be grouped as interior gateway protocols and exterior gateway protocols. Interior gateway protocols are used within an autonomous system that is designed to support fast route convergence. Exterior gateway protocols are used to route traffic between autonomous systems that are capable of holding large amounts of routes and perform routing policies. For example, if there are two internet providers, a routing policy can be defined for given prefix to prefer an ISP over another.

Fig. 1.6 Types of routing protocol category

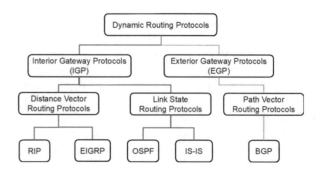

At the core of the Internet, routing protocols such as OSPF, IS-IS, and BGP facilitate how Internet routers communicate with each other to distribute information that enables them to select routes for Internet connectivity.

Recent research and investigation for the future Internet's requirements and objectives identified several technology objectives, including contextual addressing, application-aware networking, increased stability and security, faster convergence, and decreased operational costs.

Existing routing protocols are likely to be enhanced, and new routing protocols may be developed to meet emerging requirements such as application-aware networking, increased stability and security, faster convergence, and decreased operational costs.

Internet Protocol (IP) addressing facilitates how one device attached to the Internet is distinguished from every other device. It is used to direct requests to an appropriate destination (destination address) and indicate where replies should be sent to (source address). Due to the rapid growth of the Internet and exponential increase of connected devices, several short-term fixes have been developed for coping with Internet addressing demands. The continued growth and deployment of the Internet of Things and new network types such as space networking will place new requirements on existing addressing schemes.

An IP address of a host is used to identify the host and the host location. Another approach is to access data by name regardless of the origin location by using name-based addressing.

1.11 Quality of Service (QoS)

QoS in TCP/IP networks has not seen significant improvement due to the limitations of the current forwarding plane QoS functionality, which makes it difficult, if not impossible to offer scalable, low-cost QoS service differentiation. For this reason, this chapter discusses the high-speed forwarding plane aspects of QoS.

QoS is used to refer to the packet level service experience offered to traffic between an ingress and one (unicast) or more (multicast) egress point(s). QoS is also used to refer to the experience received by sequences of packets called flows and includes the aspects of throughput, congestion behavior, loss, reordering, latency, and jitter across the packets of such a flow.

This section gave an overview of the current state of QoS in TCP/IP networks, its past and present focus on congestion controlled best effort, recently with more focus on low latency, but also the revival of interest in better controlled latency, loss and throughput for deterministic and more general high-precision use-case scenarios.

The IPv4 packet header contains an 8-bit "Type of Service Octet (TOS)" to indicate to the network the QoS that the packet is requesting:

- 6-bit of "Differentiated Services Code Point" (DSCP) indicating the so-called Per-Hop-Behavior (PHB) of the desired service of the packet.
- 2 bit of "Early Congestion Notification." IPv6 [RFC8200] uses the same DSCP and ECN semantic for the TOS octet as IPv4.

IPv6 uses the same DSCP and ECN semantic for the TOS octet as IPv4. On the other hand, each MPLS label stack entry carries a 3-bit Traffic Class (TC) field (formerly known as the EXP field) that indicates (TC) of the packet indicating a PHB in the sense of the DSCP.

Integrated services (Intsrv) was the first architecture developed by the IETF in the 1990s to distinguish hop-by-hop processing of traffic requiring differentiation. It defines two services: the guaranteed service (GS) and controlled load service. The GS provides per-flow fixed bandwidth and latency guarantees. It is based on the concept of reserving bandwidth and buffer resources in advance for each flow. To maintain bandwidth guarantees, GS traffic is shaped and policed at the ingress network edge as necessary so that the flow does not consume more resources than have been reserved for it. To support latency guarantees, flows need to be reshaped on every hop to prevent loss and unpredictable variations in latency. Because of the need for upfront resource reservations, IntServ solutions are also referred to as "admission controlled" (AC).

The current Internet QoS architecture is insufficient to meet the needs of future networks for a number of reasons, including the following:

- The need for per-flow admission control makes IntServ expensive to support and scale, even if performed out of band via SDN.
- The inability to dynamically adjust the bitrate under varying network utilization makes this model too inflexible even for current and future networks.
- No mechanisms exist to support application-defined upper and lower bounds for the desired latency independent of the path round trip time (RTT).
- There are no mechanisms to slow down packets based on the desired earliest delivery time.
- Queuing cannot prioritize packets based on their desired end-to-end latency.

When mapping these evolving QoS requirements against the evolution of the Internet, it is believed that an improved QoS architecture will predominantly be required on the edge spanning topologies from industrial campus all through metropolitan/regional areas, often based on permissible path round-trip time (RTT) of control loops of applications. Today's high-speed forwarding planes could provide a lot better scalable QoS functionalities with significantly reduced operational complexity.

Today's evolving slice services in 5G/B5G only allow the network operator to parameterize existing QoS services. Per-slice programmable QoS via forwarding plane programming evolving from network forwarding plane programming languages such as P4 and programmable QoS abstractions such as PIFO (and beyond) that can enable high-precision QoS services designed by and for the actual owners of the application use cases and their partners. These directions could be a core target for networking to enable better industrial and critical infrastructures in the coming decade.

While forwarding plane performances grew by factors of 10,000 or more in the last two decades, the performance of the control plane barely rose a factor 10 or 100 in the same period. This means that on-path resource reservation via traditional approaches such as RSVP, NSIS, or similar evolving protocols in IEEE can only be adopted by investing into significantly faster control plane performance.

Any form of reservations of bandwidth resources for network 2030 should support the handling of not only fixed reservations but also those of elastic media. Programmable virtual networks are a key technology that allows future network application owners and operators to deliver their required end-to-end solution without being dependent on physical network operators or equipment vendors.

This chapter describes congestion management, integrated and differentiated services, and identifies the gaps for future networks and services. After that, the chapter describes QoS for the edge, time-sensitive networking, deterministic networking, home access, IoT for industrial applications, 5G, and beyond 5G. High-precision QoS, fine grained and path aware latency management, resilience and near-zero loss forwarding, and programmable networks and algorithms are also described.

1.12   Burst Switching

Burst forwarding is an application aware technology which optimizes both network utilization and data transmit latency where a burst is a consecutive of packets that consists an application data processing unit. The burst forwarding technology is beneficial especially for high-bandwidth and low-latency applications such as holographic type of communication.

For example, a burst can be a photo for the image processing system, or it can be a video clip in the video streaming service. The burst forwarding network uses burst as the basic transmission unit. The data source sends the entire burst at the line rate of the network interface card (NIC). End-to-end virtual channels are created for the burst transmission. In the burst forwarding network, bursts are forwarded using cut-through forwarding without congestion. In the receiver side, the application usually needs to receive the entire burst before start processing the received data. If the application data is received in packets with multiple flows, the application needs to buffer the data until the whole burst is received. In the burst forwarding network, however, the bursts are received in sequence. The application in the receiver node can immediately process the data without any further data buffering. This mechanism not only accelerates the burst data end-to-end transmission time but also optimizes the computation resource utilization of the data processing.

In a metro gate control face recognition system in Fig. 1.7, the metro gate camera takes high-resolution picture for each passenger. The average photo size generated by the camera for one passenger is around 8 MB. The cameras connect with the

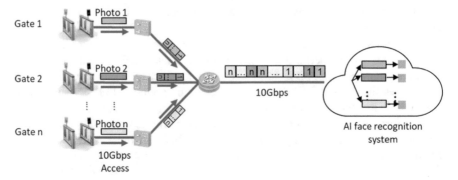

Fig. 1.7 Metro gate control face recognition system architecture

cloud AI system using 10 Gbps leased lines which can support 30 concurrent photo transmissions. The maximum end-to-end data transmission time is 193 ms.

This chapter describes use cases and issues with packet switching first. After that, the theoretical study of the burst forwarding technology associated with network throughput, host data processing performance, data transmission latency, and router buffer requirement is discussed. The burst forwarding network architecture along with data packing, data forwarding mechanism, and flow control mechanism are described in details.

1.13   Security, Privacy, and Trust

A network is considered secure if it can achieve the desired properties even in the presence of an active adversary. One prominent property is availability. The control, data, management, and configuration planes should be protected such that an adversary cannot disrupt connectivity. Another important property is trust, which is the ability of network nodes to verify origin and content authenticity of messages passed through the network. Furthermore, desirable, but difficult to achieve properties are privacy and anonymity, treated here as the ability of nodes to communicate without other network entities being able to identify the communication parties. Privacy typically refers to the secrecy of personal information, whereas anonymity is a more specific property that refers to the identity of the user or endpoint. Since personal information is usually carried within the communicated data, we focus on achieving anonymity in the network-focused context of this chapter. However, we consider the privacy of network metadata to be outside the scope of this chapter.

The chapter first states the goals of a secure inter-domain network infrastructure. Concerning the security, anonymity, privacy, and trust of a next-generation Internet, we consider the following aspects as the most critical to consider:

- **Improved trust model**: A new network trust model should be deployed to provide decentralized verifiability.

- **Transparency and control for forwarding paths**: Network paths in today's Internet lack transparency. In a first step, it would be useful to know as a sender which entities a packet traverses. In a second step, it would be useful for a receiver to achieve ingress path control for incoming traffic. Finally, in a third step, end hosts could benefit from controlling the packet's forwarding path.
- **Efficient and scalable authentication mechanisms for AS and host-level information**: Such properties will prevent IP source address spoofing attacks and enable a receiver to verify the origin of error packets.
- **Availability in the presence of an active adversary**: Communication between two endpoints should be possible, as long as a functional and connected sequence of intermediate network devices and links exists.
- **Pseudonymous sender/receiver anonymity**: Untrusted nodes (i.e., nodes under control of an adversary) in the network cannot identify the sender and/or receiver of communication without resorting to timing analysis.
- **Algorithm agility**: Cryptographic algorithms need to be replaced in case of breakthroughs in cryptanalysis or computation technology such as quantum computers.
- **Class of security level**: Not all applications or processes need the same level of security.

The chapter identifies requirements to pursue these goals. Possible pathways for achieving security and trust under these requirements are proposed.

1.14 Intent-Based Network Management

Regardless of the level of network automation, networks need human inputs for direction and guidance for how ultimately the network should be used, what services and to whom need to be provided, what operational goals to prioritize, and what other aspects to take into consideration that should affect the way the network operates. This guidance and direction is what is now commonly referred to as "intent."

Intent is defined as the ability to allow users to define management outcomes, as opposed to having to specify precise rules or algorithms that will lead to those outcomes. This requires an intent-based system to possess the necessary intelligence to identify the required steps on its own. Networks that are supported by intent-based systems that allow them to be managed using intent are referred to as "intent-based networks."

The following are some examples of intent:

- "Steer networking traffic originating from endpoints in one geography away from a second geography, unless the destination lies in that second geography." This simply states what the network should achieve without saying how.
- "Avoid routing networking traffic originating from a given set of endpoints (or associated with a given customer) through a particular vendor's equipment, even

if this occurs at the expense of reduced service levels." Again, this simply states what to achieve, not how. In addition, guidance is given for how the system should trade off between different goals when necessary.

- "Maximize network utilization even if it means trading off service levels (such as latency, loss), unless service levels have deteriorated at least 25% from their historic mean." This clearly defines a desired outcome. It also specifies a set of constraints to provide additional guidance, without specifying how to achieve any of this.
- "VPN service must have path protection at all times for all paths." Again, a desired outcome. How to precisely accommodate it is not specified.

The chapter discusses policy-based management and service management and lays out various functions of an intent-based system. How these functions are interrelated and how they can be combined into a reference architecture for intent-based networking are subsequently described.

1.15   AI-Based Network and Service Management

As mentioned in previous chapters, the complexity of networks and services is continuously increasing. In parallel to this, the complexity of management of networks and services is also increasing. Human involvement in the management processes is time-consuming and error-prone.

Network management architectures are not built to translate business requirements to services and resources supporting these services. This problem is exacerbated as the level of business abstraction increases.

Operators are also concerned about the increasing complexity of integration of different platforms in their network and operational environment. Operators need to optimize the use of networked resources and improve the use and maintenance of their networks. It is a multi-objective optimization problem, where optimal decisions need to be made, even though all objectives may not be able to be simultaneously optimized.

User needs, business goals, and environmental conditions are frequently changing. This requires improved automation and real-time closed control loops. Thus, network intelligence is needed to detect these contextual changes, determine which groups of devices and services affect each other, and manage the resulting services while maintaining SLAs.

One solution is to realize a cognitive network, where offered services can be more easily related to business needs. In this approach, intelligence is imbued into the governance of the network system and its services by making use of three important design principles: situation awareness, experiential learning, and decision-making using adaptive closed control loops.

This chapter describes current Network Management issues, translating business needs to services, incorporating dynamicity, reacting to context, and incorporating situational awareness. The translations are done on a per-pairwise-continuum basis. A high-level functional block diagram is shown in Fig. 1.8.

After that the chapter provides a cognitive architecture overview consisting of cognition, adaptive and cognitive control loop, knowledge representation, knowledge processing, dynamic command generation, and incorporating experiential and machine learning.

1.16 Quantum Computing and Its Impact

Quantum technology is a rapidly advancing field which will revolutionize computing and communications networking. Computers that perform quantum computation are known as quantum computers. Quantum computers can be used to solve previously unsolvable problems on a much larger scale and solve certain computational problems, such as integer factorization, substantially faster than classical computers. The Quantum Internet can exchange large amounts of data using quantum physics properties, thereby reducing traffic on traditional communication networks.

In classical computers, the information is represented in bits (i.e., 1s and 0s). In Quantum computers, the information is represented in quantum bits or qubit. Qubits represent the information based on the behavior of atoms, electrons, and other particles, objects governed by the rules of quantum mechanics.

Quantum particles can also be yoked together in a relationship called entanglement, such as when two photons (light particles) shine from the same source. Entangled particles can travel far from each other and maintain their connection.

It is clear that quantum computing will revolutionize networking along with Artificial Intelligence and Machine Learning techniques. With Quantum Computing,

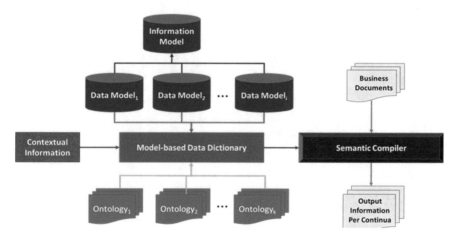

Fig. 1.8 Functional block diagram of a semantic per-continuum-level translator

we expect Network2030 to become fully automated and self-managed by being able to store and process large amounts of connectivity, application, and management information of a domain in a computer, instead of a number of networked computers in one or more data centers.

This chapter describes the Quantum technology including superposition, entanglement, teleportation, and super-dense coding. After that, the chapter describes Quantum networking landscape, current field trials, and related standard work.

Chapter 2
Future Applications and Requirements

Shen Yan and Sundeep Bhandari

2.1 Introduction

Towards the year of 2030 and beyond, many novel applications are expected to emerge as others mature, leading to increasingly intertwined human and machine communications. Application requirements and network capabilities are always spiralling. There are always some warriors trying to break through the periphery of existing capabilities, and the work of these people will further urge those that design, develop and implement networks to innovate further and faster to match these bold ideas. Therefore, an important starting point for discussing network capabilities and future development is from applications and requirements. An in-depth analysis of those requirements that are based on existing network capabilities but bring major challenges to the network will tell us where the network will develop in the future. Of course, these explorations are not without rules, but with traces to follow. Exhaustion is unrealistic. In the following content, we have selected some representative use cases for introduction and analysis. Through the description of these use cases, the authors attempt to induce readers to think about scenarios, applications, use cases, and methods to consider the needs and development paths towards implementation of the 'future network'.

Multimedia is a good example. Recently more and more of us are working online from home, streaming and casting the latest 'box-sets' and using video rather than just audio to stay in touch with family, friends and service providers. Up until recently this wasn't as common place as it is now and both industries and society

S. Yan (✉)
Principal Engineer, Huawei Technologies Co., Ltd., Beijing, China
e-mail: yanshen@huawei.com

S. Bhandari
Strategy Directorate, National Physical Laboratory, Teddington, UK
e-mail: sundeep.bhandari@npl.co.uk

© The Author(s), under exclusive license to Springer Nature
Switzerland AG 2021
M. Toy (ed.), *Future Networks, Services and Management*,
https://doi.org/10.1007/978-3-030-81961-3_2

have undergone and accepted a generational transformation in the space of c. 18 months. But what does 'generation after next' look like? Taking it one step further, we believe that when you see the pictures of your relatives far away, you must be very happy. But if you were able to touch them, sense their body temperature, and even embrace, remotely, what kind of experience would it be? Now that we can all shake hands over the network, why can't we do more? For example, meeting with your doctor via the Internet, or having a doctor sitting at home in Rome perform a minimally invasive surgery on you in Los Angeles? When we use the current network to transmit the signal for the doctor to operate the scalpel, do you dare to lie on the operating table? (At least the authors themselves dare not.) When you find that the mobile phone signal is not good, you will find a way to move closer to the base station or WiFi signal source. We always assume that people are actively chasing signals, but the base station does not move. But have you ever thought that one day, these base stations may be mobile themselves? For example, when deployed to drones, hot air balloons, or satellites, the services you use are provided to you by equipment flying in the sky. Maybe you want to ask right away, haven't we been using satellite communications for a long time? Why is it still called "future"? Yes and No. We have indeed used satellites for many years, but the current satellites are just "communication," in a term called "point-to-point communication." What we want to emphasize is that when many satellites (such as 3000 or 5000) form a network, they will serve you like a terrestrial network. Such technology is still a problem. You must have heard of IoT and AI more or less. How will these technologies develop in the future? What are the challenges to the network? Can YOU imagine that there is another you alive in the digital space? It can help YOU live and help YOU experience something in advance. For example, a virtual person you go to a virtual restaurant A for dinner, and when you come back, tell YOU that you don't like it there. This is much cooler than YOU sitting in A's chair and regretting not going to B. These scenes are what scientists try to do every day. In fact, scientists, or the scientific research work itself, also have huge demands on the Internet. So huge in fact that it is difficult to picture the enormity of the requirements from say the Square Kilometre Array telescope or the large hadron collider.

Finally, what the authors want to say is that these use cases are the results of our screening numerous research works. Cumulatively these papers and presentations offer a multitude of scenarios, possibilities, requirements and opinions. Whist some of these may fall to way side over time, appreciating the options has great value as we move forwards.

2.2 Holographic-Type Communications (HTC)

One of the most important popular services of Internet is multimedia transportation. Besides figures and voice, video is the major type of medium with its good performance for people in hearing and sight senses. By 2019, internet video traffic will account for 80% of all consumer Internet traffic [1]. It is also the reason that bandwidth requirement increasing hugely in recent years. Most of the requirements of current video are bandwidth. A typical 1080p real-time video requires stable 5 Mbps bandwidth. The corresponding numbers for 4K and 8K formats videos are 25 and 60 Mbps, respectively. Whatever the bitrate is, the current video type is still flat video and only make quantitative changes in a single dimension.

Multimedia technologies are continually evolving, with many new applications emerging in the future. One such evolution is that of holographic-type communications (HTC) which we define herein. HTC is expected to be the digital teleportation of 3D images from one or multiple sources to one or more destination nodes in an interactive manner, paving the path towards a future of fully immersive 3D interactions. This will impose heavy challenges on future networks.

Figure 2.1 illustrates a simple comparison on human-perceived visual effects for a 2D image, traditional 3D movies using binocular parallax, and a true hologram, respectively. It can be observed that the holographic display is able to satisfy all the visual cues when observing a 3D object, as we would expect a human to observe naturally [2].

In theory, holography is a method of producing a three-dimensional image of a physical object by recording, on a photographic plate or film, the pattern of interference formed by a split laser beam and then illuminating the pattern either with a laser or with ordinary light by diffraction. Such optical holograms are able to record the wavelength (colour) and intensity (amplitude) of light waves, as well as the phase of light waves (perception of depth). However, holography's technological foundation and ecosystem are not mature enough at present nor foreseeable in the coming decade, which suggests that enabling fully immersive experiences will initially be realized through the adoption of lenslet light-field 3D directly through the naked eye or realizing extreme augmented reality (AR) and virtual reality (VR) displays via head-mounted display (HMD) devices [3], in the short term.

Figure 2.2 suggests four methods of delivering 3D HTC applications. The first two, namely, true holography (which relies upon extremely large data volumes to be recorded and reconstructed) and computer-generated holograms (CGH) (where a traditional hologram is digitized), both demand bandwidth up to the Tbps level [4] for transmission. The third, lenslet light-field 3D, requires multiple parallel views to

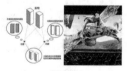

2D Image	3D Movie	Holographic Display
Flat Plane	Binocular Parallax for 3D	All 3D Cues for Objects
No Real 3D Effect	Physiological Function with	(Wave-front Reconstruction)
Can have "3D Illusion"	Eyes and Brain	

Holographic Display can satisfy all nature human observation for 3D objects.

Fig. 2.1 Comparison of human visual perception with 2D and 3D effects. (Image sources from left to right: Paris (2008 film); Avatar (2009 film); Holographic photo in ISDH2012)

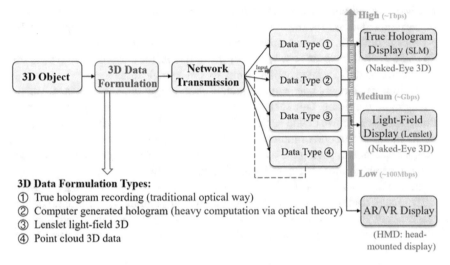

Fig. 2.2 Types of transmitting 3D data in HTC

observe 3D objects and thus typically requires a high bandwidth for data transmission, usually at the level of Gbps. As demonstrated in holoportation scenarios [3], the AR/VR-based display with HD resolution usually needs a bandwidth at tens of Mbps level, while extreme AR/VR with much higher resolution would demand bandwidth at Gbps level. Lastly, point cloud 3D data is currently the most commonly used technique for 3D imaging. Point clouds can serve as the input data of 3D models for a variety of 3D display methods including CGH. Besides the typical bandwidth requirements for these four different 3D data types used in HTC, as shown in Fig. 2.2, there exist other challenges as well, which are briefed as follows.

In the future, the HTC application may have wide scenarios especially in remote education, new interactive entertainment, or even the business online meeting may have absolutely new experience which is just like what we see in the science fiction film. **Key requirements include:**

- *High bandwidth*: Based on specific data formats that are used for different 3D holographic applications, for either naked-eye perception or HMD-assisted displaying, it needs bandwidth that vary from tens of Mbps for entry-level point cloud transmission to Gbps level for highly immersive AR/VR and light-field 3D, and it may further reach Tbps level for true hologram transmission at normal human size [5, 6].

- *Low latency*: The bottom line for visual motion effect needs the refreshing rate is larger than 24FPS. And for holographic displaying, it usually adopts 60FPS or above. Especially, under extremely immersive cases, it needs 120FPS for enjoyable 3D visual effect. Thus, converting to the network latency, it should be in the range of tens of millisecond (ms) to sub-ms. Moreover, in terms of some specific future use cases, such as holography-based tele-operations, it may further demand deterministic latency with bounded jittering. In the near future, HTC should be integrated with haptic sense data transmission [7], which further confirms the latency requirement at sub-ms level (also can refer to the next case).

- *Multi-stream synchronization*: In order to support multiparty holographic communications or multi-master and single-slave control [8], multiple transmission paths or data streams with diverse geo-locations are expected to be synchronized appropriately with limited arrival time difference, usually at the level of millisecond time interval.

- *Edge computation*: Edge computing is highly demanded near 3D data receiving endpoints, because hologram-based displaying usually needs high computation power to synthesize, render, or reconstruct 3D images before being visually shown (such as CGH).

- *Security and reliability*: For lots of future applications, such as hologram-based remote surgical control, it demands to make sure that no one is able to hack the transmission system during operation while keeping high security and reliability.

2.3   Tactile Internet for Remote Operations (TIRO)

Most of the current network applications employ file-based model which means the main service progress is triggered and promoted by file transferring. For example, when visiting website, we actually download multiple files from web servers. The similar scenarios also include music, games, or even some of online movie now separate the whole content into several pieces of files and transmit one by one to the user terminal. This kind of file-based applications has relative limited requirement of network performance which may only focus on the total complete time of the specific file in statistic. We do not care about the arriving time of each packet especially the jitter. The network plays the role of transport worker so that more people (wide bandwidth) means high performance. Fortunately, the brilliant "best-effort" design of current IP network, which is a statistic-optimal solution, perfectly matches the requirements.

However, along with the quick development and deployment of network technologies, more and more industries are beginning to use network technology in actual scenarios, and we do find some interesting cases that break the law of "bandwidth equals quality". The tactile Internet, which envisions the real-time control of remote infrastructure, will create a plethora of opportunities and open arenas for application domains, such as Industry 4.0 or tele-medical services. Immersive video streaming applications, such as aforementioned HTC 3D image streaming, will enable real-time and immersive interaction between a human operator and remote machinery.

Two typical use cases have been illustrated in Fig. 2.3. Specifically, the first case is remote industrial management that involves real-time monitoring and controlling of industrial infrastructure operations. Tactile sensors aid a remote human operator to control the machinery by means of their kinaesthetic feedbacks. A crucial component of such an operation between an operator and machine is the real-time visual

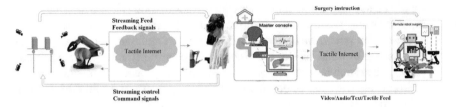

Fig. 2.3 Two typical use cases for tactile Internet

monitoring of the remote infrastructure, which is followed by haptic control. This will be provided by immersive audiovisual feeds, such as VR video streaming or other HTC-based streaming methods, together with haptic sensing data synchronization.

Another use case is remote robotic surgery. At the human-system interface (HSI), a master console is installed, where the surgeon gets a real-time audiovisual feed of the patient and operating room, as well as feeds from additional data inputs, such as diagnostics and haptic senses. The visual feed is again provided by a HTC-type streaming, depending on whether the surgeon is wearing a head-mounted device or interacting with a hologram. The surgeon then operates the haptic devices at the HSI and performs the actual surgical actions, based on the real-time visual feed and the haptic information transmitted to the robot. Meanwhile, haptic feedback is sent back to the surgeon from the patient side as well.

Generally, these two use cases require the network to have very low (near zero) latency for real-time interaction, along with guaranteed high bandwidth to support the video feed. They also necessitate strict synchronization between the various feeds to allow for a sense of interactive control. A brief description of network requirements is then given as follows.

In the future, the tactile network and related applications have wide scenarios especially in manufacturing, remote control (including surgery, car control, inter-cooperation), and so forth. This will break the physical limitation of distance and trigger more new interactions applications.

This kind of new type of services potentially brings new requirements for the future network. The **key requirements include:**

- *Ultra-low latency*: Latency is most crucial for the future high precision networks. The maximum delay that goes unnoticed by the human eyes is about 5 ms. And, for the operation to be smooth and immersive, the new paradigm even demands sub-millisecond end-to-end latency for tactile cases with instantly haptic feedback.
- *Ultra-low loss*: In such critical applications, loss of information means loss of reliability on the system. Hence, data loss should be as minimal as possible, while duplicated signalling could be enabled for higher reliability.
- *Ultra-high bandwidth*: The bandwidth is especially important in case of remote monitoring as increasing the complexity of the visual feed (from traditional 2D image to 360-degree video, and finally to holograms) makes the required band-

width grow drastically as well. For instances, a bandwidth up to 5 Gbps is required for VR feeds, and it may increase up to 1 Tbps for a large-size hologram.

- *Strict synchronization*: The human brain has different reaction times to different sensory inputs, such as tactile (1 ms), visual (10 ms), and audio (100 ms) [9]. Thus, in tactile cases, the real-time feedback from hybrid sensory inputs, which possibly emerge from different locations, must be synchronized strictly. Even in the presence of ultra-low latency, synchronization is important and needs to be on time scales that are significantly shorter than delay. Additionally, reaction time differences may allow some differentiation, regarding how to allocate networking resources. The same application may even involve multiple streams, and some of which further have different delay requirements as compared to others.
- *Differentiated prioritization levels*: The network should be capable of prioritizing streams based on their immediate relevance. Since the visual feed involves multiple views and angles for immersive media, the relevance of such different streams should be considered, and the ones with higher importance to the operator's view and current task should be given higher priority.
- *Reliable transmission*: Since reliability is the prime concern of the applications in these tactile cases, loss of packets is almost intolerable. In addition, retransmission schemes should also operate within tolerable delays.
- *Security*: During remote operations, the data transmission security should be guaranteed, without being compromised, especially for critical tactile cases associated with human lives and high-value machinery.

2.4 Space-Terrestrial Integrated Network and New Infrastructure Requirements

The aim is to leverage interconnected low Earth orbit (LEO) satellites to build a parallel Internet network that can peer with its terrestrial counterpart as the legacy infrastructure. With such integrated framework, the envisaged key benefits include (1) ubiquitous Internet access at global scale, including rural areas like oceans, deserts, as well as moving platforms such as ships and planes; (2) enriched Internet paths that may potentially lead to better data delivery performance compared to those over the terrestrial Internet determined by BGP configurations across domains; (3) ubiquitous edge caching and computing services provided by lightweight, onboard computing and storage resources on LEO satellites.

Compared to today's satellite network infrastructures, one essential aspect is that future mobile devices (e.g. smartphones, tablets, etc.) are able to directly communicate with the locally accessible LEO satellite over the head, but without necessarily relying on traditional ground station infrastructures that are constrained by geographical distributions. Figure 2.4 below provides a high-level illustration of the use case for space-terrestrial integrated network (STIN) [10].

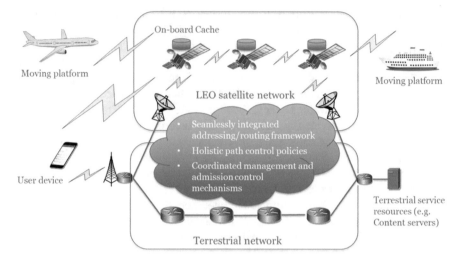

Fig. 2.4 The trend of satellite and terrestrial Internet integration

This kind of new type of infrastructure potentially brings new requirements for the future network. The **key requirements include:**

- *New addressing and routing mechanism*: Today's allocation of IP prefixes is typically done through major Regional Internet Registries (RIRs) according to specific geographical locations. Consider the IP addressing issue on potentially thousands of LEO satellites with their constellations, the interoperations with the terrestrial Internet infrastructure will incur new challenges, as the IP addresses in the space will dynamically interconnect to different domains (autonomous systems) on the ground with different IP prefixes. The new feature of allowing mobile devices to directly connect to local satellites also requires a cost-efficient addressing scheme for the mobile devices to communicate with local satellites without necessary address translation operations. The IP addressing strategy will also have direct implication to the routing mechanism both within the LEO satellite network and across the network boundaries between it and the terrestrial network infrastructure. The mobility characteristic of LEO satellite network is that the movement of the satellites is dynamic but predictable. The vast majority of network links connecting them are statically configured, while a small number of links can be established and torn down on the fly when two satellites on different orbits meet/depart from each other. Thus, an integrated routing mechanism is highly demanded, with the consideration of unique features in STIN.
- *Bandwidth capacity at the satellite side*: Compared to the high-capacity fibre optical links that constitute the traditional Internet backbone infrastructure as well as cutting edge access networks, the links connecting LEO satellites in the space and the terrestrial Internet infrastructure may become a significant bottleneck in terms of bandwidth capacity. In this scenario, the requirement is to increase the capacity in space, including peering links between satellites and also

between satellites and ground stations or user devices in order to match terrestrial capacity for future STIN-based applications.

- **Admission control by satellites:** In contrast to the traditional scenario where ground stations can be responsible for admission control on the traffic intended to be delivered through the space Internet, allowing mobile devices to directly access satellite networks will need to lift the admission control function to individual satellites which will directly interface these mobile devices. In this challenge, it is essential for each satellite (as the access point) to have necessary knowledge about the traffic load in the space network in order to make admission control decisions.

- **Edge computing and storage:** The realization of such a feature will incur challenges which in particular the hardware requirements on the LEO satellite side. For instance, the complexity of data/content procession at each satellite will be constrained by the power or battery capabilities. Lightweight edge computing tasks are still possible, which can be enabled under such constraints. Edge content caching is another application scenario that can be supported by the LEO satellite network for improving user experiences, thanks to reduced content access latency from the local cache in space. Similar to the edge computing scenario, content caching will be constrained by the data storage capacity that can be carried by each satellite.

2.5 AI-Enabled Applications

Currently, the most common use cases where AI is associated with IoT systems are those aimed at predicting future insights, detecting anomalies and taking control decisions starting from IoT streamed data. In the e-health domain, mobile personal assistants continuously monitor health data via bio-sensors and can predict critical situations like low blood sugar level and trigger alerts accordingly. Autonomous cars can feed image recognition algorithms with data provided by a multitude of on-board sensors to promptly detect obstacles and manoeuvre accordingly. In smart cities, a camera streams its data to face recognition algorithms for surveillance purposes. Predictive maintenance and condition monitoring can be performed starting from data collected from sensors embedded in a production line.

Typically, solutions for such reference applications leverage a *centralized paradigm* (commonly implemented in the remote cloud). AI algorithms (e.g. deep learning, DL) are memory- and power-hungry. Hence, most off-the-shelf IoT devices just send input raw data to the cloud which is then in charge of the model building/training as well as of the inference, whose results need to be sent back to requesting devices.

The edge will soon complement the cloud in enabling the deployment of intelligent services. Indeed, edge AI is mentioned among the top emerging technologies by Gartner in 2019 [11]. A recent IDC report [12] estimates that 45% of IoT-generated data will be stored, processed, and analysed close or at the edge of the network by 2025, with increasing market opportunity for AI-optimized processors. If deep learning (DL) services are deployed close to the requesting users, the latency and cost of sending data to the cloud for processing will be reduced, with benefits in terms of privacy preservation and offloading of the core network infrastructure.

However, a true revolution will be achieved when AI extends along the cloud-to-things continuum, embedded in IoT devices, implemented at the network edge and in the remote cloud. This would be possible, thanks to recent improvements in purpose-built AI-optimized processors and achieved advancements related to the possibility to embed AI inference in general-purpose processors. Figure 2.5 provides a high-level illustration of one potential reference architecture of this use case dealing with the **C**onnectivity and **Sharing** of pervasively distributed **AI** data, models, and knowledge (CSAI).

Such a futuristic scenario raises several daunting challenges for the design of networks of the future.

The pervasive distribution of AI capabilities to end devices, network nodes, and edge/cloud facilities is not like the placement of generic computing tasks and their subsequent connectivity and chaining. It goes well beyond since a proper understanding of AI peculiarities is required when designing networking procedures that enable a scenario where AI workloads and AI data are dynamically spread over a pervasive AI deployment, to properly match application requirements, especially in terms of accuracy and privacy.

A native AI awareness is essential in all network operations. For instance, part of the DL inference can be performed in IoT devices and heavier (training) tasks offloaded to edge and cloud facilities. As a result, according to the specific AI deployment, data of variable size (i.e. bulky raw data, intermediate data, AI models, updated model parameters, inference results) and in different formats coming from massively deployed intelligent things need to be efficiently exchanged in the network, meeting latency demands whenever real-time decisions need to be taken. Multiple AI components provided by IoT devices can be pooled together for more accurate inference results. For instance, layers of the same neural network can be split over multiple devices, according to their capabilities. The pooling procedures will benefit from decentralized networking approaches where devices have to collaborate and have capabilities to share AI-related resources and data according to

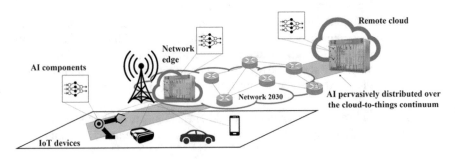

Fig. 2.5 Example reference architecture for CSAI

the needs of the environment and applications. For instance, a newly installed surveillance camera in an office building can ask a camera deployed in a different building of the same company to share the updated objects detection and tracking models, with no need to train the model from scratch.

The network has to offer the possibility of facilitating such pervasive AI deployment among intelligent things which may also need to interact autonomously. Hence, reachability of AI components needs to be ensured for the composition of an AI pipeline. More flexible network addressing schemes are required to properly name the AI components regardless of the specific position in the network where they are placed. Discovery procedures are also required to identify the most suitable things to contribute to AI-based services. Device-to-device connectivity would be required to set up autonomously and in a resilient manner, whenever privacy-sensitive data needs to be exchanged. Once properly named AI components are discovered, the network will be in charge of properly routing requests towards them. Conventional IP-based addressing and flow-based routing do not match the envisioned scenario.

For instance, groups of pervasively distributed AI components can either be simultaneously queried to perform some training tasks starting from their disjointed data subsets (e.g. in the case of federated learning) or be the simultaneous recipients of updated models. Existing network primitives cannot address such demands and novel ones (e.g. group-based push/pull) are entailed to this purpose.

Moreover, AI inference results, once computed, could be reused and serve different requests. To this aim, caching procedures could be highly relevant. Such procedures should be designed directly at the network layer to be faster and more flexibly implemented.

Nonetheless such a capillary AI deployment, more synergistically connecting the AI and IoT realms, places even more demanding requirements upon the design of future networks, entailing novel communication schemes, proper addressing solutions, and the support of strict KPIs.

This kind of AI-enabled application potentially brings new requirements for the future network. The **key requirements include:**

- *Mobility:* Intelligent things maybe either mobile (e.g., cars, smartphones carried by users) or static (e.g. smart meters, cameras). Thus, the network needs to flexibly support mobility on-demand.
- *Energy efficiency:* The decision about where to place AI components and how to interconnect them should be taken by accounting for the possible involvement of battery-constrained intelligent devices. Networking protocols are needed to ensure low energy consumption in the interactions among intelligent things to share either raw data or inferred knowledge.
- *Virtualization:* AI solutions would largely benefit from virtualization techniques able to deploy components in an agile manner. They can be packaged inside containers [13] (and even into more lightweight platforms) while reducing the deployment footprint in terms of processing and memory, to better match resource constraints of edge/IoT devices.

- *Joint network, intelligence, and computing orchestration*: The decision about how to distribute AI workloads should be performed through a synergic integration of computing, caching, and communication (3C) resources [14], to account for computing resource availability, network conditions, and popularity of requests for caching of models/inference. Moreover, it should go well beyond existing joint 3C solutions and specifically account for peculiar DL models requirements, e.g. privacy and accuracy. Such *AI awareness should be built by design* in orchestration mechanisms.
- *Bandwidth and capacity*: Massively deployed intelligent things may generate extremely large amounts of data to enable the adequate training of AI models [15]. According to the deployed AI pipeline, either large amounts of raw datasets/intermediate results to be trained or trained/updated models need to be exchanged among several entities (i.e. IoT devices, edge nodes, cloud facilities); hence, large bandwidth and capacity may be required.
- *Latency*: Data exchange among entities needs to be as fast as possible, in the order of <1 ms in the case of real-time decision-making (e.g. in an industrial plant, for an autonomous car, or for remote surgery), hence requiring extremely low-latency data transmission over both the radio interface and the core network segment.
- *AI-aware addressing*: In a pervasive AI deployment, every entity can contribute to the AI workflow. Flexible addressing capability is, thus, needed to optimally address AI components (e.g. DL models) associated with intelligent objects to facilitate discovery and composition procedures.
- *Uniform exposure*: Many purpose-built and fragmented AI solutions will be developed to serve a specific use case through proprietary APIs. To make them interoperable and flexibly chained, the network should provide uniform exposure interfaces to describe AI capabilities of intelligent things to third parties and ensure reusability.
- *Network protocol programmability*: AI components spread along the cloud-to-things continuum should be chained to ensure the exchange of data of variable sizes with low-latency and high-bandwidth demands in a flexible and dynamic manner. Moreover, it could be common that AI models (and updated ones in case of incremental deployment) need to be simultaneously spread to multiple devices (e.g. updated object/face detection models for cameras sharing the surveillance task in a smart city, or, language recognition app updates for smartphones of the same brand). Hence, proper network primitives, besides multicast and broadcast, may be required which recognize the entities to be reached and efficiently forward data to them accordingly.
- *Security and privacy*: Since most of the information used to build inference models are associated with personal devices and the way users exploit and carry them (e.g. smartphones, cars, wearables), adequate security and privacy frameworks should be conceived.

2.6 Industrial IoT and IoT Advanced Applications

The industrial networks are fundamentally different from the IT networks in terms of performance and reliability requirements. They go beyond connecting the back office to the plant floor, to integration from the device level, to enterprise business systems which results in the automatic operation and control of industrial processes without significant human intervention. These networks therefore need to deliver superior performance and mandates a real-time, secure, and reliable factory-wide connectivity, as well as for inter-factory connectivity at large scale in the future.

Factory automation and machine control applications typically demand low end-to-end latency ranging from sub-millisecond to 10 ms and small jitter (at 1 μs level) to meet the critical closed loop control requirements, as shown in Fig. 2.6. On top of that, many machine controls are multi-axis applications requiring time synchronization to manage the complex position relationships between axes. Moreover, the system reliability of the industrial network is now demanded to be 99.999999%, as any break or suspend in the production line will lead to the loss of millions of dollars. For the same reason, the security requirement of such systems stays in the high level as well.

Meanwhile under the fourth industrial revolution, referred to as Industry 4.0, OT and IT start to converge. The control functions that are traditionally carried by customized hardware platforms, such as Programmable Logic Controller (PLC), have been slowly virtualized and moved onto the edge/cloud in order to reduce the CAPEX and OPEX of the system, with increased system flexibility and capability of big data analysis. Besides the benefits it brings, the industrial cloudification puts even higher requirements on the underline networks, as the same latency, jitter, security, and reliability requirements should be implemented at a larger scale.

In addition, the application of IoT in another field may become more popular in the future. The number of objects that are reachable over the Internet is now close to ten billion, and this number is increasing rapidly [16]. These devices produce a vast amount of data and provide a remarkable number of services which need to be meshed and interconnected to extract the real value for the benefit of the society. This can be achieved through centralized approaches, where objects belonging to each platform are connected and managed by a centralized component that takes care of blending the data coming from different objects to extract the useful information. Different platforms can then be interconnected to avoid the formation of the often-criticized *silo effect* of the Intranets of Things. The control of interactions and information flows will be in the hands of the central components of each platform, which will decide what can get out of each realm and how it can be shared with the external world.

In contrast to this approach, the *Social Internet of Things* (SIoT) model intends to exploit the potential of social networking technologies to develop a decentralized approach to foster the interactions among objects that belong to communities of trillions of members. The use of social network technologies presents a different vision where objects are capable of creating and managing social-like relationships

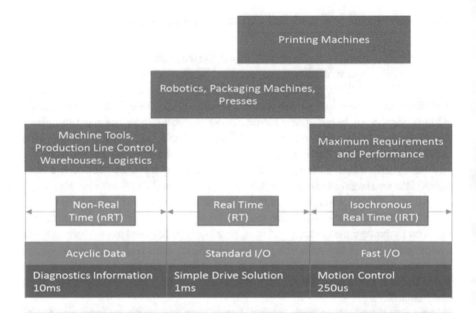

(a) IIoT latency requirements

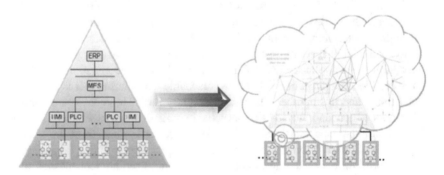

(b) The trend of industrial cloudification

Fig. 2.6 IIoT's requirements with cloudification. (**a**) IIoT latency requirements. (**b**) The trend of industrial cloudification

with each other in an (almost) autonomous way [17, 18]. In general, advantages of the SIoT are:

- By appropriately setting the rules applied to establish social relationships between objects, the resulting social graph has desirable structural characteristics, i.e. its diameter is small, and it is navigable.

- It enables new communication primitives, like *Sociocast*, which goes well beyond traditional unicast/multicast/broadcast and identifies the destinations of a given message based on their position in the social graph [19, 20].
- It simplifies the establishment of trustworthy relationships between objects so enabling differentiated level of security and, therefore, reducing its burden.
- It enables resource/service discovery across different IoT platforms.

These advantages have been demonstrated in real-world deployments for several application fields, such as transportation, energy management, and eHealth.

To implement this scenario, the network operator should take the pivotal role to support the creation and management of the social links among the objects by providing the appropriate services to augment the connected objects with the social capabilities. Accordingly, each object is supported by the network that provides the functionalities and APIs to implement a virtualized social counterpart (i.e. the virtual entity) for an object to opportunistically interact with the other virtual entities in the network.

A real scenario is one related to the delivery of parcels, where the explosion of e-commerce characterized by exponentially increasing volumes of orders has radically changed the way in which goods are delivered to customers, especially in the last mile. Such trends are expected to continue, with the recent COVID-19 pandemic giving a further boost which will not disappear at the end of the crisis. This opens an opportunity for traditional logistics operators to define new services and access untapped markets. It is however clear that to seize such opportunities, logistic operators need to address new challenges. Customers are becoming more demanding, in terms of pushing for new service models such as *same-day delivery*. Furthermore, several municipalities are closing cities to vehicle traffic, reducing the time window available for deliveries and pickups significantly. Finally, new players are entering the logistics industry and applying completely new business models, like logistics-as-a-service or on-demand logistics, thus radically changing the competitive landscape.

In such a context, it is mandatory for logistics operators to put solutions into place which minimize costs and maximize sustainability.

To achieve such objectives, it is fundamental to continuously monitor the state of all logistics assets, collect large amounts of data from the environment, and process such data for optimization purposes, by exploiting the possibilities offered by multimodality and cooperation between non-competing players. Also, there are several pilots aimed at demonstrating the effectiveness of using unmanned (both aerial and terrestrial) vehicles both for pickups and deliveries.

According to a recent study, Internet of Things (IoT) technologies will play a key role in such context [21].

A recent approach proposed in this domain is to exploit the cited Social Internet of Things (SIoT) paradigm (see http://www.cog-lo.eu).

In the context of logistics, specific advantages brought by the SIoT are:

- Logistic operations require the processing of large amounts of data generated by heterogeneous sources belonging to different organizations. The use of Sociocast helps defining the scope of each information item.
- Relationships established by social logistic objects can build links between IoT platforms belonging to different stakeholders so enabling more efficient transport and logistics services.

In Fig. 2.7, we sketch the assets of several logistic operators (identified with different colours: blue, green, black, yellow) operating in a certain area. Each of the assets is represented by a social virtual object which is a node of the SIoT graph as depicted in. Observe that assets that belong to the same logistic operators and/or are nearby are linked in the SIoT graph. Also, vehicles 1 and 13 have a relationship because they are expected to deliver parcels nearby.

In the following, we will provide a simple example in which the SIoT approach supports collaborative logistics.

Consider the case in which a vehicle fails while delivering. All parcels transported by this vehicle must be reloaded onto other vehicles for their delivery. This will require a rerouting of other vehicles. Note that collaboration with other logistic providers is likely to be needed in such a context.

Therefore, the SIoT paradigm can be exploited as follows.

Each parcel in the failed vehicle will notify the vehicles (those in the list of its "friends", and plan to pass nearby its destination and have sufficient space/capacity) that it needs a new pickup. In this scenario the 'friends' may include vehicles belonging to other logistics operators, fleets owned by non-logisitics business' or indeed domestically owned. The SIoT in fact creates relationships between objects (and thus parcels and vehicles) that are close to each other. Note that configuration policies can be defined by the owner of each logistic operator about the disclosure of information about its own fleet. In this way, the SIoT guarantees trustworthiness. The driver of the vehicle receiving the notification will decide whether to pick the parcel up or not.

Also, it might be that it is convenient to use UAVs (unmanned aerial vehicle) to transfer parcels from the failed vehicle to the new ones. In this case, if there are strong trust relationships between the major actors, it is possible to temporarily transfer the control of the UAV from one operator to the other. In this case, the advantages of exploiting a network of relationships based on trust can be extremely beneficial.

This kind of new IoT applications potentially brings new requirements for the future network. The **key requirements include:**

- *Low latency*: The IIoT systems contain many control sub-systems running at the cycle time ranging from sub-millisecond to 10 ms. In such systems, the communication typically consumes 20% of the budget. Thus, it is critical to require an extremely low latency.
- *Small and bounded jitter*: In order to recover the clock signal and reach precise time synchronization, the machine control, especially the motion control system,

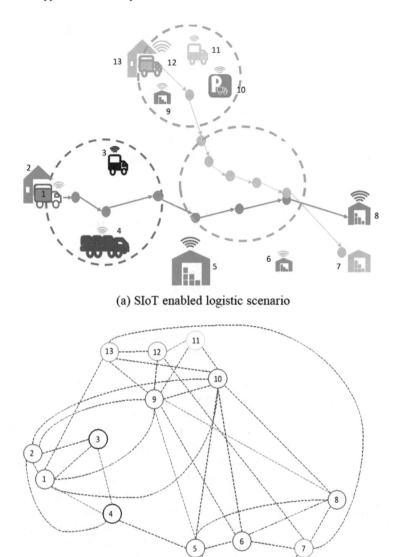

(a) SIoT enabled logistic scenario

(b) graph of the logistics scenario

Fig. 2.7 SIoT scenario. (**a**) SIoT-enabled logistic scenario. (**b**) Graph of the logistics scenario

requires very small jitter at sub-microsecond level, while such small jitter is expected to have bounded limits under some critical situations.

• *Time synchronization*: It is a fundamental requirement for multiple-axis system to have time synchronization in order to perform cooperation among various devices, sometimes remotely.

- *High reliability and security*: The IIoT system demands high reliability and high security to avoid any potential risk on interrupting the production procedure. Specifically, the reliability requirements typically range from 99.9999% to 99.999999% for IIoT applications.
- *Large-scale deterministic networking capability*: Due to the industrial cloudification, the aforementioned network requirements should be applied to large-scale deterministic networks in the near future.
- *Open network service interfaces*: This will enable new networking primitives such as Sociocast [16, 17]. To this purpose, SDN/NFV techniques might be exploited.
- *Support for friendships creation*: Social relationships among devices need to be established by monitoring device positions and the contacts among devices, e.g. through short-range connectivity or by analysing the data exchanged among them over the network. Proper APIs are required responsible for such operations and operating in a transparent manner for the end user, by allowing the user to keep control of the data.
- *Virtualization of the social objects*: The network should provide the APIs to instantiate social virtual objects associated with the objects connected to the network.
- *Security/privacy tools and infrastructure*: While the SIoT fully takes the responsibility of managing trust between smart objects, we observe that the network should provide tools to protect the SIoT from attacks. In fact, the SIoT elements contain data that can be exploited to achieve sensible information about the users.
- *Availability of computing and storage resources at the edge of the network*: In the SIoT, each object is represented by a virtual entity instantiated in a nearby server. Such servers must be "inside" the network to guarantee significant availability of bandwidth supporting the many interactions needed for SIoT relationship management. However, it should also be close to the physical object to reduce delay. It follows that effective SIoT deployment requires a dense infrastructure of computing and storage elements at the edge of the network.
- *Mobility*: Some intelligent things (e.g. smart meter, environmental sensors) are static, while others (e.g. cars, public transport means, smartphones) have high mobility or group mobility. Thus, the network needs to flexibly support mobility on-demand and track objects to update social relationships accordingly and to effectively perform location discovery of moving SIoT objects.
- *Energy efficiency*: Most "things" are battery operated, and therefore the network should put in place techniques aimed at minimizing the number of operations an object is required to execute.

2.7 Scientific Research and Big Data Applications

It is often anecdotally stated that revolutionary technological developments arise due to the demanding requirements of scientific research, the evolution of network technologies being a case in point. Looking back through history, the world's first network, ARPANET (Advanced Research Project Agency Network) was invented to support the requirements of military and scientific research. The World Wide Web, invented by Physicist Tim Berners-Lee in 1989 while working at Conseil Européen pour la Recherche Nucléaire (CERN), was initially conceived and developed to meet the demand for automated information sharing between scientists around the world. Large-scale scientific experiments produce vast amounts of data, the flows of which are increasing as rapidly as "standard" internet traffic. However, scientific data flows have different characteristics in terms of quantity and scale, orders of magnitude larger than common data flows, demanding bandwidths, and latency that the current network struggles to support.

For example, astronomical telescopes are individually configured to fulfil their specific purposes; they will be deployed globally and have varying ground communication requirements. But the data needs to be transferred synchronously and processed simultaneously, placing strain upon communication links. In another example, various particle accelerators and colliders generate massive amounts of data within very short time periods, for example, ITER [22]—"the way" fusion experiment—can generate data at 100 GB/s. The rapid collection and transmission of this present numerous challenges. It is easy to foresee that bandwidth requirements will reach Tbps in the future. Current networks cannot support such massive data transfer, so in some instances, the transfer of scientific data is still carried out in conventional ways such as physical transfer of hard disc drives.

Some examples of large-scale scientific applications include:

(a) **Astronomical Telescopes**

- **VLBI—Very Long Baseline Interferometry** [23]
- VLBI enables astronomists to observe the starry sky. A typical E-VLBI system consists of multiple distributed networked telescopes and a central correlator. Each telescope generates massive volumes of data continuously, and this data needs to be transferred to the correlator in real time. The VLBI produces 256 Mbps~16 Gbps per site and is only set to increase.
- **SKA—Square Kilometre Array** [24]
- The SKA is an array of radio telescopes made up of thousands of smaller dishes. This next generation of radio astronomy observation facility will collect and handle about 130−300 PB of data per year.
- **FAST—Five-Hundred-Meter Aperture Spherical Radio Telescope** [25]
- FAST is the largest single-dish radio telescope in the world. It is now in the early stages of exploration and has varying requirements for its operation in different modes. In the simple mode, the amount of data generated is about

6 GB/s. This telescope's annual observable time is about 2800 h, which means that the amount of data it generates will be as high as 60 PB per year. In the complex mode, data is produced at about 38 GB/s.

(b) Accelerators

- **LHC—The Large Hadron Collider** [26]
- LHC is the world's largest and most powerful particle accelerator. It consists of a 27-km ring of superconducting magnets, and its data collection rate is around 40 TB/s.

(c) **Others**

- **ITER—"The Way" in Latin** [22]
- ITER is the world's largest tokamak and has been designed to prove the feasibility of fusion as a large-scale and carbon-free source of energy based on the same principle that powers Sun and stars. The ITER front-end devices can source data at 100 GB/s.

The simple table below provides additional information about other huge scientific data applications in Table 2.1.

It is clear from the call-out box/table/figure above that today's networks are not capable of transferring the vast amounts of data produced by scientific applications at the required speed. Additionally, there are several other challenges that need to be considered when designing networks in the context of large-scale scientific applications, including:

- *Bandwidth*: We already foresee a demand for bandwidths reaching up to Tbps soon. In theory, this demand will grow exponentially as data-driven research grows.
- *Bandwidth*: We already foresee a demand for bandwidths reaching up to Tbps soon. In theory, this demand will grow exponentially as data-driven research grows.
- *Quality of service*: In a distributed workflow system, the loss of one node affects the whole system, and therefore each node requires an end-to-end guarantee. At the same time, different research applications require different scales of bandwidth for varying durations from a few minutes to a few days or even longer term. In the example above, FAST, 38 GB/s of data, is generated in a complex

Table 2.1 Requirements of applications

Huge scientific data applications	Network requirement	Store requirement	Computing requirement
High repetition frequency X-ray-free electron laser device	10 Gbps	100 PB	1–10 PF
Shanghai Light Source Phase II	1 Gbps	500 PB	20–40 PF
BESIII	100 Gbps	15 PB	10 PF
JUNO	100 Gbps	30 PB	10 PF
Major marine science and technology infrastructure	Quantum-encrypted communication	0.5 EB	100 PF

mode of operation. This is in comparison to operation in simple mode which is six times less. This variability means that the network needs to be able to dynamically allocate bandwidth and resources to be used effectively. The network may also need to pre-empt any background traffic in order to ensure adequate performance.

- *Synchronization:* Many large-scale scientific applications rely upon instruments collecting and transmitting data to a remote processing centre for real-time analysis during observation. For example, during e-VLBI observation, data is continuously collected by multiple radio telescopes distributed at different locations. The delay of one node's flow will result in the delay of the analysis result. For some observations, for example, when using e-VLBI observation for locating spacecraft, the analysis result is needed in real time. The telescopes have minimal local storage, so the data gathered has to be transferred continuously, in real time, to remote storage nodes or transferred to the processing centre in a synchronized manner.

- *Reliability*: The local storage size of scientific applications is often small if even present. This makes it challenging to retransmit lossy data, hence link reliability crucial. The transfer link of scientific data requires a high-quality guarantee, such as low packet loss rate, low latency, and low jitter. For example, the ITER nuclear fusion experiment runs 5−7 days a week and 8−16 h a day. During the experiment, the network failure time cannot exceed 1 min, requiring the network to have 99.999% availability [27]. As another example, the LHC data transmission lasts 9 months per year and can tolerate only a few hours of interruption, requiring a network availability of 99.95% [28].

- *Protocol considerations*: The traditional TCP/IP protocol suite has difficulties in supporting the timely transfer the high volume and velocity data described. Large-scale scientific applications have extremely high requirements on network quality and reliable transmission, and so in the future functions that guarantee quality of service need to be designed to overcome the shortcomings of the current best effort transmission and the domain name resolution limitations of the current internet protocol. Long-distance distributed scientific applications may be better served by changing the addressing mode of the IP protocol to content addressing. To address the challenge of congestion control in the transport layer, we could consider adding link and physical layer parameters, such as delay and buffer congestion controls, to speed up the response to changes in channel status.

In summary, scientific applications, especially for large-scale scientific projects, put forward significant challenges for networks, such as Tbps grade long-distance transfer rate, high reliability, determined delay, and intent-based provision. These are practical requirements for future networks and some of the motivations for the development of future network technologies.

2.8   Digital Twins

A digital twin (DT) is normally defined as a real-time representation of a physical entity in a digital world. Digital twins (DTs) add value on top of traditional analytical approaches by offering the ability to improve situational awareness and enable better responses for physical asset optimization and predictive maintenance. In the future, facilitated by vastly deployed DTs, the digital and physical worlds have the potential to be fully intertwined, contributing to the creation of a new norm, namely, a DT-enabled cyber-physical world.

Digital twins can be applied to various scenarios linking physical objects, including cars, building, factories, cities, environment, to process, and people. A digital twin of a city is a typical case. A city is a complex system, composed of people, things, processes, and multiple events. Creating a DT would enable all of the city's utilities, assets, and facilities (e.g. streets, communities, schools, hospitals, water supply systems, power systems) and even public events to be mapped to DT counterparts. This allows city operators to model guideline strategies and rehearse tactics to deal with multiple scenarios before they actually occur. Figure 2.8 below shows a proposed framework for smart city system based on the DT paradigm.

There are multiple technical and societal challenges associated with realizing the vision presented above, before we even consider the network requirements. However, when a DTC is realized, we foresee several key network requirements.

- *Highly diversified bandwidth on-demand*: Virtualized objects, especially dynamic ones, in a DTC generate extremely high volumes of data continuously. Meanwhile, sensory data exchanged between digitized objects or between physical and virtual objects are quite small. There the multitude of applications and data sources presented by a DTC is truly complex so a very flexible on-demand network will be critical to handling the multiple bandwidth requirements.

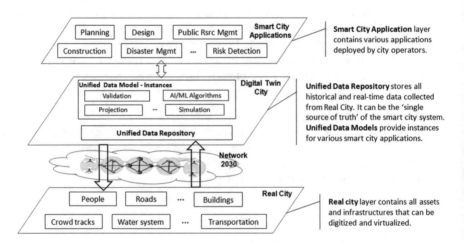

Fig. 2.8 Exemplary reference framework of a digital twin city (DTC)

- *Low latency*: In many instances, the data exchanged between the city and its DT needs to be as instant as possible in real time, sometimes to the microsecond level in order for timely responses to everything from routine resource management through to mission critical applications such as emergency response.
- *Mobility or group mobility*: In a DTC, some entities (e.g. buildings, water system, etc.) never move, while some other entities (e.g. citizens, cars, subways, etc.) are individually or even collectively highly mobile. Thus, the network must be flexible to support mobility on demand to ensure the DTC is fit for purpose.
- *Elasticity*: Different component DTs, as part of the DTC, will require disparate network resources and configurations to meet the requirements of various smart city applications. Moreover, some digital objects may request network resources for temporary tasks. This complexity will need to be handled by the network and as per the large scientific applications use case, a high degree of elasticity will be required to ensure resources can be dynamically scheduled.
- *Security and privacy*: The DTC use case involves unprecedented data and information exchange, much of which will be associated to citizens or public facilities. Data exchanges in the digital world therefore must be extremely secure, ensure privacy is fully protected, and be robust and resilient. The network of the future will of course also need to support any new security frameworks (e.g. intrinsic security, binding with digital objects) and novel privacy protection mechanisms, to achieve end-to-end security and privacy in an integrated cyber-physical world. You could envisage that the network specifically serving a DTC application might be a private network. This could be logically or physically separated from the public network but would then need to be focussed on limited applications losing some key benefits that arise from the interconnection of physical objects, people, and process.
- *Artificial intelligence (AI)*: In addition to elements described in the intelligent operations use case earlier, AI will play an important role in the DTC case and broadly across multiple DT use cases. In the DTC case, it will be required to efficiently process large-scale heterogeneous data from emerging DT platforms. As described, speed and synchronization of DTs are critical factors. The ability therefore of the network to also respond "at the speed of AI" is also required. For instance, AI could be deployed to detect network attacks and rapidly apply mitigation or recovery strategies, or AI could be deployed to increase network reliability through the introduction of an intelligent operations paradigm that enables full network automation in DTC.

2.9 Conclusion

As the reader may be aware, there are a plethora of applications emerging that will become deployable in the short, medium, and long term. Each will have their own in some cases unique, in others generic, requirements. The majority of the work embodied in this chapter was carried out as part of the ITU's Focus Group on

Network 2030 [29] where seven use cases were put forward from a diverse set of international experts. For ease of analysis and description, the members of that group agreed to cluster use cases.

The seven representative use cases described in this chapter are a representative overview. The network requirements for each use case were also clustered and evaluated in terms of five abstract network requirement dimensions, namely, bandwidth, precise-time relevant, security, artificial intelligence (AI), and ManyNets.

When clustering, the authors considered the views from the perspectives of both network operators and the end users.

Each dimension adopted in the figures below is derived by grouping several factors, as explained below through examples:

The **bandwidth dimension** is scored on a scale where a higher number denotes a stronger requirement. Factors that were considered included elements such as in 3D multimedia-related applications such as HTC truly have a high bandwidth requirement. However, in some cases, for example, in ION, it is more important to have more flexible bandwidth on demand. In addition, Industry 4.0-related applications may require asymmetric flows of high bandwidth in one direction (upstream to controller) and low bandwidth in the other (e.g. downstream command to the robots).

The **time dimension** expresses the notion of latency tolerance, which follows the principle of the lower the tolerance, the higher the score. This time dimension may refer to data transmission latency, sensitivity to jitter, and timing accuracy, or even synchronized arrival intervals. Since mobility is required from various use cases, finding the accuracy of geolocation in real time is also partially regarded as a characteristic of this dimension.

The **security dimension** envelopes privacy, trustworthiness, resiliency, lawful interception, and traceability, where larger numbers indicate more stringent demand. This dimension enables us to encompass both operator's and end user's interests. The default for security is assumed to be relatively high (i.e. 5 or above in our scoring system). In our figures, all use cases need a secure end-to-end communication infrastructure. Furthermore, it may require crossing regulatory boundaries or working with multi-domain networks, so a higher score for security in this instance means that interconnectivity between different networks preserves a user's identity and data integrity.

Artificial intelligence (AI) is currently making practical impact in all tiers of the applications. In these use cases, we primarily considered the application of AI and ML techniques that can be applied in networks, such as optimization, better predicted outcomes of network capacity planning, traffic patterns, detection of anomalies, and enhancing resilience by learning causes of past outages, and so forth. It is believed that AI will play an important role in future networks, and it uses higher scores to show more importance.

The **ManyNets** dimension represents the heterogeneity score of the networks, so that capabilities are normalized. A particular user should receive the same capability and experience through any type of number of networks it attaches to or transits through. A lower score suggests that more homogeneous networks predominate,

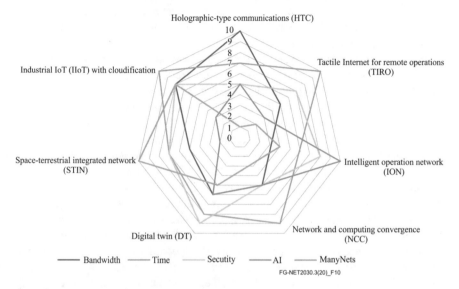

Fig. 2.9 Relative network requirement scores for seven representative use cases

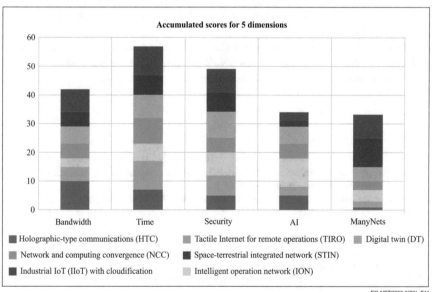

Fig. 2.10 Accumulated dimensional scores for seven representative use cases

like closed factory sites, short-range IP-based delivery, whereas higher scores imply heavier use of public infrastructure, in combination with more heterogeneous access or transit technologies.

The scores for these five dimensions range from 1 to 10,[1] and two conclusive graphs are illustrated as follows. Figure 2.9 shows the relative significance of each network requirement dimension for individual use cases, with the most prominent dimensions easily identifiable to the reader.

Figure 2.10 presents the relative importance among five network requirement dimensions, and it can be observed that the precise-time relevant dimension is the most important one and highlights that elements of this clustered dimension such as deterministic timing, bounded jittering, and time-sensitive synchronization should be the first consideration in future protocol and architecture design for networks of the future; the security dimension is second-most prominent, since new applications and services will trigger new types of security requirements, and generally, security and privacy issues should be a key priority in future networks; the bandwidth requirement is likely to be customized for specific vertical applications, for example, in AR/VR/HTC-style applications, bandwidth requirements will be significant, versus a medium-scale requirement in applications such as large-scale monitoring services via IoT sensors. The AI dimension is becoming more and more important. AI is not only applicable to improving networking operations through network status analysis but also can be widely deployed throughout whole networks to enable multiple future intelligent services. Lastly, the ManyNets dimension was originally motivated by a growing drive to integrate terrestrial networks with spatial and marine networks for completely global connectivity, but ManyNets can also extend to merge vertical industry networks, and more novel network paradigms that will emerge downstream.

The content of this chapter is introduced and expounded from the application examples. In the next chapter, readers will see that these classic cases are further examined at the network service level the network service level.

References

1. Staggering video marketing statistics for 2018, https://www.wordstream.com/blog/ws/2017/03/08/video-marketing-statistics
2. A survey on holography, http://drrajivdesaimd.com/2017/09/17/hologram/
3. Holoportation, Microsoft Research, https://www.microsoft.com/en-us/research/project/holoportation-3/
4. Network 2030: market drivers and prospects, Richard Li, FG NET-2030 Chairman, 1st Network 2030 Plenary Meeting in New York (2018), https://www.itu.int/en/ITU-T/Workshops-and-Seminars/201810/Pages/Programme.aspx
5. 3D holographic display and its data transmission requirement (2011), https://ieeexplore.ieee.org/abstract/document/6122872/
6. Holographic image transmission using blue LED visible light communications (2016), http://www.apsipa.org/proceedings_2016/HTML/paper2016/234.pdf

[1] All the scores are given for relative importance of particular network requirements, and normally 1–3 are for relatively LOW requirement; 4–6 are for MEDIUM requirement; 7–9 are for relatively HIGH requirement; and 10 means EXTREME demanding requirement.

7. HaptoClone, http://www.hapis.k.u-tokyo.ac.jp/?portfolio=haptoclone&lang=en
8. M. Shahbazi, S.F. Atashzar, R.V. Patel, A systematic review of multilateral teleoperation systems. IEEE Trans. Haptics **11**(3), 338–356 (2018)
9. The tactile internet ITU-T technology watch report (2014), https://www.itu.int/dms_pub/itu-t/opb/gen/T-GEN-TWATCH-2014-1-PDF-E.pdf
10. H. Yao, L. Wang, X. Wang, Z. Lu, Y. Liu, The Space-Terrestrial Integrated Network (STIN): an overview. IEEE Commun. Mag. **56**(9), 178–185 (2018)
11. https://www.gartner.com/en/documents/3956015/hype-cycle-for-emerging-technologies-2019
12. https://www.idc.com/getdoc.jsp?containerId=US45575419
13. R. Morabito, I. Farris, A. Iera, T. Taleb, Evaluating performance of containerized IoT services for clustered devices at the network edge. IEEE Internet Things J. **4**(4), 1019–1030 (2017)
14. M. Chen, Y. Hao, L. Hu, M.S. Hossain, A. Ghoneim, Edge-CoCaCo: toward joint optimization of computation, caching, and communication on edge cloud. IEEE Wirel. Commun. **25**(3), 21–27 (2018)
15. M. Mohammadi, A. Al-Fuqaha, S. Sorour, M. Guizani, Deep learning for IoT big data and streaming analytics: a survey. IEEE Commun. Surv. Tutor. **20**(4), 2923–2960 (2018)
16. K.L. Lueth, *State of the IoT 2018: Number of IoT Devices Now at 7B—Market Accelerating* (IoT Analytics, Hamburg, 2018)
17. L. Atzori, A. Iera, G. Morabito, The social internet of things (SIOT)—when social networks meet the internet of things: concept, architecture and network characterization. Comput. Netw. **56**(16), 3594–3608 (2012)
18. E. Papagiannakopoulou et al., The COG-LO framework: IoT-based COGnitive Logistic Operations for next generation logistics, in *IEEE WF-IoT*, 2019
19. L. Atzori, A. Iera, G. Morabito, Sociocast: a new network primitive for the IoT. IEEE Commun. Mag. **57**(6), 62–67 (2019)
20. L. Atzori, C. Campolo, A. Iera, G. Milotta, G. Morabito, S. Quattropani, Sociocast: design, implementation and experimentation of a new communication method for the internet of things, in *IEEE WF-IoT*, 2019
21. https://discover.dhl.com/content/dam/dhl/downloads/interim/full/dhl-trend-report-internet-of-things.pdf
22. ITER, https://www.iter.org/proj/inafewlines
23. VLBI, http://www.jive.nl/e-vlbi
24. SKA, https://australia.skatelescope.org/welcome/
25. FAST, https://fast.bao.ac.cn/
26. LHC, http://lhcone.web.cern.ch/
27. K. Yamanaka et al., Long distance fast data transfer experiments for the ITER remote experiment. Fusion Eng. Des. **112**, 1063–1067 (2016)
28. W.E. Johnston, ESnet4: advanced networking and services supporting the science mission of DOE's office of science (2007)
29. https://www.itu.int/en/ITU-T/focusgroups/net2030/Pages/default.aspx

Chapter 3
Future Network Services

Maria Torres Vega and Kiran Makhijani

3.1 Introduction

Network connectivity services are the built-in capabilities that network infrastructure owners provide as an interface to end users and applications to express their communication requirements. These services encapsulate many of the application delivery goals such as performance, operations, security, etc. to be met over the networks.

The previous chapter illustrated the emerging scenarios such as TIRO, HTC, and IIoT. It highlighted the need for the network to support stringent resource requirements such as ultra-low latency, ultra-low loss, ultra-high bandwidth, or strict synchronization. To enable this, not only will abundant bandwidth, time precision, and ubiquitous connectivity be necessary, but networks will also need to provide new capabilities that are not supported today. Clear examples are the ability to deliver on stringent latency guarantees or to provide precise coordination across many concurrent data streams and communication channels.

The reason for this is that current internet working infrastructure provides network services that are fundamentally built based on "best-effort". While differentiated services allow for prioritizing traffic and the reservation of resources, all of these mechanisms are associated with significant tradeoffs and limitations. Specifically, the fine granularity of accuracy with which services need to be delivered for Network 2030 applications is not possible.

M. T. Vega (✉)
Department of Information Technology, Ghent University, Ghent, Belgium
e-mail: Maria.TorresVega@ugent.be

K. Makhijani
Futurewei Technologies, Santa Clara, CA, USA
e-mail: kiranm@futurewei.com

For the development and deployment of Network 2030 applications, new services need to evolve beyond best effort and support the new concept of "high precision" in terms of quantifiable latency guarantees, of synchronization of a packet, of flows across multiple communication channels and communicating parties, and also in terms of behavior when faced with congestion and resource contention.

This chapter focuses on the definition of new network layer services, or "network connectivity services" (Chap. 4) in order to support emerging applications and vertical industries in the year 2030 and beyond. The term 'network services' will be used in this chapter, since that is the common short-hand terminology used in the industry, however, a broad range of services are covered in the book and connectivity differentiates this from other definitions.

It is emphasized that the services which are already supported today are expected to continue to (co-)exist as illustrated in Fig. 3.1. Thus, the high-precision services defined here will not necessarily replace today's network services, nor will the network needed to support these new services necessarily replace today's network. Instead, it should be anticipated that new services will be added and provided in addition to the existing ones, which will continue to be offered in many cases. Essentially, each network service of today or network 2030 serves a purpose and addresses application delivery requirements.

In this chapter, we start by briefly defining a set of fundamental concepts that apply to any network services, followed by state of the art. Furthermore, this chapters dwells on formalizing network 2030 services in greater detail.

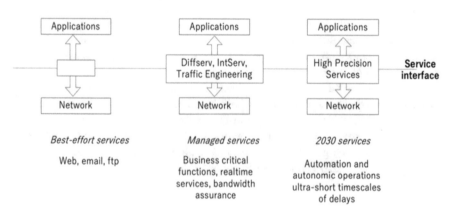

Fig. 3.1 Evolution of network services

3.2 Background

In order to provide a background to define the Network 2030 services, this section first defines current network service capabilities. Subsequently, it provides an overview of performance of services and current Quality of Service (QoS) technologies.

3.2.1 Network Service Capabilities

Network services are a collection of capabilities from the network that provide value to application delivery logic. Such capabilities offer appropriate network resources to enable application functionality anywhere over the network without modifying their core logic.

The scope of network services covers a wide variety of network functions, ranging from basic connectivity, QoS (preferential traffic treatment), path control, security, telemetry, resiliency, redundancy, and performance monitoring. The delivery of services requires equal part efforts at operations and management, control, and user planes. To illustrate this, consider a managed video streaming service. It is not as trivial as delivering packets between the content server and the end-user. Instead, it needs to be managed, access controlled, delivered with expected QoS, and monitored for performance. It requires expertise in several aspects of the network infrastructure. The management functions need to identify where the content is located, planning for how many users will use the service, and periodically monitor for performance and failures. The control functions are responsible for distributing the service characteristics over the networks to treat video service with a uniform quality. It also supports the realization of management tasks. The user plane (or data plane) needs to carry sufficient hints, metadata, or markers in the packets to classify the video service. The routers then process the video streaming packets based on control plane supplied policies or configurations.

Figure 3.1 presents an evolution diagram of network services. First, **best effort services** (left side in Fig. 3.1) do not provide any guarantees. This means that if the resources are available, and the destination is reachable, the packets get delivered. This type of service is suitable for applications with high to average tolerance to delays and bandwidth. The middle part of Fig. 3.1 shows managed services suitable for business-critical operations utilizing specific network capabilities. They are classified into the **core differentiated services** as in Fig. 3.2, supported by today's networks. An application can request bounded guarantees of bandwidth and latency, minimizing delay variation, and traffic engineering (for path management). This type of services is covered in Sect. 3.2.3.

These differentiated services will not be sufficient for holographic-type communications, tactile Internet, or machine-type communications. Therefore, additional core capabilities are introduced, shown in service evolution, Fig. 3.1.

Fig. 3.2 Core network
capabilities

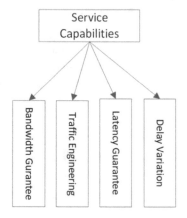

The text in this chapter does not differentiate management, control, or data plane functions since they require a reference implementation framework. The "capability" aspect of new service capabilities is generically discussed to develop a thorough understanding of new services. How the procedures are developed and distributed is an implementation detail, not the purpose of this chapter. Only later in Sect. 3.9, a short discussion of this will be covered.

3.2.2 Service Levels and Assurances

In this chapter, the terms service-level objective (SLO), service-level agreements (SLA), and key performance indicators (KPI) will be used when defining future services and their behavior. These terms are briefly described for the sake of completeness and within the context of this chapter.

The performance of network services is often described through "service levels." Other associated terminology with the service levels and used among different stakeholders are the SLA, the SLO, and the KPI. No service description is complete without describing its performance measurement characteristics. Simply put, the service level is the measurement of the performance of a service. Network service providers (NSPs) deliver services at different levels of quality based on the cost of that level of service.

The terms of service are called service-level agreements. **SLAs** [1] are defined as a formal contract between the provider and a customer at the business layer. SLAs are an explicit statement of what quality of service a customer can expect. The performance of the deployed service is measured relative to this SLA. Thus, they can also define penalties on violations. SLAs are composed of a set of **SLOs** [2], where these are concrete target values or range of values (upper and lower bounds) of a particular metric (e.g., bandwidth, one-way response time, jitter, etc.) that a service level aims to achieve. Finally, the KPI is a direct performance measure of a metric at a given time or a sample period. If the value of KPI is within the SLO bounds it measured, then the service is said to be performing well, and there are no violations to the SLA.

While "service levels" are formal descriptions in a business layer at BSS/OSS, they get realized in the networks as quality-of-service mechanisms, i.e., when a service is provisioned in the network those service levels are translated to QoS metrics such as SLO parameters or thresholds.

Different types of QoS available today for application use are discussed next.

3.2.3   Quality of Service

A large percentage of traffic in the networks is forwarded over packet-based networks in a best-effort manner. This type of traffic forwarding means a unit of data (packet) delivered independently as long as the network has resources available to process those packets. When the network is busy or an error in transmission occurs, packets are dropped. Thus, it is up to the application to discover such losses and implement recovery mechanisms suitable for their needs. While many applications tolerate such disruptions, others, especially those in the enterprise networks (including both local area and wide area networks), are far more sensitive to network disruptions. Such applications require network service providers to meet a certain level of performance in the networks. An application needing preferential treatments over the best-effort service uses differentiated quality of service (QoS). It describes its network resource constraints using QoS parameters such as bandwidth, delay, jitter, packet loss, etc.

In current networks, there are a number of ways to support QoS. As the traffic flows through the network, it needs to identify the service associated with that traffic and then reserve or provision network resources as necessary. In order to implement QoS guarantees, network nodes are required to implement different types of queuing, scheduling, and shaping techniques. The QoS support exists in data-link layer technologies such as Ethernet, Frame Relay, and ATM or in network layer which is primarily IP based.

QoS in Ethernet networks is marked using the 802.1p priority bits (incorporated as a part of 801.1Q standard [3] standard), and it can support up to eight priorities. In addition, Ethernet also supports more advanced QoS schemes described under time-sensitive networks (TSN). The TSN task force [4] has developed a suite of protocols to support low-latency provisioning, scheduling, and corresponding resource reservations in the Ethernet networks. Ethernet technologies are suitable for local-area networks (LAN), but for large-scale systems using higher-layer (TCP, HTTP, etc.) applications, maintaining L2-based QoS consistency requires complex and careful provisioning. Additionally, as networks cross boundaries, say from LAN to WAN, those internal QoS must be mapped to WAN QoS. Thus, it should be taken into consideration that QoS translations from one technology to the other are necessary.

Differentiated Services (DiffServ) is the earliest and most widely deployed IP QoS architecture. It requires the marking of packets as an indication of what kind of service they are mapped to and thus providing treatments accordingly. There are two limitations with the DiffServ model: (a) we are limited by the number of type of services supported in a network, since DiffServ uses differentiated services code point (DSCP) bits in the IP packet header; (b) a more critical issue is the way it deals with end-to-end service guarantees. The problem with DiffServ is that each network (or for that matter even a router) has a different interpretation of DiffServ bits, because each network nodes may deploy different algorithms for scheduling and queuing.

Integrated Services (IntServ) provide QoS support to real-time applications using Guaranteed Services as described in RFC2212 [5]. Guaranteed Services reserves resources in advance for a given flow, which are for exclusive use by packets of that flow. This ensures that the packets are delivered within the requested time to the receiver. The IntServ traffic is shaped at the ingress network edge and at each hop so that the flow does not consume more resources than have been reserved. Without shaping, collisions and resource contention between packets would occur, which would lead to the possibility of loss and unpredictable variations in latency. The other controlled load service (RFC2211) [6] maintains bandwidth guarantees by providing queuing delays minimization for the contracted volume of traffic. The easiest way to understand load controlled traffic is that an application continues to receive same round-trip times (RTT) as they would have under a network with no congestion. The Resource Reservation Protocol (RSVP) (RFC1633) [7] is used to signal these requirements across the nodes in the network.

3.3   Future Services Classification

In order to support emerging applications and vertical industries in the year 2030 and beyond, there is a need for defining new network-layer services, dubbed "network services." While the previous section has described currently used network services, this section will focus on the classification of new services based on the requirements of future applications and networks. It is important to point out that the new services defined will not necessarily replace today's network services nor will the network need to support these new services necessarily replace today's network. Instead, it is envisioned for the new services to be provided in addition to the existing ones, therefore, enhancing their capabilities.

In this work, we split the services into foundational and compound services, depending on where the service is placed (Fig. 3.3). A foundational network service requires dedicated support from some or all network system nodes to be able to deliver the service between two or more end users (application system nodes). One clear example of a foundational network service already installed in network infrastructure is the IP packet routing and forwarding. On the other hand, a compound (or composite) network service is composed by one or more foundational services.

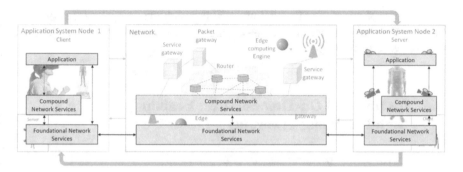

Fig. 3.3 Service classification overlap in a future network end-to-end system

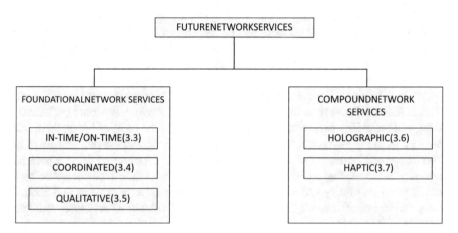

Fig. 3.4 Classification detail and chapter distribution

In this chapter, we consider compound network services as those requiring at least one next-generation foundational service together with a number of pre-existing foundational network services. Therefore, a compound network service in itself does not necessarily introduce new network service or requirements into the network system nodes.

In the remainder of this chapter, new foundational network services as well as a number of compound services needed to enable future applications are discussed. In particular, this chapter defines and describes the following new network layer services (Fig. 3.4). The most relevant foundational services are presented in Sects. 3.4, 3.5, and 3.6. First, Sect. 3.4 deals with in-time/on-time services, which deal with service-level objectives related to packet travel time and latency. Second, Sect. 3.5 presents the coordinated network services, which aim at providing coordination among multiple flows with interdependencies. Third, qualitative communications, described in Sect. 3.6 suppress retransmissions of less relevant portions of the payload in order to meet requirements on latency if the application is tolerant to quality degradation (such as interactive applications). After the definition of the

foundational services, Sections 3.7 and 3.8 deal with the compound network services related to holographic communication, haptic communication, and high-speed
data delivery, respectively.

3.4   In-Time and On-Time Services: Enabling
High-Precision Networking

3.4.1   Introduction and Rationale

The first important category of Network 2030 services concerns communication
services that adhere to stringent quantifiable latency objectives. First, haptic applications require end-to-end networking latency with an upper bound on the order of
5 ms. This low latency is needed in order to allow for round-trip control loops to
communicate haptic feedback in well under 10 ms, even as low as 1 ms in some
cases [8]. If such guarantees cannot be met, not only does the quality of experience
for users deteriorate, but the applications themselves may not become usable at all.
This is the case because to the end user, the illusion of remotely "touching" something and the ability to remotely operate machinery on the basis of haptic feedback
are lost. In another example, the industrial and robotic automation requires not only
"not-to-exceed" latency, but latency that is in effect "deterministic," with packets
not only not exceeding a certain latency. Furthermore, an industrial controller may
require very precise synchronization and spacing of telemetry streams and control
data, facilitating, for example, precise operation of robotic effect along multiple
degrees of freedom.

    Network 2030 services therefore need to support "high-precision" communica-
tions services, where "high precision" refers to a precise latency that packets may
incur, which is explicitly specified. We refer to those services also as "in-time" and
"on-time" services, with respect to the latency objectives that are imposed on the
packets that deliver those services. Contrary to existing technology, in which networks can be engineered and optimized for "low" latency, but the actual latency that
is obtained still needs to be measured, latency objectives in Network 2030 should be
provided as a specific parameter for the service.

3.4.2   Technical Definition

Latency is defined as the time that elapses from when the transmission of the first
bit of a packet is started until the last bit of the packet is received by a receiver across
the network [9]. Putting this definition in the context of an end-to-end system, the
sending/receiving point is considered a host or a border router of a domain (usually
an autonomous system). Thus, end-to-end latency is the aggregate of multiple

latency components, including latency that is incurred by physical propagation and processing of packets along individual hops for queuing, packet serialization, and packet processing. Applications such as haptics, autonomic mission-critical infra- structures, or interactive applications need end-to-end latency in order to allow for round-trip control loops (in the order of 1–10 ms). Furthermore, these applications need for the latency not only "not-to-exceed" certain value, but also it is of utmost importance that its effect is deterministic, with packets not only not exceeding a certain latency but also not delivered any sooner. For instance, an industrial control- ler may require very precise synchronization and spacing of telemetry streams and control data, facilitating precise operation of robots along multiple degrees of free- dom. These future network services therefore need to support high-precision latency communication services. Depending on the latency objectives imposed on the pack- ets delivered, we can classify the network service as "in-time" and "on-time" (Fig. 3.5).

In-time services aim to deliver packets within a required latency. This means that packets may be delivered at any time before or on the latency deadline. If a client application requests an in-time service, it will specify the required maximum latency as well as optional constraints in terms of the expected bandwidth, and the maxi- mum packet loss rate. Applications such as immersive video on demand supporting buffering capabilities would be the typical applications to ask for this type of services.

On the other hand, on-time services ensure the arrival of data within a specific time window. In the same manner as in-time services, they will impose a maximum latency. Moreover, they will indicate a minimum latency. Therefore, a packet must be delivered no later than upper bound of the time window, but also no earlier than the lower bound of the time window. In the extreme case scenario, time window size will nominally be set to zero, resulting in latency which is deterministic within the bounds of the clock uncertainty. A client application requesting an on-time service will specify the required latency (specified, e.g., using a target latency midpoint and time window, or lower and upper latency bounds, or even a target delivery time which is converted into latency by the network). An example of this type of services

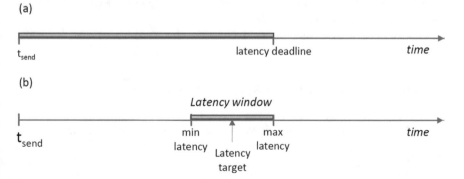

Fig. 3.5 High Precision Communications. **(a)** In-time services **(b)** on-time services

would be an interactive immersive conferencing, in which users at different locations need to be able to listen and talk with one another. Another example would be a remote surgery case, where the surgeon needs to perform certain actions at an exact time by means of the robotic arm on location.

3.4.3 Gap Analysis

As mentioned before, in-time services must provide the ability to support a quantifiable end-to-end latency for packet delivery across a network that must not be exceeded, given a set of constraints (which include a rate at which packets can be sent, and a loss rate that would be acceptable). Therefore, these are the requirements:

Services may be required to ensure that clients adhere to agreed constraints (e.g., a not-to-exceed packet rate). For this purpose, they may perform admission control or rate limiting as needed. Alternatively, services may simply monitor agreed-to-constraints to warn users in case violations occur. Any high-precision commitments given by the provider of a high-precision network service will no longer apply in case of violation of constraints—while the network may still deliver the demanded latency if it is possible to do so, it is not committed to do so.

"Miss rate" is defined as the ratio between the number of packets that do not meet the latency objective (including packets that are lost) and the total number of packets. In other words, a low-precision packet which misses its latency objective is considered the same as a "lost" packet. In an extreme case, it is possible that a miss rate could be specified as zero, in which case no misses would be acceptable. However, it should be noted that in reality, a miss rate (and loss rate) of zero will be impossible to achieve at all times and under all circumstances (e.g., in case of occurrence of a cosmic event), although it can be asymptotically approached and guarantees can be given that in the presence of, e.g., single device and single link failures, no misses will occur.

On-time service, on the other hand, must additionally support a quantifiable end-to-end latency that must be met within a given window. The window boundaries define a latency that is not to be exceeded (as in the case of an in-time service), as well as a minimum acceptable latency.

Accordingly, an in-time service request is characterized by the following parameters:

1. Required latency
2. Packet rate (possibly refined further, e.g., sustained vs. burst)
3. Miss rate

An on-time service request is characterized by the following parameters:

1. Required latency
2. Latency window size
3. Packet rate

4. Miss rate

Traditional networks support multiple mechanisms to reduce and optimize latency. However, these mechanisms do not support latency objectives that can be quantified in advance. The gap that needs to be addressed for Network 2030 services concerns the ability to deliver on latency objectives that are precisely quantified as part of the service request.

As introduced in the background section, the internet QoS architecture has two complementary high-level QoS architectures: *Integrated Services (IntServ) and Differentiated Services (Diffserv).*

Another option is the *Time-Sensitive Networking (TSN).* TSN is a set of updates to the IEEE Ethernet standard that aims to empower standard Ethernet with time synchronization and deterministic network communication capabilities.[1,2] TSN can be best understood as an Ethernet layer 2 variation of the IntServ service, where it supports a model for deterministic shaping that does not require perflow state on transit nodes. Additionally, it adds Frame Replication and Elimination for Reliability where packets are replicated $n + 1$ times on ingress, sent across failure disjoint paths and then the replicas are eliminated on egress. This so-called proactive path protection supports close-to-zero loss in the face of link or equipment (node, linecard) failure. TSN does not target to provide on-time guaranteed service over large-scale networks and long distances.

Finally, the *Deterministic Networking Architecture (DetNet)* [10] is an architecture that has been proposed by the IETF DetNet Working Group in order to ensure a bounded latency and low data loss rates within a single network domain. The DetNet architecture intends to provide perflow service guarantees in terms of (1) the maximum end-to-end latency (called bounded delay in DetNet) and bounded jitter, (2) packet loss ratio, and (3) an upper bound on out-of-order packet delivery. Some options considered in DetNet may in the future also be able to provide bounded delay variation between packets of a flows. Although DetNet provides efficient techniques to ensure deterministic latency, scalability remains a challenge. In particular, implementing the DetNet techniques requires the data plane to keep track of perflow state and to implement advanced traffic shaping and packet scheduling schemes at every hop, which is not scalable because core routers can receive millions of flows simultaneously. In the control plane, if resource reservation protocol (RSVP) is used, every hop needs to maintain perflow resource reservation state, which is also not scalable.

Even if current solutions have been brought forward, the Internet QoS Architecture, TNS, and DetNet are not sufficient for Network 2030 for a number of reasons, including the following:

1. The need for perflow admission control makes IntServ expensive to support and scale, even if performed out-of-band via SDN.

---

[1] https://1.ieee802.org/tsn/

[2] https://support.industry.siemens.com/cs/ww/en/view/109757263

2. The inability to dynamically adjust the bitrate under varying network utilization makes this model too inflexible even for current, let alone future networks. Originally built for non-IP voice/video applications that required fixed bandwidth and support for constant bit rates (CBR), Network 2030 applications require support for variable bit rates and elastic bandwidth. This is not adequately supported.
3. No mechanisms exist to support application-defined upper and lower bounds for the desired latency independent of the path round-trip time (RTT).
4. There are no mechanisms to slow down packets based on the desired minimum delay.
5. Queuing cannot prioritize packets based on their desired end-to-end latency.

3.4.4 Performance Design Targets

Network 2030 applications may impose required latency as low as 5 ms, for example, for tactile Internet applications. Granularity that is specified and measured in microseconds (for end-to-end latency) may need to be supported (e.g., for certain Industrial Internet applications). Likewise, accuracy on the order of 1 ms may need to be supported.

3.5 Coordinated Communication Services

The Internet is a spatiotemporal heterogeneous environment, yielding different content delivery behaviors in time and space due to variable endpoint capabilities and network conditions. Actually, no two paths (or even different flows on the same path) can be assumed to have identical properties in terms of delay, jitter, and bandwidth. It is imperative that the immersive holographic media will be the basis of next level of advancements in remote teleconferencing, social meetings, and even remote commercial performances. These applications involve data that is either produced or consumed by multiple end users which must be rendered coherently for each end user. The role of the networks is to preserve this coherency by coordinating delivery of interdependent data, so that each user has same view of the entire application.

3.5.1 Multiparty Coordination

In this section, two emerging scenarios are presented to explain the coordination or dependency between different participating streams that are part of the same application.

3.5.1.1 Virtual Orchestra

Imagine an instrument ensemble in which the holographic life size 3D projections of musicians, each in a different place in the world come together and perform live on the stage in front of you. Assume a conductor on the stage shaping the sound of the ensemble with his gestures. These gestures must be received at the same time by the remote musicians at different locations to play their instruments at a particular time with a specified tempo. Similarly, the music transmitted from those locations to the stage must be played together with the same beats and tempo.

Any delay or early arrival of a particular instrument can cause the ensemble to go out of tune and destroy the entire performance. Therefore, the network shall support the coordination of rules from conductor to all musicians and audio/visuals from musicians to the stage. In particular, in a large-scale ensemble when a large number of instruments are involved, to preserve the integrity of performance, it may be necessary to allow dropping sound and hologram streams of a musician that cannot arrive at the same time with others and to provide mechanisms for subsequent fast synchronization.

A virtual orchestra [11] seeks to provide a large, diverse group of participants the same concert experience as they would have if all of the orchestra members and participants were colocated. This type of application has not only stringent low-latency and high-throughput requirements but also need for coordination of the interrelated multiple streams involved. Since the network paths for each participating stream may be different, for a perfectly harmonious concert, the application will have to monitor each stream's network conditions continuously and synchronize the streams accordingly.

3.5.1.2 Multiparty Holographic Collaborations

A collaborative 3D environment [12], or online interactive immersive games, also requires a very close coordination among multiparties. For example, in a holographic collaborative environment, the exact placement of a virtual object must always be known to all the receivers. Any changes to the position of objects should be rendered in other locations at the same time for all parties to have a consistent view; otherwise, the receivers will operate on different views of the digital scenario as they are completely unaware of each other's behavior. Such different views of positions of objects will certainly happen because the end-to-end path latencies will vary for each sender-receiver pair. To enable a fully synchronized operation and cope with heterogeneity of delivery paths in the network, we require some mechanisms which provide all parties with related information at about the same time.

3.5.2   Understanding Coordinated Services

Current network technologies at best can support pseudo-coordination of media streams because human beings can tolerate intermittent disruptions caused by packet drops or delays. Virtual orchestra or similar remote immersive collaborations are more complex since they require coordination in the order of sub-milliseconds among all the participating streams. Furthermore, an application may want to express other interdependent factors such as ordering arrival of packets across multiple streams, interdependency between audio, video, and sensory streams, so they are processed together as a group. To be able to coordinate transmission of such applications with ease in networks, coordinated communication service as a generalized network communication service is described. It expresses interrelated dependency requirements between multiple streams from sender or receiver endpoints which are then supported by the network.

At the time of writing this chapter, multiparty media streaming is mainly supported at the application level using WebRTC platforms. WebRTC is a server-based architecture that uses either multipoint control unit (MCU) or selective forwarding unit (SFU) as described in RFC7667 [13] for the distribution of streams to and from multiple participants. MCUs cope with synchronization or dependency between the multiple streams by processing streams from one or more participants, decode them individually, and finally compose them in a single stream which is then sent to all the participants. Lately, SFUs have emerged as a preferred approach to simply relay received streams from different senders to all the participants. The difference in two approaches is in the way intelligence about the media stream is managed; MCUs maintain all media-specific intelligence and complexity, whereas with SFUs, end devices are required to manage encoding, rendering, etc. media functions themselves. Both the solutions are opaque to the network and do not respond fast enough to changing network conditions. Moreover, applications do not assure that dependency constraints, such as all the sourced streams, must arrive simultaneously at all locations regardless of the heterogeneity of network conditions.

This is explained in Fig. 3.6, three collaborators working on the same 3D scene are at different remote locations served by the networks with different slow, average, and high link capacities. A change in visual scene triggered by an average link capacity collaborator (User C) will be received relatively sooner by a high capacity user (User B) as compared to the other one with the low bandwidth (User A). As a result, each will have a different or an uncoordinated view of the scene.

Alternately, coordination can be specified selectively for the part of the co-flow. If two observers in Fig. 3.6 are looking at two different point cloud objects in a 3D scene, they will ask for different information. They need not be coordinated as changes in an independent view of the scene do not affect the experience of the other observer. Thus, not all packets need to carry dependency information.

Coordinated services are more advanced than the traditional multicast service. Since multicast service does not address the overall behavior of different streams in

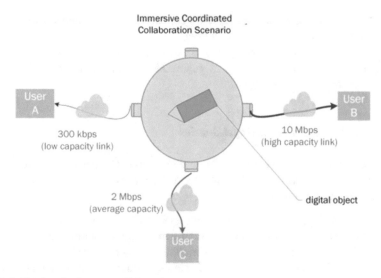

Fig. 3.6 Synchronized remote collaboration

a group which are co-dependent on each other in more than one way, synchronization is the most important dependency to provide a realistic interactive experience.

3.5.3 Characteristics for Coordination

Coordinated services provide a guarantee of delivery of multiple flows in a dependent manner (called co-dependent flows or co-flows for short). The co-flows may express different kinds of dependencies or relationships. Such a coordinated service shall be able to coordinate delivery of co-flows over different categories of group communications. The mechanisms to support coordinated service require new capabilities and coordinated network functions.

3.5.3.1 Application Supplied Metadata

Applications are expected to identify and describe their coordination requirements to the network. It should include knowledge of co-flows and dependency parameters, therefore, the following minimal information must be specified.

- *Indication of a coordinated service*, to inform the network that a particular packet or stream is requesting coordination with related traffic during transit.
- *Co-flow identification (co-flow id)*, a unique application-supplied identifier which determines which flows (source, destination, application tuples) are co-dependent. Depending upon the type of application, this may be same as multicast

group addresses (e.g., large-scale IPTV) or can also be an independent identifier embedded in the control fields of a flow.

• *Dependency parameters*, to determine which parts of the co-flows are related or dependent as the service needs to guarantee only the coordinated delivery of dependent/related information.

3.5.3.2 Different Types of Dependencies

The need for coordination across multiple flows arises when there exists some relationship between them. A few examples of such relationships in terms of coordination dependencies are as follows:

• Time-based dependency: This type of dependency guarantees that the co-flows will meet time-related guarantees. Such dependencies include the same arrival time of the same flow at different destinations, the exact arrival time from various sources at the same destination, and relative sooner or later time dependency. These are the guarantees of coordination and not always concern with meeting hard real-time dependencies like on-time guaranteed services.

• Ordering dependency: This type of dependency guarantees that user-specified relative order between co-flows is met in spatiotemporal terms. It includes conditions in which co-flows need not arrive at the same time but must follow a particular time gap or may consist of conditions such that specific bytes of a number of packets are from one member flow which must be received before another member flow. For example, if a particular task failed or slowed down in an automated assembly line, the subsequent operations must also halt or slow down accordingly.

• QoS fate sharing: This is a relationship constraint that specifies that member-flows of co-flows all need to be consistent with the type of QoS requested by the client. For example, if one member flow experiences quality degradation, then (and only then) it might be acceptable for other members of co-flows to be subjected to the same reduced service level. It may be possible for applications to describe other kinds of dependencies where networks can facilitate by providing coordination guarantees without having applications to deal with the management of such dependencies.

3.5.4 Co-flow Communication Groups

Coordinated services encapsulate all types of group communications. Different types of groups should be clearly specified by the applications since it assists with proper and reliable processing of received flows. For example, a receiver needs to know how many media streams to expect before rendering them together and what recovery procedures to trigger if some of the streams are too early or too late. The following scenarios are possible and are shown in Fig. 3.7.

- One-to-one communication: Co-flows, materialized from a single source to a single destination, may go through the same path or multiple paths (Fig. 3.7a, b). This is explained in Example 1; separate streams of multisensory holographic communications can be sent independently over the network and then combined at the destination.
- Many-to-one and one-to-many communication: Co-flows may originate from a single source to different destinations as shown in Example 2 in Fig. 3.7 from conductor to musicians, or from multiple sources to a single receiver in the same example from musicians to stage. Also shown in Example 3 is movement of a virtual object synchronized through one to many, and multiple sources modifying the same virtual scene at a particular destination as many-to-one communication. These two scenarios, referred to as many-to-one (from A, B, C to D in Fig. 3.7c) and one-to-many communications (from D to A, B, C in Fig. 3.7c), are typical incasting and multicasting, respectively.
- Many-to-many communication: All the flows need to be coordinated in multi-party communications such as immersive teleconferencing with AR/VR or holographic-type coordination (Example 3 above) or an advanced case of virtual orchestra when musicians may need to coordinate among themselves in Example 2. This is not simply a combination of incast and multicast, but a multisource-multidestination, fully cooperative environment (Fig. 3.7d).

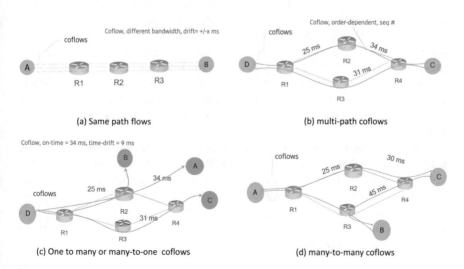

Fig. 3.7 Types of co-flows in coordinated services

3.5.5   In-Network Coordination Operation

Coordinated service delivery is achieved by managing time dependency over heterogeneous links/paths between different pairs of senders and receivers to avoid different arrival times. The latencies of co-flows are coordinated with the slowest flow (i.e., that with the longest arrival time). An application may need to create a high-level coordinated session that involves all the endpoints in order to use or instantiate coordinated service. The session then performs management of all connectivity operations with support from the network.

1. Determining paths and relative delays
2. Adapt to subsequent changes to path and latencies
3. Dynamic changes to co-flow memberships
4. Distribution of coordination dependencies along the network

These functions may be performed on the end host itself or delegated to the router gateway connected to end host.

3.5.6   Coordinated Service Network

Similar to building a multicast network or an SFU-based media streaming networks, coordinated service networks are type of an overlay over existing network infrastructure, in which an overlay coordinated network topology of nodes with coordinated service capability. Such nodes are called coordination points.

Figure 3.8 illustrates coordinated services support in the network through two scenarios of coordination for multi-sender and multi-receiver coordination, respectively. Co-flows originating from multiple senders U4 and U2 are to be received by destination U1 with the dependency constraint, for example, "together" without any hard limits of time. A member flow of co-flows from U1 may also need to be received by destinations U2 and U3 with user-defined constraints, e.g., at the same time.

The coordinated network will perform functions to monitor changes in path delays, group membership changes, and additionally have to be capable of providing high-precision services (Sect. 1.4) and sufficient storage. Specific mechanisms to distribute co-flow information are not covered in this chapter since several approaches are possible. In one approach where multicast PIM (RFC 7761) tree is viable, co-flow ids, dependency information is distributed along the PIM tree between the different hosts.

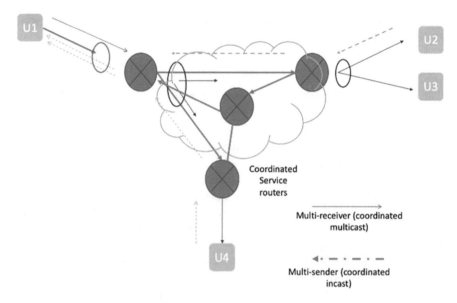

Fig. 3.8 Realizing coordinated communication networks

3.5.7   Summary

Having an explicit support for coordinated service to an application offloads coordination to the network. By focusing only on application's own logic, its performance is expected to improve.

To support coordinated service type of functionality, today's applications need to form and manage groups of endpoints and determine and monitor the characteristics of the paths between them for themselves. These applications also need to accurately measure the time of delivery from sender to receiver at every endpoint for all sources and destinations in co-flows. If coordinated services were to be implemented in hosts, then each host would need to keep track of run-time network state with respect to the dependency constraints. This would further require that the senders would need to manage complex scheduling when transmitting information to different receivers in order to manage transmission times to each receiver corresponding to end-to-end latency over each path, i.e., sending on slower links sooner than the faster ones. The receiver side also would need to provide complex buffer management to buffer received data at the receiver until it is ready to consume it. This would lead to the sub-optimal use of memory at endpoints while waiting for other member flow data to arrive. The fact that many dynamic changes occur inside the network compounds those challenges and creates numerous difficulties for applications to manage coordinated services at the endpoints. An ability to support coordinated network services within the network itself, managing the delivery of member flows according to their interdependencies and coordination requirements as a function of the network, therefore becomes critically important to support applications that depend on co-flows.

3.6 Qualitative Communication Services

So far, the network is blind to the semantics associated with the packets. The payload semantics may represent the (a) opaque boundary or grouping of data, (b) importance of different parts of the payload, (c) a relationship between different parts in the same or the other packets, (d) or may even be a functionally computed output from the transmitted data and the network conditions. If the network could perceive such semantics of the packet payload, it could then perform contextual actions and operations as per those semantics. The data sent will differ from the data received, yet the supplied data will be useful to the receiver. This novel concept is referred to as qualitative communications and the corresponding service offered by the network as *qualitative services*.

Such different type of "quality of data in transit" can utilize different types of functionalities, such as qualitative-based (mark priorities to the different parts with in one packet [14]), semantic-based (associate with prior knowledge of objects), entropy-based (redundant data detection [15]), and random linear network coding-based (for reliability) [16] communications. All these approaches will associate a differentiating property within the packet payload, potentially divided into multiple parts with a degree of importance, entropy, semantic values, etc., for each part or a group of parts. This approach is highly desirable for holographic type communications [12] with very low tolerance of losses and delays but could render media just fine with lesser data received. Qualitative communications become even more significant topic of research in 6G networks [17].

Data communication network treats a packet as the minimal, independent, and self-sufficient unit that gets classified, forwarded, or dropped entirely by a network node, according to the local configuration and congestion condition. The network protocols always ensure that the data sent matches the data received exactly bit-by-bit at the receiving end. This is referred to as data integrity. A packet gets dropped or entirely lost if the transmission media is faulty or the congestion has occurred, which must be retransmitted from the sender. Despite all kinds of congestion control mechanisms (based on the end-to-end principle), the congestion events remain unpredictable. The current network mechanisms to achieve data integrity include reliability, error detection, and corrections.

Instead of dealing with bit-by-bit loss or corruption, *Qualitative Communications* perceive the semantics of the packet payload. Then the unit of action taken by the network does not need to be on the entire packet but based on the semantics. This may help eliminate retransmissions in the networks and yet being able to supply data to receiver with a tolerable quality. The semantics are a certain value understood by the applications, a relative importance of the payload, or even relationship to the data received in past. Thus allowing the payload to be serviced not as a single raw stream of bits but as differentiating relevance or semantics to different chunks within the payload.

Qualitative communication service works at sub-payload level as shown in Fig. 3.9b. Intuitively, qualitative services attach different attributes within the payload and provide means to drop, repair, and recover user data in the network without compromising data integrity. The qualitative metadata (in Fig. 3.9b) carries them as sub-payload semantic attributes such as priorities p0, p1, and p2. This is clearly different from traditional packet Fig. 3.9a which provides no interpretation of packetized data.

3.6.1 From Quantitative to Qualitative Communications

Whenever network conditions are favorable, receiving complete data as sender intended is the most preferred mode of communication. However, in certain cases, when the data is transiting through a network under stress, receiving some information in timely manner is more important than receiving all of it late.

For example, volumetric media applications cannot afford retransmission delays since it can cause degraded user experience, and additional load on an already constrained network. The problem is amplified with larger packet sizes since a more significant amount of data is lost. Similarly, the emergency infrastructure services such as first responders' network are often constrained by their network capacity, when there's need to deliver information very fast, critical parts of conversations, or visual feeds should neither be delayed nor lost.

Packet losses are expensive and happen in a network for three reasons:

- Congestion discard
- Equipment (including link) error or failure
- Bit errors on the links

Of these three causes, congestion discard is by far the most common, but as applications demand extreme reliability, the other two causes become significant.

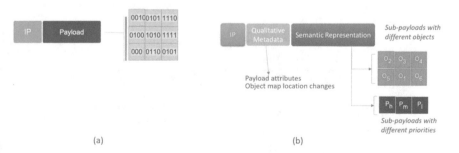

(a)   (b)

Fig. 3.9 Packetization in traditional and qualitative services. (**a**) Traditional raw packetization, (**b**) quantitative packetization

As we move to very high bandwidths, there is a tendency to move to larger packets to provide line-rate data transmission. Whether a packet is small or large, its header needs to be processed; hence, larger packets offer certain efficiency gains. However, as a result, each discarded packet results in a larger quantum of data that needs to be retransmitted than in the past.

Alternatively, in some circumstances, packets can be fragmented into parts to avoid maximum transmission unit (MTU) issues. This process considers all fragments of equal value, and all of the fragments are forwarded to the destination. If one or more of the fragments fails to reach the fragment reassembly point, then the whole of the packet is discarded. In either case, effort made by the hosts and routers is wasted as data needs to be retransmitted that otherwise wouldn't have to. This violates the guiding principle of "work conservation," which states that systems should perform any work only once if at all possible, ensuring that any outcomes are preserved once they have been achieved to avoid having to redo the same processing steps multiple times.

In all the current approaches to congestion avoidance through packet discard, an assumption is made that all portions of the packet are of equal relevance. However, in practice, some payload portions may be more important to applications than others. The qualitative networking approach exploits this fact by allowing senders to group payload within a packet by relative priority, then allowing the network to selectively discard portions of lesser priority when needed.

Specifically, qualitative services allow applications to differentiate between different portions of packet payload (Fig. 3.10), referred to as "chunks," and describe their relative priority to the network. Packets carry the necessary metadata needed to describe those chunks. If needed, a lower priority chunk can be dropped from the packet payload, while the higher priority chunk can be preserved to continue to their destination. This way, congestion can be reduced and continuity of delivery of critical data to the application, while minimizing the need for retransmission, can be ensured.

Qualitative communications thereby addresses both the latency and work conservation issues associated with the approach taken by the established reliable transport protocols.

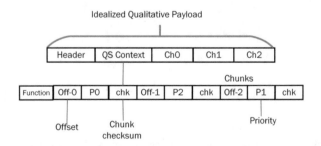

Fig. 3.10 Idealized qualitative packet

3.6.2   Understanding Qualitative Services

The qualitative communication service allows the network to deliver more important information in the packets to the destinations, by preventing whole packets from being completely discarded in the face of congestion.

To enable and implement qualitative communication service, support from both the application and the network is required.

3.6.2.1   New Packetization

Qualitative service requires a new packetization method in which the payload is constructed as a series of chunks (see Fig. 3.10), and the information needed to extract, prioritize, and process the chucks is carried in the packet header. A qualitative packet may carry metadata such as priority levels that allow the network nodes to know which chunks to drop, and the threshold beyond which a packet must be discarded rather than be further degraded. It may also carry a function instead that when applied to the chunk, we result in new data which is understood by the receiver.

The packet may carry enough error detection and correction information such that the useful chunks may be extracted from a packet that is partially corrupted and would otherwise be discarded due to a CRC error. The error rates in optical networks are such that this would rarely be required, but the higher error rates other types of transmission media might cause such a capability to be of use.

Figure 3.10 shows such a packetization method in which three chunks are arranged as high, mid, and low level. Both the offsets and checksum of each chunk are part of the metadata which is referred to as QS header.

3.6.2.2   Application Level Support

The end-to-end qualitative control method on source and destination is explained as below.

- Source application function: The characterization of what information is qualitatively more significant (i.e., qualitative context) is decided and assigned by the source application. It is necessary that the application understands the encoding of the user data in the payload, so that the qualitative context of the chunks can be indicated in the packet. A qualitative context (QS context in Fig. 3.10) includes a function selected by the application which can be used to identify the relationship, degree of significance of chunks, and/or to help the recovery of lost chunks. This context allows the network to operate on a qualitative packet without needing to look inside the chunk payload. It should be possible for the source application to further rearrange the positions of the chunks in the payload according to the qualitative context, e.g., significance of the chunks, which helps the network nodes to run the qualitative communication service by, for example, always dropping the chunks from the tail of the packet payload when it is necessary to reduce the size of the packet.

- Destination application function: Upon receiving a qualitatively treated packet, the destination application needs to decide whether to accept/acknowledge the packet qualitatively, an indication to the sender node of the qualitative outcome, and if/how to recover the original packet through the available metadata and payload chunks. The receiver may send the feedback about its satisfaction level concerning the received packet, whether more information is needed, and which piece of information needs to be fetched from the source application or from the caching locations.

3.6.3 In-Network Qualitative Service Support

As part of qualitative operation upon identifying congestion, forwarding nodes perform selective trimming of a payload from less to higher priority chunks. Until the network conditions improve, the receivers receive poor quality streams.

One possible high-level network treatment is explained through Fig. 3.11. It shows a network of three nodes, managed through AQM rules with different priority queues (best effort, mid priority, and high priority).

A packet arrives on the left node, sees no congestion, and forwards the packet to the middle node. On the middle forwarding node, congestion is experienced (such as the egress queue is 80% full), and the corresponding rule to drop low priority applies to each chunk in the payload. As a result, the lower priority chunks are dropped.

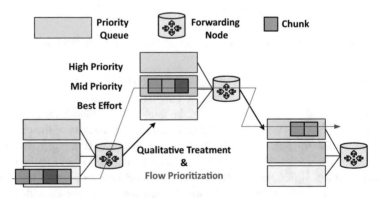

Fig. 3.11 In-network transport prioritization

3.6.3.1 Adaptive Rate Control

In order to ease the network load and reduce congestion, the receiver can check the QS context trigger an adaptive congestion control by notifying the sender about the level of congestion in the network. It can send an acknowledgment with the quality of packet value, which the sender uses to alter its transmission rate. This adaptive rate control utilizes network resources more effectively. Moreover, it significantly reduces data delivery delay by partially delivering the packets and dynamically managing packet sizes, which is critically important in emerging real-time applications.

3.6.3.2 Flow Fairness

This use of quality of packet improves network efficiency and fairness among users. In particular, if a packet that has been qualitatively treated already is in contention for buffer space with a packet that has not experienced chunk drops yet (all other priorities being equal), the forwarding element should treat the intact packet. Moreover, since a packet with a reduced payload is more congestion friendly, the forwarding layer should give it a higher priority.

Figure 3.11 shows how the priority of qualitative packet changes while forwarding through the network nodes.

3.6.4 Summary

The existing transport solutions only operate on full packets and use retransmission mechanisms to maintain the completeness of a data stream discarded due to data loss due to congestion discard. Qualitative services minimize such discards leading to a higher effective throughput.

The congestion discard is quite problematic to emerging applications. Transport protocols such as TCP and QUIC work on the assumption that the occasional congestion discard is worth the price in exchange for operating the network as a whole at its maximum capacity. Some of the demanding applications that Network 2030 considers find even this loss unacceptable. A fundamental characteristic of the qualitative communication service is to avoid retransmissions by delivering partial yet useful packet fragments to the end user.

A cardinal rule of networking is to never look at the contents in the packet and only perform forwarding functions based on packet headers. Qualitative communications must operate with this level of necessary payload opacity. Qualitative communication service comes at the cost of adding metadata information in the packet to identify the chunks and to assist the network nodes in deciding how to edit the packet to reduce its size when this is needed. Thus, on the one hand, the qualitative communications approach reduces the payload size to deal with congestion.

However, on the other hand, it is necessary to add additional information in the header to enable this network service. This results in a trade-off between bandwidth utilization in the normal case vs. the error/congestion case. Although a complete analysis of this is an area for research, this overhead is likely to be minimal in future networks where the packet payload size is expected to significantly increase.

3.7 Holographic-Type Services

3.7.1 Introduction and Rationale

The use of holograms as a means for users to interact with computing systems has long captured people's imagination, as evidenced in movies such as "Star Wars" or "Minority Report." As holographic display technology has made significant advances, holographic applications are well on their way to becoming a reality. Many such applications will involve network aspects, specifically the ability to transmit and stream holographic data from remote locations across the network to render it on a local holographic display. Examples of such applications abound. For example, holographic telepresence will project remote participants as a hologram to local meeting participants in a room. Remote troubleshooting and repair applications will allow technicians to interact with holographic renderings of artifacts located in a remote location. Training and education can provide users the ability to dynamically interact from remote with ultrarealistic holographic objects for teaching purposes. Audiovisual feeds for robotic tele-surgery, as mentioned in Sect. 3.4.1, can involve holograms as well. Then there is immersive entertainment, gaming, sports, and much more. It is easy to foresee that the vast majority of those applications will involve holographic-type communications (HTC), i.e., the ability to transmit and stream holographic data across networks. Rather than representing simply yet another media type, there are several unique aspects about holographic data that pose significant challenges to networks. The following background is intended to help appreciate some of these challenges. In a hologram, the same image is captured from different viewpoints, tilts, and angles. Depending on the position of the viewer relative to the image, a different "field" in an array of images is seen, with each image depicting the same "object" or "scene" from a slightly different viewpoint. For smooth holographic representations, differentiated images for roughly every 0.3° difference in angle are needed, implying that a hologram able to accommodate 20° differences in viewing angle and 10° of tilt, a two-dimensional array of 1800 separate images, are needed. The raw amount of bandwidth required is enormous; however, clever compression/decompression schemes across the image array allow the encoding and rendering systems to exploit the fact that individual images in the array include only-minute differences. Another option of representing holographic data is through the use of point clouds consisting of volumetric data. In this case, objects are represented as "point clouds," i.e., sets of three-dimensional "volume pixels," or voxels, in a conceptual three-dimensional box. Instead of streaming arrays of images, volumetric media data is streamed. The actual image can then be dynamically rendered from any viewing angle at the local endpoint, placing the point cloud object into a scene or even rendering multiple point cloud objects simultaneously. In order to reduce the amount of holographic data to be streamed, applications are expected to take advantage of techniques such as user interactivity prediction schemes.

The goal is to minimize the volume of data that needs to be transmitted while maintaining acceptable quality. This occurs by focusing on the data that will likely have the highest effect on quality first, for example, transmitting image data of fields that are in focus at the highest quality, while transmitting other images at lesser quality (e.g., reducing resolution, frame rate) or not at all (e.g., dropping certain tilts and angles). Since the user may change viewpoint or position, supporting such schemes requires highly adaptive and ultralow latency control schemes to be able to adapt streamed holographic contents as needed.

3.7.2 Technical Definition

HTC services will provide a set of network services used to transmit streams with holographic data, i.e., data that can be used to render holographic images. There are different flavors of holographic data streams that have to be supported:

1. **Point cloud based**, i.e., the sender sends volumetric data objects from which holograms are rendered at the receiver side. In many cases, a volumetric data object can be decomposed into multiple, smaller volumetric data objects, e.g., "3D tiles." Depending on viewpoint and position of the end user on the receiver side, some data objects may be obstructed at any one point in time. HTC services need to support rapid "switching" between different data objects as they come in and out of view in order to preserve bandwidth while maintaining high image quality, respectively, allowing the pool of available bandwidth to be preferably applied to those data objects that will be in view and in focus.
2. **Image array based**, i.e., the sender sends an array of images instead of a point cloud. Analogous to point clouds, depending on viewpoint and position of the end user on the receiver side, different fields in the image array may be able to be prioritized. HTC services need to support rapid "switching" between different fields in the array, prioritizing the image quality of some feeds over that of others as they come in and out of user view.
3. **Multiple camera feeds**, i.e., a set of senders send a series of two-dimensional images, possibly coupled with depth information. In that case, HTC data is sent "raw," not preprocessed, and feeds get combined at the receiver side to result in one holographic image/point cloud.

Depending on the type of content, the communication service will be tailored according to the following guidelines:

1. It can involve multiple channels of holographic data (e.g., one per component point cloud or 3D tile in case of volumetric data, one per field in an image array, one per camera feed). Each of these channels may map to a separate flow with stringent in-time requirements to ensure an internally consistent/synchronized holographic rendering. Some channels may have differing resilience requirements—a drop of some data, while not desirable, may result in a slight degradation in quality of experience for users but still yield an acceptable result. In some cases, a drop of data in one channel may lead to the data in that channel to be deprioritized completely—it may be preferable to deprioritize one channel versus other channels (or drop it completely) instead of having uniform slight degradation across channels. (However, multidimensional compression across different fields in the image array can occur. In such a case, resilience requirements may be dramatically increased, and different prioritization schemes may apply.) Aggregate resources for the totality of holographic data may be shared (resulting, e.g., in a requirement for "aggregate bandwidth") and may need to be continuously reallocated among the channels (as optimization schemes continuously adapt which contents to stream based on user interactivity and parts of holographic images coming into and going out of user focus).
2. It may involve an additional channel of "manifest data" that indicates how to compose the holographic image from the multiple feeds. This data needs to be especially protected, as any corruption of data may render other holographic data useless.
3. It will involve a "back channel" to control transmission and prioritization between 3D tiles or image array fields, as end user viewpoints shift and different parts of the holographic data come into and out of view.

An HTC Network Service will allow clients to specify some of their parameters in order to allocate resources. Some of these are the following:

1. The number of channels for the holographic
2. The aggregate bandwidth that can be allocated among the channels
3. The acceptable end-to-end latency, specified as an in-time requirement that must be met by all holographic channels as well as any manifest channel.
4. The latency that is needed for the back channel (which determines how much in advance user interactivity and changes in user viewpoint need to be predicted and adjustments of individual channel feeds needs to occur).

The HTC Network Service can be composed from a set of coordinated services, consisting of:

1. A set of channels to carry holographic data from holographic source to destination/rendering endpoint. Each of those channels will share the same in-time requirement. In addition, the aggregate bandwidth of each channel must not exceed the overall bandwidth allocated for the coordinated service.
2. A channel to carry manifest data from source to rendering endpoint. The latency of this channel must not exceed the latency of any of the holographic data channels.

3. A control channel in the opposite direction (from rendering endpoint to holographic source) to adjust manifest and streamed data as needed. This channel can be provided through a separate instance of an in-time service and does not need to be included as part of the coordinated service.

3.7.3 Gap Analysis

The main gap that exists concerns the ability of networks to support foundational services with sufficiently low latency (in-time services with quantifiable latency) and sufficiently high bandwidth. In addition, existing technology does not facilitate the notion of aggregate bandwidth shared across and dynamically reallocated among a set of flows. Similar to tactile networking services, an HTC Networking Service is an example of a composite service that would be reasonably straightforward to provide once foundational services for Network 2030 become a reality, but that cannot be provided by networks today due to lack of such services. Another challenge lies in the ability to deliver holographic-type data with very low latency. A lack of low latency can partially be traded off against an increase of bandwidth: higher latency implies that the time horizon of user interactivity prediction needs to be longer, respectively, that there is enough additional data provided to "tide the user" over, while adaptations among the data channels occur (so that different fields in the array or different 3D tiles in the point clouds can be transmitted in higher quality).

3.7.4 Performance Design Target

1. **Low latency:** Latency requirement on the order of 10 ms for allowing instant viewer position adaptation at 60 frames/s. However, latency requirement can be relaxed, e.g., for lower frame rates and at the expense of higher bandwidth, can become as low as conventional interactive video (on the order of 100 ms, gated by latency requirement for interacting with the remote party, not by latency requirement regarding viewpoint prediction).
2. **Ultrahigh bandwidth:** Required bandwidth may start from roughly 1 Gbps and increase up to 1 Tbps [10] (Fig. 3.12) but depends heavily on encoding and trade-offs regarding bandwidth and compute for optimization Schemes. A feed from a current commodity RGB-D sensor like Intel Real Sense or Microsoft Kinect generates roughly 2 Gbps of raw data (for 512 * 424 pixels with 2 Bytes of depth data) but can be compressed further.
3. **Strict synchronization:** At 60 fps, latency variation across channels should not exceed 7 ms (duration for half a frame).
4. **Support for concurrent flows:** Depending on point cloud and image array dimensions, the order of 1000 concurrent flows may need to be supported.

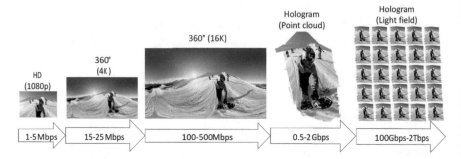

Fig. 3.12 Evolution of content with their bandwidth requirements

5. **Ultralow miss rates:** Specifically in the presence of strong compression techniques.
6. **Dynamic prioritization of streams:** The network should be capable of prioritizing streams based on dynamic and varying criteria (related to viewing position and user focus).

3.8 Haptic Communication

3.8.1 Introduction and Rationale

ITU defines the tactile Internet as the network that combines ultralow latency with extremely high availability, reliability, and security.[3] The tactile Internet envisions real-time monitoring, management, and control of remotely located infrastructure and devices involving haptics. In some sense, the term "tactile Internet" may be a slight misnomer, as tactile is only one of two types of haptic feedback, referring to things that one can feel when touching a surface, such as pressure, texture, vibration, and temperature. The other type of haptic feedback is kinesthetic, referring to forces (e.g., gravity, pull) that act on muscles, joints, and tendons in an "actuator" such as an arm, contributing to (among other things) a sense of positioning awareness. Both types of haptic feedback are important for tactile networking applications. We refer to communications involving one or both types of haptic feedback accordingly as "haptic communications." Haptic communications are expected to form the backbone of the Industry 4.0[4] along with other application domains such as tele-health, online immersive gaming, remote collaboration, etc. The tactile Internet envisions the creation of a paradigm shift from content delivery to skill set/labor-delivery networks. While traditional networks support audiovisual

---

[3] https://www.itu.int/en/ITU-T/techwatch/Pages/tactile-internet.aspx

[4] https://www.forbes.com/sites/bernardmarr/2018/09/02/what-is-industry-4-0-heres-a-super-easy-explanation-foranyone/495df4d29788

communications, the tactile Internet will enable haptic communication, i.e., providing a medium to transport the sense of touch (tactile) and actuation (kinesthetic) in real time. Haptic communications accentuate true immersive steering and control in remote environments along with novel immersive audio/video feeds. The three previous revolutions in manufacturing were all triggered by technical innovations— mechanization powered by water and steam in the first revolution to mass production and assembling using electricity in the second to adoption of programmable logic controllers for automation in the third. The next revolution will be triggered by networks that facilitate communication between humans and machines in Cyber Physical Systems (CPS) over substantially large networks. Industry 4.0 envisions communication between connected systems, thereby making decisions without human intervention. In order to bring that vision of a "smart" factory into reality, collaboration among the CPS, the Internet of Things (IoT), and the Internet of Systems (IoS) is necessary. The tactile Internet forms the core of such collaboration (Fig. 3.13).

The stringent ultralow latency required by haptic communications coupled with novel immersive audiovisual feeds, opens avenues for a plethora of application domains. One example use case involves remote industrial management. Remote industrial management involves real-time monitoring and control of the industrial infrastructure operation. This will allow a human operator to monitor a remote machine aided by immersive audiovisual feeds, such as virtual reality (VR) video streaming or holographic-type communication (HTC), and to control the machinery by means of their kinesthetic feedback involving haptic devices, as depicted in the figure below.

Common to each of these use cases is the need for communication channels that are characterized by extremely low latency [19]. There is a strict time budget for the round-trip time from when an actuator is operated by a human until the tactile feedback is provided. This is on the order of 5 ms or even less. Anything longer and the ability to confidently operate the machinery from remote breaks down rapidly.

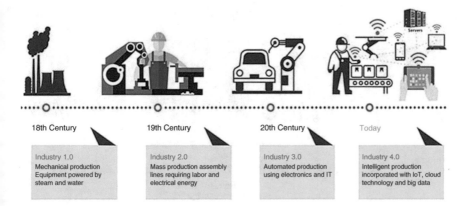

18th Century	19th Century	20th Century	Today
Industry 1.0 Mechanical production Equipment powered by steam and water	Industry 2.0 Mass production assembly lines requiring labor and electrical energy	Industry 3.0 Automated production using electronics and IT	Industry 4.0 Intelligent production incorporated with IoT, cloud technology and big data

Fig. 3.13 Industrial evolution [18]

While time budgets are slightly longer for audiovisual feedback, the same is true there. Furthermore, because applications are mission critical and retransmission of packets is not an option due to latency concerns, packet loss is not tolerable. As multiple data feeds are involved for data that needs to be rendered and acted on in unison, there is also a need for precise synchronization [20] (Fig. 3.14).

3.8.2   Technical Description

A haptic networking application in general involves two channels that provide a tactile control loop (Fig. 3.15):

1. A haptic feedback channel, used to communicate haptic data from one or more remote haptic sensors (e.g., sensors in a robotic arm) to a haptic effector (e.g., a "data glove" rendering tactile sensations to a user). Haptic data includes tactile data, such as surface texture and pressure points, and kinesthetic data, such as force feedback and location/positioning awareness.
2. A control channel, used to operate a remote actuator (e.g., a robotic arm).
3. (Optional) Live visual feed(s) from the remote location (e.g., high-resolution video, immersive video/VR, holograms)
4. (Optional) Live audio feed(s) from the remote location,
5. (Optional) Live telemetry feed(s) from the remote location.

Each channel by itself can be mapped to individual communication flows, e.g., instances of an in-time/on-time service. To improve synchronization among those channels, instances of a coordinated service can be used. However, instead of requiring applications to manage and orchestrate those channels themselves, a network service composition function can be offered, providing a haptic networking service.

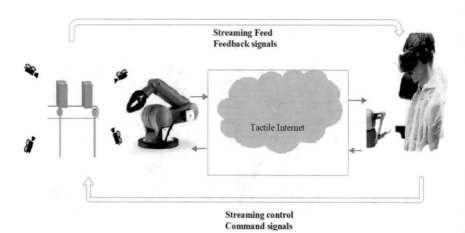

Fig. 3.14 Remote management and control of an assembly line

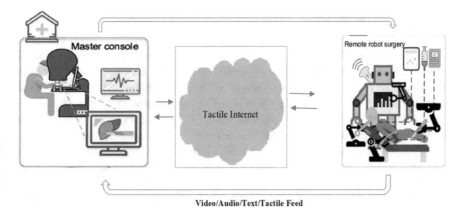

Fig. 3.15 Remote surgery example

A haptic networking service provides a set of coordinated services, consisting of the following:

1. A haptic feedback channel.
2. A control channel.
3. (Optional) Channels for additional feeds that are bundled with haptic feedback and need to be synchronized with it.

The haptic networking service needs to ensure that the requirements of a corresponding haptic networking application are met. Therefore, it should allow clients to specify different network and service parameters. For instance, it should specify the required round-trip haptic control latency. Moreover, the haptic codecs used and bandwidth and miss rate requirements would help the service to adapt the network provision. Finally, information about additional channels/feeds to bundle with the haptic control loop needs to be provided in order to synchronize them with the haptic feedback.

Based on the description, this composite service could be mapped onto the following foundational network services:

1. Automatic selection of proper latency parameters for in-time service instances for haptic control and haptic feedback channels. Round-trip latency requirements for the tactile application as a whole are broken down into individual one-way latency requirements for control and feedback channels. Typically, one-way latency requirements for both channels will be the same and together add up to the round-trip latency requirement, but other mappings are conceivable.
2. Automatic selection of parameters such as packet rate and acceptable miss ratio for those channels tuned to the needs of the specific haptic codecs and encodings.
3. Additional instances of in-time/on-time services for additional channel feeds. Note that some of those instances can themselves be instances of composite services, as in the case of a holographic feedback.

3.8.3   Gap Analysis

The main gap that exists today concerns the ability of networks to support foundational services with sufficiently low latency to realize a tactile feedback channel, coupled with extremely low miss rates to account for the high reliability of tactile applications. In other words, the gap consists in foundational services that meet the performance design targets as specified below. A haptic networking service is an example of a composite service that would be reasonably straightforward to provide once foundational services for Network 2030 become a reality, but which cannot be provided by networks today due to lack of such services. Over short distances, haptic applications can be supported today. The challenge, and gap, lies in enabling haptic communications across larger networks and across longer distances in geography. Performance design targets for ultralow latency of very few milliseconds, as specified below, quickly run into physical limitations due to signal propagation that cannot exceed the speed of light (300 km one way in 1 ms, or 200 km in an optical fiber). For those reasons, in reality, the distances across which haptic communications services can be offered will still be bounded, and tactile networking applications may need to leverage additional techniques (such as bringing intelligence and compute close to the network edge) to mitigate those limitations.

3.8.4   Performance Design Targets

The following design targets will be required in order to enable haptic communication in future networks.

1. **Ultralow latency:** Latency is most crucial for the future high-precision networks. The maximum latency that goes unnoticed by the human eyes is 5 ms [8]. For the operation to be smooth and immersive, the new paradigm even proposes sub-millisecond end-to-end latency for tactile feedback, which should be sufficient to cover the most general use cases envisioned by the Tactile Internet.
2. **Ultralow packet loss:** In such critical applications, loss of information means loss of reliability on the system. In addition, retransmission is generally not an option due to latency concerns. Hence loss should be as close to zero as is practical.
3. **Ultrahigh bandwidth:** The bandwidth requirement is especially important in case of remote monitoring as increasing the complexity of the visual feed (from 360° video to holograms) makes the required bandwidth grow drastically as well. A bandwidth up to 5 Gbps is required for VR feeds, and it increases up to 1 Tbps for holograms [9].
4. **Strict synchronization:** The human brain has different reaction times to different sensory inputs (tactile (1 ms), visual (10 ms), or audio (100 ms)). By themselves, some streams (e.g., audio) might thus allow for slightly higher latency than others (e.g., tactile). Nonetheless, synchronization is important, even in the

presence of ultralow latency, as synchronization needs to be on time scales still significantly shorter than latency. This means that tolerable latency for, e.g., video might actually be lower in scenarios when the visual information needs to be synchronized with tactile feedback than in other scenarios.

5. **Prioritization of streams:** The network should be capable of prioritizing streams based on their immediate relevance. Since the visual feed involves multiple views and angles for immersive media, the relevance of such different streams should be considered, and the ones with higher importance to the operator's view and current task should be given higher priority.

3.9 Future Services Design Considerations

Network 2030 services are essential for bringing emerging applications commercially to the mainstream. This section discusses overall architectural, technical, design, and deployment challenges and a logical way forward to adopting these services. A small discussion focuses on a unified approach and cooperation between the network operators and application developers to successfully deploy these services.

The Internet has become a fundamental resource, and it is at the center of all future innovations. Therefore, introducing any new services has to seamlessly work with the existing infrastructure without disrupting current operations and capability. Therefore, the design of future network services needs to evolve incrementally along with the current services. A thorough understanding of all aspects of networking is required and should be discussed from the following perspectives. Even though marker drivers are mentioned here, bringing these services requires an assessment of value to the end users, cost, and complexity associated with these services' deployment.

3.9.1 Technological Challenges

High-precision services are at the core of future applications. In spite of the unavoidable limitations of time, distance, and capacity, latency-sensitive applications are quite diverse. The "Tactile Internet" applications need 1 ms of latency, and the distance will be limited to 200 km. Industrial automation and control require different ranges of latency for different types of control and feedback control loops. Performance targets have been identified as the following:

1. <1 ms, also called "Isochronous Real-Time Applications" or tactile Internet
2. 1–10 ms, e.g., general real-time industrial control and automation
3. 10–20 ms, e.g., high-voltage power grids
4. <30 ms, for example, intelligent transportation.

3.9.1.1 Algorithmic Advancements

HPC describe different types of time guarantees and include applications that are not limited to low latency (called in-time guarantee) but include precise latency (called on-time guarantee) which concerns itself with exact time when to deliver data to the destination end point. For example, 50 ms latency on-time guarantee, 100 ms latency on-time guarantee, etc., such precise guarantees will be used to support "coordinated guarantees." Support for such diverse latency is only feasible through enhancements in packet queuing and scheduling mechanisms with the ability to process application-specified latency constraints. In addition, these capabilities need to be addressed in a hardware-friendly manner without prohibitive performance penalties.

3.9.1.2 In-Network Support for Transport Protocols

Future networks also include volumetric media such as holograms. Achieving consistent high throughput for holographic-type communications is extremely challenging. The networks today do not support predictable throughput (keep probability of packet loss constant), instead focus on fairness (L4S [21]) of throughput between different flows. The design of transport protocols needs a more coordination with in-network media processing for constant throughput. The volumetric media devices are trending toward light-weight, low-power wearable. Many of them will usually have small storage, which makes it difficult to retransmit. Therefore, traditional congestion and flow control algorithms do not apply, hence requiring forwarding nodes to participate actively in flow and congestion control mechanisms.

It is anticipated that QUIC is the next new transport protocol. It is designed to bring a modular approach to transport-related features, with focus on security and fast session setup. Future services should continue to leverage modular functionality, customizing transport on the basis of application requirements. In-network support, to provide continuous instantaneous network state, will allow future transport protocols to respond quickly to those changing conditions.

3.9.2 Architectural and Infrastructure Enhancements

The growth of the Internet has been ad hoc. Thus, today's network remains best effort. Most of the problems are solved by adding more capacity; however, this will not be sufficient. Therefore, network operators will need to ensure that applications meet their communication constraints.

In order to support future services, the fundamental architecture of the network needs not be changed as shown in Fig. 3.16; the transport and network layers remain with enhancement for the future services support.

3.9.2.1 Application to Network Interfaces

Many applications will demand a closer interaction with the network. Thus one key enhancement to architecture is the need for an application to network interface. These network service interfaces will be used by senders to specify service-level objectives (SLOs), such as required end-to-end latency targets, for in-time and on-time services. This could involve the ability to indicate an SLO parameter when a socket is opened or created or the ability to select between different socket types that are associated with different SLOs.

Furthermore, facilities to negotiate SLO should be provided between application and network. For example, an indication whether a requested target will indeed be supported by the network, if not then applications could "negotiate down" to identify a target that can be supported.

With these interfaces, rapid customization of networking services should be possible. Rapid customization may be seen analogous to serverless lambda programming paradigm; these interfaces need to be secure and that does not enable novel attack vectors, for example, that cannot be used to compromise network provider infrastructure or traffic by other users.

3.9.2.2 Protocol Innovations

Protocols provide a common and standard set of rules to communicate between one or more parties (end nodes or forwarding nodes). Protocol enhancements should continue to leverage the current programmability trends. Per packet service customization is important for high-precision, coordinated, and qualitative services. Protocols supporting these services will also need to be designed with considerations to trust, privacy, and security.

Traditional Internet protocols evolved to provide reachability at the global scale. The past decade has seen requirements shifting toward customizing transit through networks using source routing, service chaining, and policy-based forwarding and routing protocols. Innovations in Network 2030 protocols will further this trend by encompassing service-centric functions in forwarding, control, and management planes.

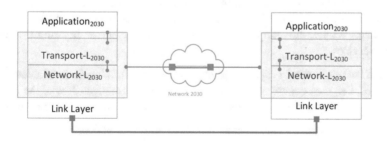

Fig. 3.16 Future service stack

3.9.3   Mindset for Service Assurance

Most Network 2030 services are associated with guarantees of requested service level objectives. Guarantees demand their price, making it increasingly important to be able to validate that promised service levels were delivered on. This will require advances in accounting technology. First of all, measurements of parameters corresponding to service level objective will need to be accurate to ensure that the performance targets are met. Another aspect of service assurance is providing proof-of-service delivery to the consumers particularly for mission critical applications. The best effort charging and accounting may no longer be sufficient for 2030 networks services; network service providers will rely on incentive-based schemes for high-precision or holographic-type service offerings. With such schemes, network providers will be able to allocate their resources more effectively than today in ways that best support economic goals.

Today's accounting technology largely relies on interface statistics and flow records. Those statistics and records involve sampling and are thus subject to sampling inaccuracies. In addition, this data largely accounts for volume but not so much for actual service objectives (e.g., latencies, let alone coordination across flows) that are delivered. For Network 2030 services, fine-grained service performance measurements will rely heavily on active measurement techniques, but there is a significant overhead, including the consumption of network bandwidth as well as additional processing on edge nodes. Techniques that rely on passive measurements are unfeasible in many network deployments and hampered by encryption, as well as issues relating to privacy, the concerns for which are expected to increase further.

3.9.4   Encapsulating Network 2030 Services

It is obvious that high-precision services can be associated with a higher cost of service delivery and strict guarantees of resources. The service levels are more stringent, and the cost of violations is a failure. Imagine the cost of failure to deliver a tactile Internet service—it can cause entire factory floor to come to a halt or even worse may lead to machines crashing into each other if one machine waited for too long in an assembly line. Thus, we need to consider leaner vertical integration approaches without breaking the layering principle, i.e., cross-layering is a key to future services.

- Each layer has certain capabilities to provide resiliency, bandwidth latency guarantees, i.e., the concept of service needs to be uniform regardless of the layer.
- Cross-layering will enable using the service requests from one layer to the other.

    Service levels need to get directly translated into user planes.

3.10   Conclusion

Network services are the built-in capabilities that network infrastructure owners provide as an interface to end users and applications. Therefore, enabling the applications envisioned by Network 2030 will require to devise new services able to cope with the novel requirements. These services will not substitute the ones already in place, but rather they will complement them and improve the overall capabilities. This chapter deals with the definition of future services section. First, it covers limitations in the current network services and their inability to deploy emerging applications. Then, we provide a classification of the envisioned future services. The remainder of the chapter deals with the description of the different services. The chapter is finalized by a discussion of open questions and challenges.

References

1. D.C. Verma, Service level agreements on IP networks. Proc. IEEE **92**(9), 1382–1388 (2004)
2. Hasan, B. Stiller, SLO auditing task analysis, decomposition, and specification. IEEE Trans. Netw. Serv. Manag. **8**(1), 15–25 (2011)
3. IEEE standard for local and metropolitan area networks—bridges and bridged networks. *IEEE Std 802.1Q-2014 (Revision of IEEE Std 802.1Q-2011)* (IEEE, Piscataway, 2014), pp. 1–1832
4. Time-Sensitive Networking (TSN) Task Group, IEEE802.org
5. S. Shenker, C. Partridge, R. Guerin. RFC2212: specification of guaranteed quality of service (1997)
6. J. Wroclawski, RFC2211: specification of the controlled-load network element service (1997)
7. R. Braden, D. Clark, S. Shenker, RFC1633: integrated services in the internet architecture: an overview (1994)
8. G.P. Fettweis, The tactile internet: applications and challenges. IEEE Veh. Technol. Mag. **9**(1), 64–70 (2014)
9. A. Clemm, M. Torres Vega, H.K. Ravuri, T. Wauters, F.D. Turck, Toward truly immersive holographic-type communication: challenges and solutions. IEEE Commun. Mag. **58**(1), 93–99 (2020)
10. B. Varga, N. Finn, P. Thubert, J. Farkas, Deterministic networking architecture (DetNet), in *Internet Draft-draft-ietf-detnet-architecture-13*, May 2019
11. K. Makhijani, H. Yousefi, K.K. Ramakrishnan, R. Li, Extended abstract: coordinated communications for next-generation networks, in *2019 IEEE 27th International Conference on Network Protocols (ICNP)*, 2019, pp. 1–2
12. Holographic type communication, delivering the promise of future media by 2030, in *ITU-T, Fifth ITU Workshop on Network 2030*
13. M. Westerlund, S. Wenger, RTP topologies. RFC 7667 (2015)
14. Packet trimming to reduce buffer sizes and improve round-trip times—extended abstract
15. L. Dong, K. Makhijani, R. Li, Qualitative communication via network coding and new IP: invited paper, in *2020 IEEE 21st International Conference on High Performance Switching and Routing (HPSR)*, 2020, pp. 1–5
16. L. Dong, R. Li, In-packet network coding for effective packet wash and packet enrichment, in *2019 IEEE Globecom Workshops (GC Wkshps)*, 2019, pp. 1–6
17. T. Tarik et al., *White Paper on 6G Networking, 6G Research Visions* (University of Oulu, Oulu, 2020)
18. What is industry 4.0? https://www.blueoceands.com/home/events/news/industry4-0

19. A. Aijaz, M. Dohler, A.H. Aghvami, V. Friderikos, M. Frodigh, Realizing the tactile internet: haptic communications over next generation 5G cellular networks. IEEE Wirel. Commun. **24**(2), 82–89 (2017)
20. J. van der Hooft, T. Wauters, M. Torres Vega, T. Mehmli, F. De Turck, Enabling virtual reality for the tactile internet: hurdles and opportunities, in *1st International Workshop on High-Precision Networks Operations and Control (HiPNet)*, Rome, Italy, Nov 2018
21. B. Briscoe (ed.), *Low Latency, Low Loss, Scalable Throughput (L4s) Internet Service: Architecture*, draft-ietf-tsvwg-l4s-arch-09

Chapter 4
Overall Network and Service Architecture

Mehmet Toy and Atilla Toy

4.1 Introduction

Future applications and services that are expected to emerge in this decade and beyond and their requirements are described in Chap. 2 and 3. This chapter describes an overall architecture for the network and services to support these applications.

The future network architecture is an end-to-end integrated, automated, intelligent, and dynamic architecture that combines connectivity, applications, and computation and storage resources [1]. This architecture is driven mainly by proliferation of virtualization, artificial intelligence (AI)/machine learning (ML) techniques, Application Programing Interfaces (APIs) for automation, advances in computing, and current and expected future applications requiring enormous bandwidth, the end-to-end delay of a couple milliseconds, and near-zero packet loss.

Future applications and services are expected to be used by various end devices including robots, self-driven cars, and drones as depicted in Fig. 4.1.

In order to support various applications driven by devices in Fig. 4.1 and devices to be developed in the future, the network infrastructure is expected to include fixed and wireless networks, cloud and space communications infrastructures as depicted in Fig. 4.2. Virtualization, memory, and computing technologies in addition to AI/ML techniques will continue to impact networks and services to achieve highly flexible, automated, high bandwidth, and intelligent networks and services.

Due to stringent future application performance (i.e., delay, jitter, and loss) requirements, it is clear that the resources will be moved to the networking edge. The networking core will play its traditional role of controlling and supporting the

M. Toy (✉)
Verizon Communications, Inc., Allendale, NJ, USA
e-mail: mehmet.toy@verizon.com

A. Toy
365 Operating Company LLC, Norwalk, CT, USA

© The Author(s), under exclusive license to Springer Nature   93
Switzerland AG 2021
M. Toy (ed.), *Future Networks, Services and Management*,
https://doi.org/10.1007/978-3-030-81961-3_4

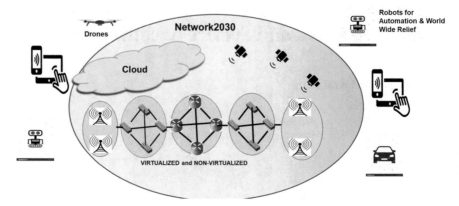

Fig. 4.1 An example of future network infrastructure and end devices [1]

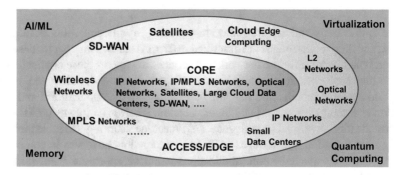

Fig. 4.2 Expected infrastructure of future networks [1]

edge. At the edge, in addition to Service Providers, the users will have a control of resources and services offered at the edge. Federation among the edges supporting similar applications and federation among Service Providers providing similar services are expected to grow by time for efficient utilization of resources and improve user experiences.

Networks and services may be grouped into two categories as underlay networks and services, and overlay networks and services that are built over the underlay networks and services. The following sections will describe these concepts and an overall architecture for future networks and services. We will use future networks and Network2030 interchangeably throughout this chapter.

4.2   Underlay and Overlay Networks

Underlay network is a physical infrastructure that transmits frames or packets over network devices like switches and routers. An overlay network is a virtual network that is built on top of this underlay physical infrastructure.

In the underlay networks, the control and data plane are within the same physical boxes, while they are separated in the overlay networks.

An overlay network is formed on top of the underlay to construct a virtualized network. The data plane traffic and control plane signaling are controlled within each virtualized network, providing isolation among the networks and freedom from the underlay network. The overlay network is used to decouple a network service from the underlying infrastructure where the underlying physical infrastructure of the network has no or little visibility of the actual services offered. This layering approach enables the underlay network such as the core network to scale and evolve independently of the offered services.

For example, Internet itself is nothing more than an overlay network on top of an optical infrastructure. Most of the Internet paths are formed over a dense wavelength division multiplexing (DWDM) infrastructure that creates a virtual (wavelength based) topology between routers and utilizes several forms of switching to interconnect routers together [2]. On the other hand, some of the Internet paths are overlaid over SONET/SDH time division multiplexing (TDM) networks that provide TDM paths to interconnect routers. Therefore, pretty much every router path in the Internet is an overlaid path.

In most cases, routers are interconnected together through Ethernet switched networks. They are essentially overlaid on top of a Layer-2 infrastructure. A router has absolutely no visibility in the Layer-2 paths. If communication between routers is lost because of a loop in the underlying Ethernet network, the routers will never be able to recover. A packet loss in the Ethernet network is not visible in the routers, unless it is correlated through a management system.

In wireless networks, wireless services are mainly provided through an overlay network of General Packet Radio Service (GPRS) Tunnelling Protocol (GTP) tunnels. Essentially, traffic from each mobile Subscriber is encapsulated in a tunnel and routed to the serving GPRS support node (SGSN) and pilot GSSN (PGSN) gateways of the wireless network.

Similarly, an MPLS L2/L3 VPNs is an overlay network of services on top of an MPLS transport network. A transport label identifies how the packet should be forwarded through the core MPLS network. Packets are routed through one or more overlays.

Overlay networks maintain state information at the edges of the network and hide individual services from the core network. This approach decouples the services from the underlying physical infrastructure whether there is a need for a tight coupling between overlay and underlay networks or not. Very often the right network design demands a strong correlation. Ignoring physical networking and treating as a black box can lead to severe issues.

For the IP/optical convergence, the underlay optical network is built and managed without considering the overlay IP network. Contrary to optical networks though, MPLS VPNs rely in a relatively tight coupling of overlays and L1 paths. MPLS paths are setup through a careful understanding of the physical topology, and VPN services are overlaid over transport paths with full visibility and correlation between transport and services. The reason for these selections is to allow features like traffic management, fast traffic restoration, etc. These networks utilize control plane protocols that will quickly detect impairments and try to restore traffic paths for the overlay network.

A comparison of underlay and overlay networks is given in Table 4.1 [4].

4.2.1   Underlay Network Architectures

In this section, we describe optical and IP/MPLS backbone and wireless networks as examples for underlay networks.

Table 4.1 Underlay and overlay comparison

Capabilities	Underlay network	Overlay network
Traffic flow	Transmits packets which traverse over network devices like switches and routers	Transmits packets only along the virtual links between the overlay nodes
Deployment time	Less scalable and time-consuming activity to set up new services and functions	Ability to rapidly and incrementally deploy new functions through edge-centric innovations
Frame/packet control	Hardware oriented	Software oriented
Frame/packet encapsulation and overhead	Frame delivery occurs at Layer-2, and packet delivery and reliability occurs at Layer-3 and Layer-4	Needs to encapsulate frames or packets across source and destination, hence incurs additional overhead
Multipath forwarding	Less scalable options of multipath forwarding. In fact using multiple paths can have associated overhead and complexity	Support for multipath forwarding within virtual networks
Managing multitenancy	NAT- or VRF-based segregation required which may face challenge in big environments	Ability to manage overlapping IP addresses between multiple tenants
Scalability	Less scalable due to technology limitation	Designed to provide more scalability than underlay network. For example, VLAN (underlay network) provides 4096 VLAN support while VXLAN (overlay network) provides up to 16 million identifiers
Frame/packet delivery	Responsible for delivery of frames/packets	Offloaded from delivery of frames/packets
Protocols	Underlay protocols include Ethernet Switching, VLAN, routing, etc.	Overlay network protocols include Virtual Extensible LAN (VXLAN), Ethernet Virtual Connection (EVC), Network Virtualization using Generic Encapsulation (NVGRE), Stateless Transport Tuning (SST), Generic Routing Encapsulation (GRE), IP multicast and Network Virtualization overlays 3 (NVO3) [3]

4.2.1.1 Optical Networks

Internet and non-Internet services, such as mobile backhaul, wholesale and transit, wavelength switching, and private-line business services share the same underlying optical transport infrastructure with the foundation for Internet and IP services. They rely on the backbone network's available capacity to operate. In parallel to increasing traffic demand, Service Providers increase the backbone capacity to support this growth and to create a reserve that can absorb the spikes in traffic that might occur.

An IP/optical network consists of optical fiber, the IP nodes where fibers meet, the optical nodes stationed intermittently along fiber segments, and the edge nodes that serve as the sources and destinations of traffic.

There are two common core architectures, hollow core and lean core [5]. Hollow core architectures attempt to eliminate the costs associated with core backbone routers by replacing them with an optical transport network (OTN) switching layer performing transport switching function which offers a lower overall cost per bit for a given interface speed. In the hollow core model, these switches create a dense mesh of circuits between each of the edge and peering nodes. As with the full IP core, the packet, and optical TDM and DWDM layers are managed separately. There is little control-plane integration. This lack of control-plane integration prevents topology information from being shared between the packet, optical TDM, and DWDM layers. Routers know only the routing topology, and the DWDM and TDM switches know only the optical topology.

DWDM is the process of multiplexing signal of different wavelength onto a single fiber. Through this operation, it creates many channels (i.e., virtual fibers) each capable of carrying a different signal where all sharing a single transmission medium.

Lean core architectures are an adaptation of full IP architectures in which backbone routers have reduced network processing unit (NPU) functionality or memory. With reduced NPU memory, the router can only learn internal routes, which forces Operators to use Multiprotocol Label Switching (MPLS) to forward Subscriber traffic instead of IP.

This model still manages packet, optical TDM, and DWDM layers separately, and control-plane integration remains limited as in full IP core architectures. As with conventional architectures, topology information is effectively isolated within the packet, optical TDM, and DWDM layers which limits operational efficiency.

In order to overcome these challenges, including scale, flexibility, and cost, without adopting solutions that sacrifice one in favor of the other, a new architecture is needed. This architecture is a Converged Transport Architecture (CTA) consists of packet routers and DWDM reconfigurable optical add-drop multiplexers (ROADMs). This architecture also eliminates boundaries between the packet and DWDM layers. Eliminating these boundaries through control plane and Network Management System (NMS)/Operation Support System (OSS) capabilities can dramatically improve the efficient sharing of information within the network and across organizational boundaries, resulting in a more dynamic, efficient backbone network.

The first option for CTA is building the core with bigger IP routers. This is the "safest" path to take in many respects and involves the simplest topology (Fig. 4.3).

The potential implications of bigger IP routers in the same architecture could include an increasingly cumbersome and expensive infrastructure, as well as mismatched functionality of IP for simply and economically moving large numbers of packets. This complexity can compromise capabilities, escalate power and space consumption, and make convergence difficult.

Second option is to have OTN in the core and routing at the edge. An alternative architecture to deal with issues with the first option consisting of OTN in the core and routing at the edge of the network is depicted in Fig. 4.4.

A core architecture based on OTN means that one must provision for peak traffic. For rapid provisioning and automated restoration of channels, both DWDM and OTN domains are dynamically configurable. ROADMs (reconfigurable optical add/drop multiplexers) route wavelengths from one end of the DWDM network to the other under software control.

The core of OTN circuits is not suited for the bursty and dynamic traffic. The complexity shifts to the edge and increases costs even more.

A third option is to continue to maintain multiple layers and to bypass core routers whenever possible.as depicted in Fig. 4.5.

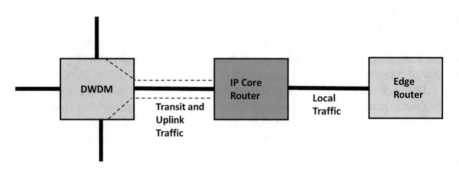

Fig. 4.3 Building the core with bigger IP routers

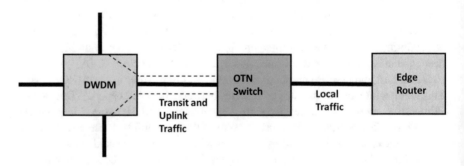

Fig. 4.4 OTN in the core and routing at the edge

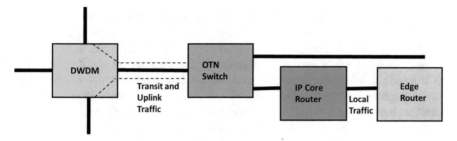

Fig. 4.5 Bypassing core routers

Packet Services	Legacy Services
OTN Switching	IP or IP/MPLS Switching
OTN Transport	
Lambda Switching (Photonic Switching or Wavelength Switching)	
Fiber Channels	

Fig. 4.6 Packet optical architecture

The core network infrastructure is expected to be flexible enough to accommodate changes in traffic characteristics. A detailed modeling [6] demonstrates that the core of the network needs to be packet-based to be most efficient. Aggregation and multiplexing need to happen at the packet level. Integration of the packet layer and optical transport layer (Fig. 4.6) will improve the economics of the core network, both CapEx and OpEx.

Seamless integration of the IP/MPLS and optical control planes will allow for coordinated provisioning, management, and restoration of these multiservice networks. A unified packet transport network consolidates service delivery infrastructure by carrying circuit-based services as well as packet-based services.

OTN transport consisting of long-haul transmission, switching, and multiplexing consolidates L1, L2, and L3 networks over wavelengths. It improves operation, administration, and management (OAM).

On the other hand, IP or IP/MPLS handles bursty variable-rate traffic and dynamically aggregating traffic. Thousands of variable-rate traffic flows are processed at the packet level. MPLS also provides protection for these flows. The degree of mixing OTN and MPLS depends on operational and financial constraints.

The state-of-the art DWDM system supports 2 Tbps [7] on a single mode fiber. With ROADMs, effective optical routing and switching is performed. With software-defined elastic optical network (SD-EON), the optical networking became more flexible in assigning bandwidth to connections compared to fixed-grid WDM networks. This capability decouples intelligent control layer from optical layer.

SD-EONs divide available spectral resources into narrow frequency slots that are assigned to connections according to their bandwidth requirements [8]. However, single-mode fibers used in SD-EONs reach its upper bound due to nonlinear Shannon limit. In order to overcome this limit, space division multiplexing (SDM) technique is proposed. SDM enables parallel optical signal transmission through spatial modes co-propagating in suitable designed fibers. It is possible to apply EON and SDM simultaneously to achieve spectrally spatially flexible optical network.

SDN is enabled a centralized and programmable network control and management for EONs that uses rule-based policies for service provisioning. As network traffic and topology change, these rules may require changes. Application of machine learning techniques such as deep reinforcement learning to fault management [9] and provisioning of EON and SDM should improve the flexibility of optical networking and utilize the optical networking resources more efficiently. From the abstracted optical network, virtual slices can be established.

As part of an end-to-end network slicing to support dedicated 5G network slices in meeting diverse performance requirements such as bandwidth and delay constraints of users, optical network slicing is necessary. Abstracted optical network resources are sliced with isolation techniques [10].

4.2.1.1.1   Future Expectations

As move into 2030 and beyond, we expect to see increase in capacities of fiber, optical amplifiers, and optical switches. With super dense coding of quantum computing, two bits worth of data is carried in one qubit. In addition, quantum entanglement can transfer information between endpoints using very little bandwidth. As a result, a quantum-based network can transmit large amounts of data between endpoints with little communication resources. The data efficiency of networks will increase dramatically.

4.2.1.2   Wireless Networks

In this section, we will briefly describe a global system for mobile communication (GSM) architecture first. After that, we will describe 5G and 6G architectures, and the future of wireless networks.

GSM Architecture

Global system for mobile communication (GSM) is a digital cellular technology used for transmitting mobile voice and data services. The concept of GSM emerged from a cell-based mobile radio system at Bell Laboratories in the early 1970s. It is a circuit-switched system that divides each 200 kHz channel into eight 25 kHz time

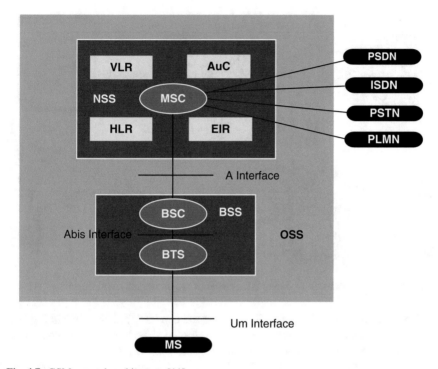

Fig. 4.7 GSM network architecture [11]

slots and operates on the mobile communication bands 900 and 1800 MHz in most parts of the world. In the USA, GSM operates in the bands 850 and 1900 MHz [11].

As depicted in Fig. 4.7, a GSM network comprises of the following:

- Mobile Station (MS)
- Base Station Subsystem (BSS)
- Network Switching Subsystem (NSS)
- Operation Support Subsystem (OSS)

The additional components of the GSM architecture depicted in Fig. 4.8 are:

- Home Location Register (HLR)
- Visitor Location Register (VLR)
- Equipment Identity Register (EIR)
- Authentication Center (AuC)
- SMS Serving Center (SMS SC)
- Gateway Mobile Switching Center (GMSC)
- Chargeback Center (CBC)
- Transcoder and Adaptation Unit (TRAU)

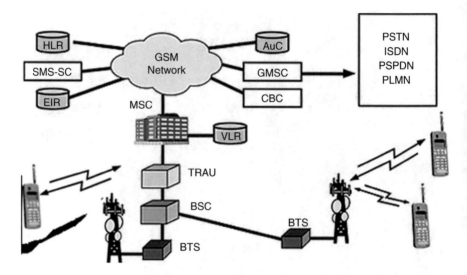

Fig. 4.8 GSM network architecture with its databases and messaging system functions [11]

The MS and the BSS communicate across the Um interface as shown in Fig. 4.7. It is also known as the *air interface* or the *radio link*. The BSS communicates with the Network Service Switching (NSS) center across the *A* interface.

A Mobile Switching Center (MSC) acts as a control center of a Network Switching Subsystem (NSS). The MSC connects calls between Subscribers and also provides information needed to support mobile service Subscribers. Based on the size of the mobile operator, multiple MSC can be implemented.

The MSC is stationed between the base station and the Public Switched Telephone Network (PTSN). All mobile communications are routed from the base station through the MSC. The MSC is responsible for handling voice calls and SMS including other services like FAX. The MSC initiates call setup between Subscribers and is also responsible for real-time prepaid billing and account monitoring. The MSC is responsible for inter-Base Station Controller (BSC) handovers and inter-MSC handover.

A BSC initiates an inter-BSC handover from the MSC when it notices a cellphone approaching the edge of its cell. After the request is made by the BSC, the MSC scans through a list to determine adjacent BSCs and then proceeds to hand over the mobile device to the appropriate BSC. The MSC also works with the HLR by using the database of the HLR to determine the location of each mobile device in order to provide proper routing of calls.

In a GSM network, the following areas are defined:

- **Cell** is the basic service area where one base transceiver station (BTS) covers one cell. A BTS is a fixed radio transceiver that connects mobile devices to the network. It sends and receives radio signals to mobile devices and converts them to digital signals that it passes on the network to route to other terminals in the network or to the Internet.

- Each cell is given a Cell Global Identity (CGI), a number that uniquely identifies the cell.
- **Location Area** (LA) is formed by a group of cells. This is the area that is paged when a Subscriber gets an incoming call. Each LA is assigned a Location Area Identity (LAI). Each LA is served by one or more base station controllers (BSCs).
- **MSC/VLR Service Area** is the area covered by one Mobile Switching Center (MSC).
- **PLMN** is the area covered by one network operator that contains one or more MSCs.

GSM network types and Core Network Operator are defined in [11]:

- **Conventional Network**: A PLMN consisting of radio access network and core network, by which only one serving operator provides services to its Subscriber. Subscribers of other Operators may receive services by national or international roaming.
- **Common PLMN:** The PLMN-id indicated in the system broadcast information as defined for conventional networks, which non-supporting UEs understand as the serving operator.
- **Core Network Operator:** An operator that provides services to Subscribers as one of multiple serving operators that share at least a radio access network. Each core network operator may provide services to Subscriber of other Operators by national or international roaming.
- **Gateway Core Network**: A network sharing configuration in which parts of the core network (MSCs/SGSNs/MMEs) are also shared.
- **Multi-Operator Core Network:** A network-sharing configuration in which only the RAN is shared.
- **Non-supporting UE:** A UE that does not support network sharing in the sense that it ignores the additional broadcast system information that is specific for network sharing for 3GPP UTRAN and GERAN. In other specifications, the term "network sharing non-supporting UE" may be used.

An MSC Server or MSS is a 2G/3G core network element which controls the network switching subsystem elements. MSS can be used in GSM networks as well, if the manufacturer has implemented support for GSM networks in the MSS. In fact, MSS along with other 3G network elements such as media gateway (MGW) can be configured to support GSM network exclusively.

MSC server functionality enables split between control plane signaling and user plane bearer in network element called a media gateway (MGW), which guarantees better placement of network elements within the network. The server communicates with other distributed elements using industry open standards such as media gateway control protocol, megaco/H.248, session initiation protocol, Message Transfer Part 2 (MTP2) User Adaptation Layer (M2UA), and Message Transfer Part 3 User Adaptation Layer (M3UA).

Traditional cellular, or Radio Access Networks (RAN), consist of many stand-alone base stations (BTS). Each BTS covers a small area, whereas a group BTS provides coverage over a continuous area. Each BTS processes and transmits its own signal to and from the mobile terminal and forwards the data payload to and from the mobile terminal and out to the core network via the backhaul. Each BTS has its own cooling, back-haul transportation, backup battery, monitoring system, and so on. Because of limited spectral resources, network operators "reuse" the frequency among different base stations, which can cause interference between neighboring cells.

There are several limitations in the traditional cellular architecture:

- Each BTS is costly to build and operate.
- When more BTS are added to a system to improve its capacity, interference among BTS is more severe as BTS are closer to each other and more of them are using the same frequency.
- Because users are mobile, the traffic of each BTS fluctuates which is called as "tide effect," and as a result, the average utilization rate of individual BTS is pretty low. However, these processing resources cannot be shared with other BTS. Therefore, all BTS are designed to handle the maximum traffic, not average traffic, resulting in a waste of processing resources and power at idle times.

In the 1G and 2G cellular networks, base stations had an all-in-one architecture. The RF signal is generated by the base station RF unit and propagates through pairs of RF cables up to the antennas on the top of a base station tower or other mounting points. This all-in-one architecture was mostly found in macro cell deployments.

For 3G, a distributed base station architecture was introduced. In this architecture, the radio function unit, also known as the remote radio head (RRH), is separated from the digital function unit, or baseband unit (BBU) by fiber. Digital baseband signals are carried over fiber, using the Open Base Station Architecture Initiative (OBSAI) or Common Public Radio Interface (CPRI) standard. The RRH can be installed on the top of tower close to the antenna, reducing the loss compared to the traditional base station where the RF signal has to travel through a few kilometers long cable from the base station cabinet to the antenna at the top of the tower.

The 3G wireless systems were proposed to provide voice and paging services to provide interactive multimedia including teleconferencing and internet access and variety of other services. However, these systems offer wide-area network (WAN) coverage of 384 kbps peak rate and limited coverage for 2 Mbps. Hence providing broadband services is one of the major goals of the 4G Wireless systems.

4G supports interactive multimedia, voice, video, wireless internet, and other broadband services. It supports up to 20 Mbps and operates in 2–8 GHz frequency band. 4G is a packet-based network and carries voice as well as data by providing different levels of quality of service (QoS). Location registration, paging, and handover are supported for mobile devices. The global roaming can be achieved with the help of multi-hop networks that can include wireless local area networks (WLANs) or the satellite coverage in remote areas.

Evolved Packet Core (EPC) is a framework for providing converged voice and data on a 4G Long-Term Evolution (LTE). EPC is the core component of Service Architecture Evolution (SAE), 3GPP's flat LTE architecture. Its key components are:

- Mobility Management Entity (MME) which manages session states and authenticates and tracks a user across the network
- Serving Gateway (S-gateway) which routes data packets through the access network
- Packet Data Node Gateway (PGW) which acts as the interface between the LTE network and other packet data networks; manages QoS and provides deep packet inspection (DPI)
- Policy and Charging Rules Function (PCRF) which supports service data flow detection, policy enforcement

C-RAN (centralized RAN or cloud RAN) may be viewed as an architectural evolution of the distributed base station system. It uses the CPRI standard, low-cost coarse or dense wavelength division multiplexing (CWDM/DWDM) technology, and mm wave to allow transmission of baseband signal over long distance thus achieving large-scale centralized base station deployment. C-RAN applies data center network technology to allow a low cost, high reliability, low latency and high bandwidth interconnect network in the BBU pool. It utilizes open platforms and real-time virtualization technology rooted in cloud computing to achieve dynamic shared resource allocation and support multi-vendor, multi-technology environments.

5G Architecture

The primary goal of previous generations of mobile networks has been to simply offer fast, reliable mobile data services to network users. 5G utilizes a more intelligent architecture, with Radio Access Networks (*RANs*) no longer constrained by base station proximity or complex infrastructure. 5G leads the way toward disaggregated, flexible, and virtual RAN with new interfaces creating additional data access points.

5G is the new generation of radio systems and network architecture that will deliver extreme broadband, ultra-robust low latency connectivity, and massive networking for human beings and the Internet of Things. It will combine existing Radio Access Technologies (RATs) in both currently licensed and unlicensed bands and add novel RATs optimized for specific bands and deployments, and scenarios and use cases.

Programmability is central to achieving the super-flexibility that mobile network operators need to support the new communication demands that come from a wide array of devices and users. In order to manage all these diverse demands, 5G networks need to be programmable, flexible, modular, and software-driven.

In 3GPP 5G architecture [12], services are provided via a common framework to network functions that are permitted to make use of these services. Modularity, reusability, and self-containment of network functions are additional design considerations for a 5G network architecture described by the 3GPP specifications.

Multiple frequency ranges are now being dedicated to 5G new radio (NR). The portion of the radio spectrum with frequencies between 30 and 300 GHz is known as the millimeter wave, since wavelengths range from 1 to 10 mm. Frequencies between 24 and 100 GHz are now being allocated to 5G in multiple regions worldwide.

In addition to the millimeter wave, underutilized UHF frequencies between 300 MHz and 3 GHz are also being repurposed for 5G. The diversity of frequencies employed can be tailored to the unique applications considering the higher frequencies are characterized by higher bandwidth, albeit shorter range. The millimeter wave frequencies are ideal for densely populated areas, but ineffective for long-distance communication. Within these high and lower frequency bands dedicated to 5G, each carrier has begun to carve out their own discrete individual portions of the 5G spectrum.

Network Function Virtualization (NFV) enables the 5G infrastructure by virtualizing appliances within the 5G network. This includes the network slicing technology that enables multiple virtual networks to run simultaneously. NFV can address other 5G challenges through virtualized computing, storage, and network resources that are customized based on the applications and customer segments.

NFV extends to the RAN through, for example, network disaggregation promoted by alliances such as Open Radio Access Network (O-RAN). This enables flexibility and provides open interfaces and open source development, ultimately to ease the deployment of new features and technology with scale. The O-RAN ALLIANCE objective is to allow multi-vendor deployment with off-the-shelf hardware for the purposes of easier and faster interoperability. Network disaggregation also allows components of the network to be virtualized, providing a means to scale and improve user experience as capacity grows. The benefits of virtualizing components of the RAN provide a means to be more cost effective from a hardware and software viewpoint especially for IoT applications where the number of devices is in the millions.

Currently Common Public Radio Interface (CPRI) is used as fronthaul connection in 4G. Remote radio heads are distributed in towers across cities or suburban areas, for good coverage in built-up areas. They are then connected to baseband units in centralized locations. This connectivity is known as "fronthaul."

Enhanced CPRI (eCPRI) is used to support 5G by enabling increased efficiency. It provides flexible radio data transmission through a packet-based fronthaul network, for example, IP or Ethernet. As 5G becomes increasingly present in our lives, the fiber between the radio units and baseband units should see an increase in traffic. eCPRI is introduced with 5G to efficiently utilize the bandwidth and make radio interface more flexible. CPRI is a point-to-point interface, meaning that vendors will be exclusive within their own network. eCPRI works on an open interface, allowing more convergence in the industry as carriers complement networks with shared equipment.

It will improve bandwidth efficiency and increase capabilities and lower latencies. The interface will allow for carriers to move from the baseband units to the radio. This allows for simpler deployments of massive MIMO and increased flexibility.

The interface can be framed within Ethernet to take advantage of the networks already in place. Ethernet can carry eCPRI from several system vendors at the same time. eCPRI is expected to reduce the required bandwidth by up to ten times.

Another breakthrough technology integral to the success of 5G is beamforming. Conventional base stations have transmitted signals in multiple directions without regard to the position of targeted users or devices. Through the use of multiple-input, multiple-output (MIMO) arrays featuring dozens of small antennas combined in a single formation, signal processing algorithms can be used to determine the most efficient transmission path to each user, while individual packets can be sent in multiple directions then choreographed to reach the end user in a predetermined sequence.

With 5G data transmission occupying the millimeter wave, free space propagation loss, proportional to the smaller antenna size, and diffraction loss, inherent to higher frequencies and lack of wall penetration, are significantly greater. On the other hand, the smaller antenna size also enables much larger arrays to occupy the same physical space. With each of these smaller antennas potentially reassigning beam direction several times per millisecond, massive beamforming to support the challenges of 5G bandwidth becomes more feasible. With a larger antenna density in the same physical space, narrower beams can be achieved with massive MIMO, thereby providing a means to achieve high throughput with more effective user tracking.

The 5G core, as defined by 3GPP, utilizes cloud-aligned virtualized functions, service-based architecture (SBA) that spans across all 5G functions and interactions including authentication, security, session management, and aggregation of traffic from end devices.

The 4G Evolved Packet Core (EPC) is significantly different from the 5G core. 5G supports millimeter wave, massive MIMO, network slicing, and essentially every other discrete element of the diverse 5G ecosystem leveraging virtualization and cloud native software design at unprecedented levels.

Among the other changes that differentiate the 5G core from its 4G predecessor are user plane function (UPF) to decouple packet gateway control and user plane functions, and access and mobility management function (AMF) to segregate session management functions from connection and mobility management tasks.

In order to bridge the gap between 4G and 5G, there are two 5G architecture options, non-standalone (NSA) mode and standalone (SA) mode 5G. The 5G non--standalone architecture utilizes existing LTE RAN and core networks as an anchor, with the addition of a 5G component carrier. Despite the reliance on existing architecture, non-standalone mode will increase bandwidth by tapping into millimeter wave frequencies. 5G standalone mode is essentially 5G deployment from the ground up with the new core architecture and full deployment of all 5G hardware, features, and functionality.

3GPP has defined several options for using LTE and 5G NR radios together with LTE core, Evolved Packet Core (EPC), and with the new 5G Core (5GC). Among these options, Options 2 and Option 3X are to be preferred by the industry [13]. These two options are as illustrated in Fig. 4.9 where solid lines indicate user plane traffic and dashed lines indicate control plane traffic.

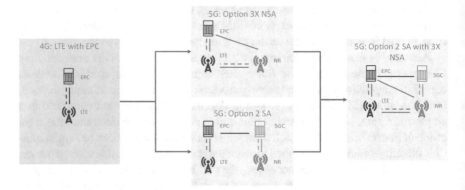

Fig. 4.9 4G-to-5G migration options [13]

Early 5G devices and networks might use NSA Option 3X (i.e., 5G NR connected to the existing EPC) for deployment speed and service continuity, leveraging LTE coverage. Option 3X uses dual connectivity between LTE and 5G radios to achieve data rate aggregation. 5G NR is connected as a secondary node to the EPC which allows enhanced mobile broadband services on 5G-enabled devices while maintaining 4G EPC services. Option 3X also relies on the LTE radio for all control plane signaling and wide-area radio frequency coverage.

The next distinct phase of 5G deployment will be using the Option 2 architecture which attaches 5G NR to the new 5G Core (5GC). Option 2 (with or without 3X NSA support) is therefore widely seen as the long-term target architecture where user data traffic directly flows to the 5G gNB part of the base station in Option 3X.

The Option 2 solution uses 5G NR for data and for control plane signaling and requires a 5G low-frequency band for wide-area coverage as it cannot utilize the LTE layer as in Option 3X.

The 5G Base Station's Baseband processing block has been decomposed into Centralized and Distributed Units (CU and DU) that need not be co-located with the radio unit (RU). This optional connectivity between RU and DU is called "fronthaul," which can utilize bridged Ethernet networks (Fig. 4.10).

The 5G system architecture [12] separates the user plane (UP) functions from the control plane (CP) functions, allowing independent scalability, evolution, and flexible deployments. The architecture enables each network function (NF) and its network function services (NFSs) to interact with other NF and its NFSs directly or indirectly via a Service Communication Proxy if required. It minimizes dependencies between the Access Network (AN) and the Core Network (CN) with a common AN-CN interface which integrates different access types such as 3GPP access and non-3GPP access.

The authentication framework is unified. Roaming with both home-routed traffic as well as local breakout traffic in the visited PLMN is supported.

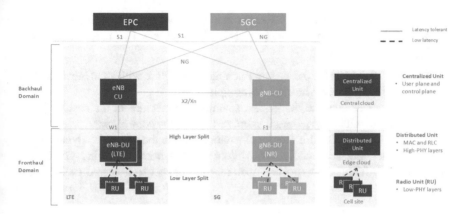

Fig. 4.10   High-level RAN architecture [14]

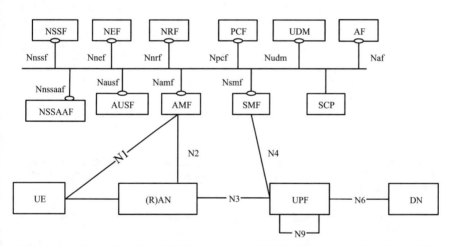

Authentication Server Function (AUSF); Access and Mobility Management Function (AMF); Data Network (DN), e.g., operator services, Internet access or third-party services; Network Exposure Function (NEF); Network Repository Function (NRF); Policy Control Function (PCF); Session Management Function (SMF); User Equipment (UE); Unified Data Management (UDM); User Plane Function (UPF); Network Slice Selection Function (NSSF); Service Communication Proxy (SCP); Application Function (AF); (Radio) Access Network ((R)AN); Network Slice-Specific Authentication and Authorization Function (NSSAAF)

Fig. 4.11   5G System non-roaming architecture [12]

The 5G system architecture consists of various network functions depicted in Fig. 4.11. Network functions within the 5GC control plane uses service-based interfaces for their interactions:

- Network functions (e.g., AMF) within the control plane enable other authorized network functions to access their services.

- The interaction between the NF services in the network functions is described by point-to-point reference point (e.g., N11) between any two network functions (e.g., AMF and SMF).

Figure 4.12 depicts the 5G system roaming architecture with local breakout with service-based interfaces within the control plane.

Figure 4.13 depicts the 5G system roaming architecture in the case of home routed scenario with service-based interfaces within the control plane.

Network sharing [12] is a way for Operators to share the heavy deployment costs for mobile networks, especially in the rollout phase. A network sharing architecture shall allow different core network operators to connect to a shared radio access

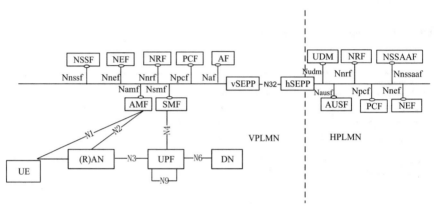

Visited Public Land Mobile Network (VPLMN); Home Public Land Mobile Network (HPLMN); Visited Security Edge Protection Proxy (vSEPP); Home Security Edge Protection Proxy (hSEPP); reference point between two UPFs (N9)

Fig. 4.12 Roaming 5G system architecture-local breakout scenario in service-based interface representation [12]

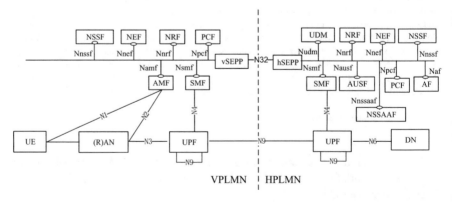

Fig. 4.13 Roaming 5G system architecture—home-routed scenario in service-based interface representation [12]

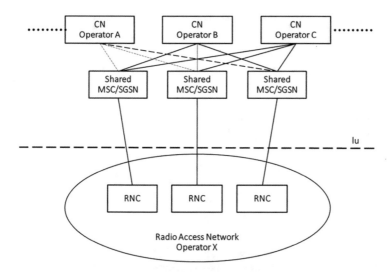

Mobile Switching Center (MSC); Serving GPRS Support Node (SGSN)

Fig. 4.14 A Gateway Core Network (GWCN) configuration for network sharing [15]

network including radio resources. There are two identified architectures to be supported by network sharing. They are shown in Figs. 4.14 and 4.15.

In both architectures, the radio access network is shared. Figure 4.14 shows reference architecture for network sharing where access network and core network nodes, MSCs and SGSNs, are also shared. This configuration is referred to as a Gateway Core Network (GWCN) configuration.

Figure 4.15 shows the reference architecture for network sharing in which only the radio access network is shared, the Multi-Operator Core Network (MOCN) configuration.

The UE behavior in both of these configurations is the same.

The goal of next-generation communication systems is to achieve high spectral and energy efficiency, low latency, and massive connectivity because of extensive growth in the number of Internet-of-Things (IoT) devices. These IoT devices will realize advanced services such as environment monitoring, and control, virtual reality (VR)/virtual navigation, telemedicine, digital sensing, high definition (HD), and full HD video transmission in connected drones and robots.

5G use cases are grouped as massive machine-type communication (mMTC), ultrareliable and low latency communication (URLLC), and enhanced mobile broadband (eMBB) [16]. The GSMA expects 5G to deliver high-speed, low-latency, reliable, and secure enhanced mobile broadband which is expected to be delivered in its early deployments.

eMBB provides greater data bandwidth complemented by moderate latency improvements on both 5G NR and 4G LTE. It will bring the benefits of 5G to the wider public and help to develop use cases such as emerging AR/VR media and applications, ultra HD or 360-degree streaming video.

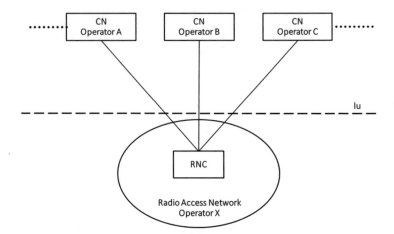

Fig. 4.15 A Multi-Operator Core Network (MOCN) in which multiple CN nodes are connected to the same RNC and the CN nodes are operated by different Operators [15]

URLLC is a primary enabler for a number of unique use cases in the areas of manufacturing, energy transmission, transportation, and health care.

mMTC can support extremely high connection density of online devices. Service providers can combine mMTC technology together with a MEC (multi-access edge computing) to build the infrastructure needed to support this massive IoT network.

6G Architecture

Forecasts suggest that by 2030, around 50 billion of these IoT devices will be in use around the world, creating a massive web of interconnected devices [17]. It is very challenging for the existing multiple access techniques to accommodate such a massive number of devices. Future networks need to support this massive access in beyond 5G (B5G)/6G communication systems [18].

Using millimeter-wave (mmWave) and terahertz (THz) frequency bands, massive bandwidth, and highly directive antennas is expected to be available to the 6G mobile devices to enable new applications and seamless coverage. Ultrahigh-precise positioning will become available with 6G due to high-end imaging and direction-finding sensors.

6G capabilities may be summarized as [19]:

- Very high data rates, up to 1 Tbps
- Very high energy efficiency, with the ability to support battery-free IoT devices
- Trusted global connectivity
- Massive low-latency control (less than 1 ms end-to-end latency)
- Very broad frequency bands (e.g., 73–140 GHz and 1–3 THz)
- Ubiquitous always-on broadband global network coverage by integrating terrestrial wireless with satellite systems

- Connected intelligence with machine learning capability and AI networking hierarchy

6G communications is also expected to significantly improve security, privacy, and confidentiality. It is estimated that the 6G system will have 1000 times higher simultaneous wireless connectivity than the 5G system. URLLC, which is a key 5G feature, will be a key driver again in 6G communication by providing end-to-end delay of less than 1 ms [19]. Volume spectral efficiency, as opposed to the often-used area spectral efficiency, will be much better in 6G [20]. The 6G system will provide ultra-long battery life and advanced battery technology for energy harvesting. In 6G systems, mobile devices will not need to be separately charged. A comparison of 5G and 6G is given in Table 4.2.

6G is expected to support of three new service types beyond the eMBB, uRLLC, and mMTC services supported by 5G [19]:

- **Computation-Oriented Communications (COC):** The AI-empowered 6G will allow smart devices to take advantage of federated learning and edge intelligence. Instead of targeting just rate and latency provisioning, CoC might choose an operating point in the rate-latency-reliability space depending on the availability of communication resources to achieve a certain computational accuracy.
- **Contextually Agile eMBB Communications (CAeC):** The provision of 6G eMBB services is expected to be more agile and adaptive to the network context such as link congestion and network topology, physical environment context such as surrounding location and mobility, and social network context such as social neighborhood and sentiments.
- **Event-Defined uRLLC (EDuRLLC):** In contrast to the 5G uRLLC application scenario such as industrial automation where redundant resources are in place to recover from failures, 6G will need to support uRLLC application scenarios with spatially and temporally changing device densities, traffic patterns, and spectrum and infrastructure availability.

As we mentioned above, by 2030, we expect massive number of connected devices. The huge amount of data produced by these massive numbers of devices will require very high-performance processing units and robust backhauling links.

Table 4.2 Comparison of 5G and 6G

Key performance indicator (KPI)	5G	6G
Traffic size	10 Mb/s/m^2	~ 1–10 Gb/s/m^3
Downlink data rate	20 Gb/s	1 Tb/s
Uplink data rate	10 Gb/s	1 Tb/s
Uniform user experience	50 Mb/s, 2D everywhere	10 Gb/s, 3D everywhere
Latency (radio interface)	1 ms	0.1 ms
Jitter	Not specified	1 μs
Reliability (frame error rate)	1–10^{-5}	1–10^{-9}
Energy/bit	NS	1 pJ/b
Localization accuracy	10 cm in 2D	1 cm in 3D

Base stations (BSs), access points (APs), and mostly the central processing units may utilize ML and AI algorithms, and the backhauling links may utilize optical fiber and/or photonic communications.

By intelligent networking, all the end devices would be aware of the location and features of BSs/APs in their vicinity, and all of the BSs/APs would be aware of the locations, features, and QoS requirements of devices in their vicinity. Robust interference management/optimization techniques can be applied to maximize the efficiency of the wireless network. Central processing units will be fast enough to manage and switch the resources (bandwidth, time, power) among multiple end users, and data processing will be conducted at the baseband processing units (BPUs).

Some of 6G architectural components may be summarized as below [18]:

- Air Interface: 6G will concentrate on the current terahertz frequency range with extremely wide available bandwidth. The availability of extremely wide bandwidth would change the emphasis from spectrally optimized solutions to improved coverage solutions. In these new frequency spectrums, the tradeoff between spectrum performance, power efficiency, and coverage will play a key role in developing devices.

- The non-orthogonal multiple access (NOMA) is considered for the B5G/6G mobile networks. In NOMA, all of the users are allowed to access the complete resource (frequency band) simultaneously. The rate-splitting multiple access (RSMA) is also considered as a new access technology for 6G communication systems. Both NOMA and RSMA rely on the successive interference cancellations (SIC) to decode the information for the user. RSMA uses the SICs to decode the common message firstly and then decode the private message.

- New Spectrum: mmWave is a candidate for 5G. In this case, personal BSs and satellite connectivity can get merged into cellular communication. Therefore, using an unlicensed spectrum is proposed, to use the mmWave, THz band, and visible light spectrum, simultaneously. In this higher frequency band, the signal is attenuated very rapidly with the distance traveled. For example, a 3G or 4G BS can have a coverage of about several miles, whereas a 5G or 6G BS coverage may limit to only a few hundreds of meters. To resolve this issue in mmWave and THz communications, the idea of using massive multiple inputs and multiple outputs (MIMO) and beamforming emerged.

- Since 6G will accommodate a wide range of communication devices ranging from IoTs to live HD video transmission, 6G will need to be in line with all previous technologies. Therefore, a flexible and multi-radio access technologies (RAT) system architecture will be an essential component in the 6G network.

- AI/ML: B5G/6G wireless networks have increased complexity, requiring smarter methods for handling network features, detecting anomalies, and understanding performance trends. In order to preserve a certain level of performance, AI/ML will boost the decision-making process. The operation and implementation of RAN for 6G need a new strategy.

- Incorporating AI in wireless algorithms (e.g., for channel estimation, for channel state information (CSI) feedback, decoding, etc.) may bring a change in the direction of these algorithms [18]. By applying ML, deep learning, and AI algorithms to the communication network, we can instantly manage the resources as per the user requirements. The probability of choosing the best solution is improved in this way, and the network can maintain its optimum state.

- AI and ML will play important roles in the self-organization, self-healing, and self-configuration of 6G wireless systems.
- Advanced Beamforming with Very Large Scale Antenna (VLSA): The beamforming is to steer the beam to only the desired direction or user. Since energy is not spread in all directions, the transmission range is thus improved by concentrating the beam in one direction.
- Intelligent Reflecting Surfaces (IRSs): IRSs can be the potential area for beamforming in 6G. IRSs are composed of thin electromagnetic materials, which can reflect/configure the incoming electromagnetic rays in an intelligent way by configuring the phase of reflected rays by a software. A large number of low-power and low-cost passive elements reflect the incident signals with configurable phase shifts without the requirements of additional power, encoding, decoding, modulation, and demodulation requirements. They can be installed at locations such as high-rise buildings, advertising panels, vehicles such as cars, airplanes, unmanned aerial vehicles, and even the clothes of the pedestrians. They can enhance the signal-to-interference-plus-noise-ratio (SINR) with no change in the infrastructure or the hardware of the communication network.
- IRS can reduce the hardware complexity at the receiver and the transmitter by reducing the number of antennae installed at them, thereby, reducing the radiofrequency (RF) chains at the transmitter and the receiver.
- Orbital Angular Momentum (OAM)-aided MIMO: A new dimensional property of the electromagnetic waves is called as the orbital angular momentum (OAM) which is the transmission of multiple data streams over the same spatial channel. An electromagnetic wave carrying the OAM has the phase rotation factor of $\exp(-jl\Omega)$ where l is OAM state number represented in integer and Ω is transverse azimuth angle. The OAM can have an unlimited number of orthogonal modes, which allows the electromagnetic waves to multiplex multiple data streams over the same spatial channel, thereby, enhancing the spectral efficiency and transmission capacity. OAM support a high number of user in mode division multiple access (MDMA) scheme without utilizing extra resources (i.e., frequency, time, and power.
- OAM-based MIMO systems have advantages over the conventional MIMO systems in terms of capacity and long-distance line-of-sight (LoS) coverage. Therefore, OAM has great potential for applications in 6G wireless networks.
- Coexistence of variable radio access technologies: 6G can lead to a ubiquitous networking infrastructure. Each node in this network would be intelligent enough to sense the conditions of the channel and the specifications of QoS at any other node. For example, the use case and the network availability will decide the network as cellular, wireless LAN, Bluetooth, ultra-wideband (UWB), etc.
- 6G is to be designed in such a way that it will converge all of the wireless technologies.

Future Expectations

As we move toward 2030, we expect 6G to be implemented. Currently, government organizations are monitoring the spectrum and allocating the spectrum to the Operators. The owner of the spectrum has the full right to use that spectrum. Any other operator cannot use the spectrum allocated to some other Operator due to not having efficient spectrum monitoring and managing techniques. We expect free spectrum sharing will become a reality in 6G with robust spectrum monitoring using AI and block chain technologies.

The Quantum Computing and the Quantum Machine Learning will play key roles in channel capacity allocation, channel estimation, channel coding, localization, load balancing, routing, and multiuser transmissions, fast and optimum path selection for data packets.

The demand for massive connectivity in wireless networks has triggered the network resource management such as power distribution, spectrum sharing, and computational resources distribution. Blockchain should help in managing the relationship between Operators and users with the application of smart contracts [18]. Furthermore, blockchain should help management of energy and unlicensed spectrum management, seamless environmental protection and monitoring, smart healthcare, and cyber-crime rate reduction. Blockchain is also expected to be used for complex transactions initiated and trigger massive machine-to-machine (M2M) transactions.

One of the key network abilities that will allow us to build a flexible network and services on top of the common physical infrastructure is network slicing. As 5G becomes widely deployed, network slicing should become the fundamental technology to enable a wide range of use cases to cost-effectively deliver multiple logical networks over the same common physical infrastructure. It is expected that network slicing capabilities are to be enhanced as we move into 6G and beyond.

4.2.2 Overlay Network Architectures

An overlay network is built on top of another network. Overlay networks build the foundation of virtual networks, and they run as independent virtual networks on top of a physical network infrastructure. These virtual network overlays allow resource providers, such as cloud providers, to provision and orchestrate networks alongside other virtual resources. They also offer a new path to converged networks and programmability.

The most common examples of network virtualization are Virtual LANs (VLANs) and Virtual Private Networks (VPNs), Virtual Private LAN Services (VPLS), and virtual networks connecting Virtual Network Functions (VNFs).

VLANs are logical local area networks (LANs) overlaid on physical LANs. A VLAN can be created by partitioning a physical LAN into multiple logical LANs using a VLAN ID. Alternatively, several physical LANs can function as a single logical LAN. The partitioned network can be on a single router, or multiple VLANs can be on multiple routers just as multiple physical LANs would be.

An example Carrier Ethernet Service defined in [21, 22], E-Line service, is depicted in Fig. 4.16. The E-Line service is an overlay service consisting of Ethernet Virtual Connections (EVCs), EVC End Points (EPs), User Network Interfaces (UNIs), and External Network Interfaces (ENNIs) that can ride over VLANs, Label-Switched Paths (LSPs) of MPLS, etc.

A VPN consists of multiple remote endpoints (typically routers) joined by some sort of tunnel over another network. Two such endpoints constitute a point-to-point Virtual Private Network (VPN) as depicted in Fig. 4.17. Connecting more than two endpoints by putting in place a mesh of tunnels creates a multipoint VPN.

In recent years, the introduction of virtualization has driven development of new virtual networks and services (i.e., overlay networks and services). Most of the virtualized systems can be represented as in Fig. 4.18 [23] where a virtualized hardware and operating system via a virtualization layer supports virtual machine (VM)

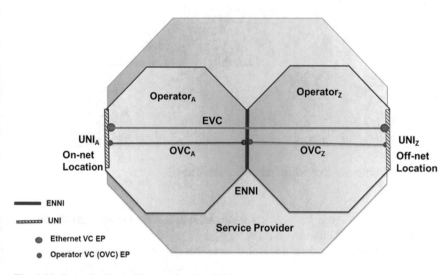

Fig. 4.16 Example Carrier Ethernet Service, E-Line

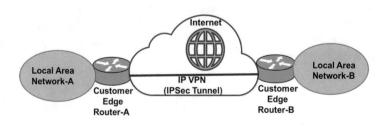

Fig. 4.17 A point-to-point IP VPN

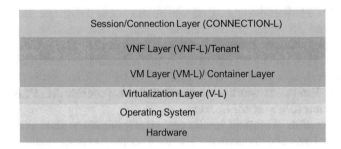

Fig. 4.18 A virtual system [23]

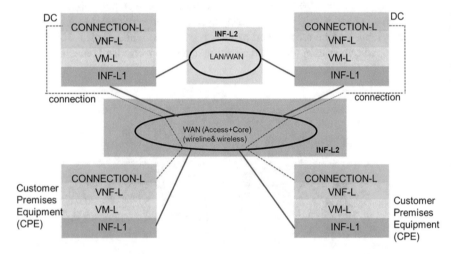

Fig. 4.19 An example of virtualized network and services [23]

and/or container layer. In turn, the VM/container layer supports virtual network function (VNF) and connection layer.

A virtual network and service architecture that employ virtual systems described in Fig. 4.18 are depicted in Fig. 4.19. The virtual system layers below VM/container layer are called infrastructure Layer-1 (INF-L1), while the segment of the network formed by legacy monolithic devices is called the infrastructure Layer-2 (INF-L2). The virtual services (i.e., Cloud Services) ride over connection layer.

An example of virtual services is Software-defined wide-area network (SD-WAN) services that have been deployed in recent years in the industry. We will describe that in the following section.

4.2.2.1 SD-WAN

Growth in public cloud applications drives a cloud-friendly enterprise network model. Enterprise customers require cloud-ready, inexpensive, and easily managed network service. SD-WAN is designed to allow enterprises not only to connect their branch offices using traditional wireline networks such as MPLS WAN but also to provide connectivity to public cloud applications.

Virtual Private Network (VPN) provides WAN connections over the public network infrastructure. However, it has several disadvantages:

- Configuring VPN services is a time-consuming and complicated process. During this process, customers make an order for their required services at the SP's business portals. Then, Operators at the SPs take days to handle the requests including manual authorization and careful network configurations. After that, VPN services can be launched and provided to customers.
- Although VPN provides end-to-end connections through manual installation, it is difficult to establish diverse connections such as interconnecting multiple data centers in short time.
- Achieving on-demand bandwidth allocation requires frequent changes of network configurations. In the VPN, such changes may take days or weeks to activate. Similarly, adjusting class of service (CoS) types of MPLS VPN typically takes days.
- Network policies are required in VPN services, and configuring them often needs network operators to manually install rules into corresponding devices (e.g., firewall). Such complicated and error-prone process makes it impossible to provide customized policies.

Cloud applications require on-demand WAN services to provide desired connections, on-demand bandwidth, and customized network policies. As an overlay network, SD-WAN uses the SDN principle that separates the control plane from the data plane and allows users to run the service over various underlay networks including Internet. This provides great flexibility to users and SD-WAN Service Providers.

SD-WAN is capable of utilizing both the internet and the existing MPLS network in order to offer the best possible WAN optimization for the business. Therefore, SD-WAN and MPLS are able to coexist if there is sensitive traffic that can justify the MPLS cost.

An example SD-WAN network is shown in Fig. 4.20 that has a centrally enforced security and application flow policy which would manage and secure all MPLS, broadband, and wireless links. With zero-touch provisioning (ZTP), one can simply ship the SD-WAN Edge device to the required site for automatic SD-WAN access. The task of setting up policies per site, per tenant, and per department becomes quick and simple with the aid of an intuitive user interface (UI) and automated workflows.

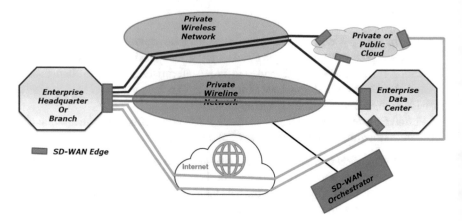

Fig. 4.20 An example SD-WAN network

SD-WAN aims to improve the user experience at the application level by constantly monitoring the quality of service (QoS) parameters (e.g., delay, jitter, loss) to ensure that the application data is sent over the available service-level agreement (SLA) compliant path.

SD-WAN enables the abstraction of configuration into business policy definitions that span multiple data plane components and also remain stable over time, even as the underlay network changes. Policy definitions can refer to users and groups, applications, and service levels (i.e., SLAs) that user should receive.

The control plane provides the programming flexibility and centralization over a diverse and distributed data plane. The underlay networks are abstracted allowing self-provisioning delivery model.

SD-WAN provides consolidated monitoring and visibility across the variety of underlay physical transports and Service Providers, as well as across all remote sites. This monitoring capability offers business-level visibility, such as application usage and network resource utilization.

Detailed performance monitoring across all components of the data plane that is coupled with the business policies make intelligent steering of application traffic across different paths and resources within the virtual WAN network.

MEF defined SD-WAN services [24]. An SD-WAN service provides a virtual overlay network that enables application-aware, policy-driven, and orchestrated connectivity between SD-WAN User Network Interfaces (UNIs) and provides the logical construct of a L3 Virtual Private Routed Network for the Subscriber that conveys IP Packets between Subscriber sites.

An SD-WAN Service operates over one or more Underlay Connectivity Services. Since the SD-WAN service can use multiple disparate Underlay Connectivity Services, it can offer more differentiated service delivery capabilities than connectivity services based on a single transport facility.

An SD-WAN service is aware of, and forwards traffic based on, Application Flows. The Service agreement includes specification of Application Flows and Policies that describe rules and constraints on the forwarding of the Application Flows.

SD-WAN benefits can be manifested in the ability to adjust aspects of the service in near real time to meet business needs. This is done by the Subscriber by specifying desired behaviors at the level of familiar business concepts, such as applications and locations and by the Service Provider by monitoring the performance of the service and modifying how packets in each Application Flow are forwarded based on the assignment of policy and the real-time telemetry from the underlying network components.

MEF SD-WAN Services have three components as shown in Fig. 4.21:

- SD-WAN Virtual Connection (SWVC)
- SD-WAN Virtual Connection End Point
- SD-WAN UNI

Figure 4.21 depicts one of the Subscriber sites connected to a Private or Virtual Private Cloud which may not be located at the Subscriber's physical location where SD-WAN Edge is usually a virtual network function (VNF), whereas for the other

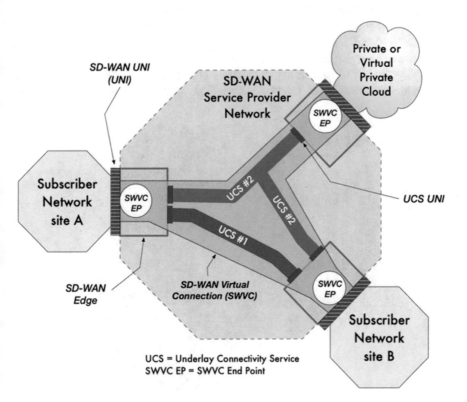

Fig. 4.21 SD-WAN service components [24]

sites, the SD-WAN Edge could be either a physical network function (PNF) or a VNF.

The SD-WAN Edge is a physical or virtual network function that includes the SWVC End Point. The SD-WAN Edge has the UNI on one side (left) and the Underlay Connectivity Services on the other side (right).

Underlay Connectivity Services can include a variety of services such as Ethernet Services [25], IP Services [26], L1 Connectivity Services [27], and public Internet Services. Each Underlay Connectivity Service terminates at its own UNI as shown in Fig. 4.22.

Figure 4.22 also shows Tunnel Virtual Connections (TVCs) across the Underlay Connectivity Services. An SD-WAN Service Provider typically builds point-to-point paths called TVCs across the various Underlay Connectivity Services that compose an SD-WAN Service. The SD-WAN Edge selects a TVC over which to forward each ingress IP Packet.

Each TVC built over an Underlay Connectivity Service can be private or public with performance and bandwidth constraints. It can be *encrypted* or *unencrypted.* By building point-to-point TVCs, a Service Provider creates a virtual topology that can be different from the physical topology of the Underlay Connectivity Service. For example, if one of the Underlay Connectivity Services is an EP-LAN service connecting all of the SD-WAN Edges, but the Service Provider only builds TVCs from the Headquarters site to each remote site (and not between the remote sites), then the SD-WAN Service is, effectively, a hub and spoke even though the Underlay Connectivity Service provides a full mesh.

Forwarding of IP Packets across different Underlay Connectivity Services with different attributes based on Policies applied to Application Flows is a key

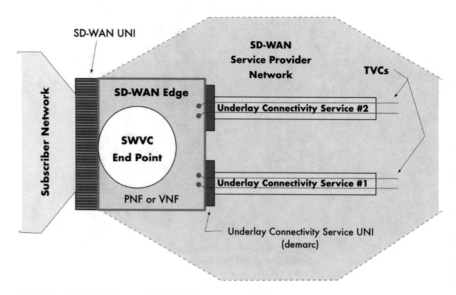

Fig. 4.22 SD-WAN Edge and TVCs [24]

characteristic of SD-WAN Services. The Subscriber and the Service Provider agree on the Application Flows that are identified at the SD-WAN Edges. For each of the agreed-on Application Flows, a Policy is assigned, which defines how IP Packets in the Application Flow are handled. For example, an Application Flow can be described by a broad set of characteristics of the packet stream identified at the UNI such as addresses, ports, and protocols. Also packets belong to an application such as voice or video can form an application flow.

Forwarding of an Application Flow is based both on the Policy assigned to the flow and IP forwarding requirements, which together determine the best TVC for forwarding each IP Packet in the Application Flow.

A Policy is a list of Policy Criteria. For example, IP Packets forwarded over the SWVC can be encrypted. An Encryption Policy Criterion provides a mechanism to specify whether or not encryption is required. Another Policy Criterion provides control over whether or not an Application Flow can traverse a public Internet Underlay Connectivity Service.

Policies are assigned to Application Flows and Application Flow Groups at each SWVC End Point. A policy provides details on how Ingress IP Packets associated with each Application Flow are handled by the SD-WAN Service, providing rules concerning forwarding, security, rate limits, etc.. Policies only apply to ingress IP packets. Packets that arrive at the SD-WAN Edge from other sites are forwarded to the UNI regardless of policies that are associated with their Application Flow.

4.2.2.2 Future Expectations

As we mentioned before, the overlay network is used to decouple a network service from the underlying infrastructure. This layering approach enables the underlay network to scale and evolve independently of the offered services. Growth in underlay wireless and wireline network capacities and network slicing techniques should accelerate the growth of overlay networks. We expect this trend to continue by 2030 and beyond.

4.3 Service Architecture

Section 4.2 described underlay and overlay network architectures. This section describes an architecture for services that can ride over these networks. The architecture is based on Cloud Services Architecture defined in [28–31].

Key actors of future services as depicted in Fig. 4.23 are:

- Network2030 User: A person or organization or a machine that maintains a business relationship with and uses service from a Service Provider.

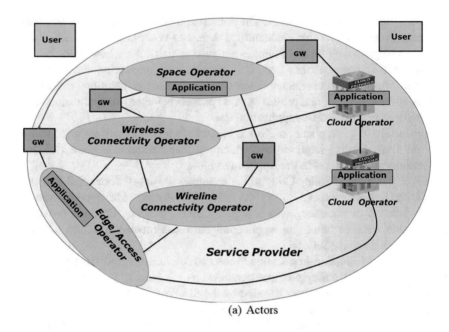

(a) Actors

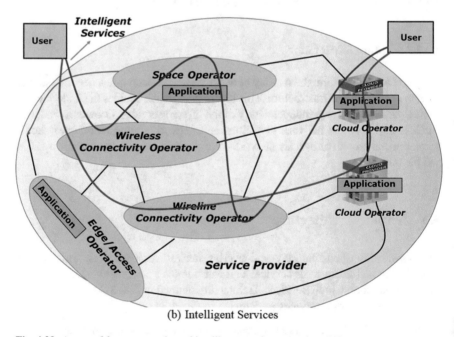

(b) Intelligent Services

Fig. 4.23 Actors of future networks and intelligent services over them [1]

- Operator: An organization with administrative control over connectivity and/or applications, and which provides services to other Operators or to Service Providers.
- Connectivity Operator: An organization that provides connectivity services to a Service Provider. In case of Internet, the connectivity operator is an Internet Service Provider (ISP).
- Space Operator: An Operator that may provide connectivity as well applications in the space.
- Edge/Access Operator: An operator that provides edge computing and/or access networking.
- Cloud Operator: An entity that is responsible for providing applications available to users or Service Providers. It can be public or private.
- Network2030 Service Provider (SP): An entity that is responsible for the creation, delivery, and billing of services and negotiates relationships among Cloud Operators, Connectivity Operators, Space Operators, and Users. It is the single point of contact for the user.

Today, a Service Provider which is responsible from a service over Internet end-to-end does not exist. Whether Internet Service Providers (ISPs) will act as a Service Provider by 2030 or not remain to be seen. However, this concept will allow us to define relationships among entities involved in providing an Internet service and automate the processes to support best effort as well as high-quality services dynamically for a user without manual coordination among Internet providers in advance. The services supported by Network2030 are called as Intelligent Services due to their expected end-to-end automation and dynamicity.

4.3.1 Characteristics of Future Services

Network2030 includes connectivity and application functionalities with greater flexibility and automation in service order, provisioning, monitoring, and billing. Some of its characteristics are [28–30]:

- Consisting of virtualized components such as virtual network functions (VNFs) and cloud-native network functions (CNFs), and non-virtualized components such as Physical Network Functions (PNFs).
- Consisting of network functions with just non-virtualized components (PNFs) or both virtualized components (VNFs and CNFs) and non-virtualized components (PNFs).
- Consisting of applications built with virtualized components (VNFs and CNFs).
- Consisting of connections provided by one or more Public Cloud Operator (s), Private Cloud Operator (s), Fixed and Wireless Connectivity Operator (s), Edge/Access Operators, and Space Operator (s).
- Consisting of applications provided by one or more Cloud Providers, Space Operators, and Edge/Access Operators.

- Supporting best effort as well as highly available (i.e., higher than five of nine availability), highly secure, and high-precision services requiring bandwidth from Gbps to Tbps.
- Supporting elasticity for dynamic service configurations by users and locations of the service functionality.
- Supporting service monitoring and usage-tracking by users.
- Supporting programmability, self-service by users, and collaboration among Operators.
- Supporting scalability of resources dynamically.
- Supporting various high availability options from physical layer to application layer.
- Supporting "pay as you use" (i.e., usage based billing).

It is expected that Network2030 Service Providers strike a balance among programmability, self-service by users, and the Service Provider control of resources to ensure integrity, security, and availability of Network2030. Service providers may need to place appropriate controls for the self-service and programmability to avoid possible unintended failures.

4.3.2 Interfaces

A Network2030 User interfaces to a Service Provider's facilities via a User Interface, **Network2030 User Network Interface (Network2030 UNI)**, as depicted in Fig. 4.24 which is implemented over a bidirectional link that provides various data, control, and management capabilities required by the Service Provider to clearly demarcate two different domains involved in the operational, administrative, maintenance, and provisioning aspects of the service.

The user in Fig. 4.24 can be an enterprise with multiple users sharing the same Network2030 UNI.

Depending on the service offering, the protocol stack for Network2030 UNI can be from Layer-1 to Layer-7 (Fig. 4.25). For example, if the service offering is a connectivity service, then the Network2030 UNI is a Layer-2 interface for Carrier Ethernet Services and a Layer-3 interface for IP services. If the service offering is a multimedia service, then Network2030 UNI is a Layer 7 interface..

In all layers of interfaces depicted in Fig. 4.25, we expect to see artificial intelligence (AI)/machine learning (ML) as a common capability.

Figures 4.24 and 4.25 describe an interface between a Network2030 User and a Network2030 Service Provider. Standard interfaces between users and private Service Providers have been defined for services of Layer-1 [27], Layer-2 [25], Layer-3 [26], SD-WAN [24], and Cloud [29]. However, there is no Service Provider for services provided over Internet. We expect to use Network2030 UNI for end-to-end services provided over Internet by ISPs, Cloud Operators, etc. that we call as Network2030 Service Provider.

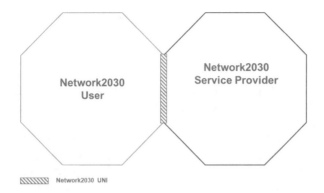

Fig. 4.24   Network2030 UNI

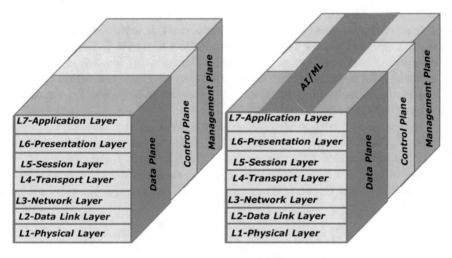

Fig. 4.25   Network2030 UNI Protocol Stack with and without AI/ML functionalities

It is clear that attributes for the Network2030 UNI will be different for different connectivity and application layers. They are expected to be defined and standardized in parallel to the evolution of Network2030.

Some of the attributes of the Network2030 UNI are described in [31]:

- Physical layer attributes such as the attributes of Ethernet.
- Connectivity attributes such as number of V-LANs and Ethernet Virtual Connections supported.
- Application attributes such as Virtual Machine (VM) and Virtual Network Function (VNF) attributes.
- Traffic management attributes such as bandwidth.
- Resiliency attributes such as access link redundancy.
- Fault management attributes such as alarms/events associated with the interface failures.

- Performance management attributes associated with measurements at the interface.
- Security attributes for securing user access to Network2030 services.
- Billing attributes.

The Network2030 SP establishes connections, Network2030 Virtual Connections (Network2030 VCs), among Network2030 users and Network2030 Applications as depicted in Fig. 4.26.

The Network2030 Application has an interface called Network2030 Application Interface. This interface marks the boundary between Cloud Applications and elements of connectivity for the Network2030 Service. Network2030 Application Interface can be an interface to a VNF, CNF, VM, Container, or Application. Fig. 4.27 depicts the interface for VNF and CNF.

The protocol stack for Network2030 Application Interface can be from Layer-2 to Layer-7 as depicted in Fig. 4.28.

In providing services to a Network2030 user, two Operators interface each other via an Operator-Operator Interface, **Network2030 External Network Network Interface (ENNI)**, as depicted in Fig. 4.29. Network2030 ENNI is defined as a reference point representing the boundary between two Operators that are operated as separate administrative domains. This reference point provides demarcation between two Operators for services.

Network2030 ENNI protocol stack is the same as the protocol stack for Network2030 UNI in Fig. 4.16. Depending on the service offering, the protocol stack for Network2030 ENNI can be from Layer-1 to Layer-7. For example, if the service offering is a connectivity service, then the Network2030 UNI is a Layer-2 interface for Carrier Ethernet Services and a Layer-3 interface for IP services. If the

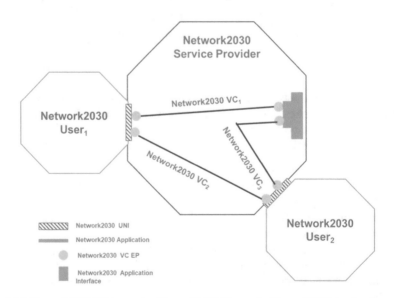

Fig. 4.26 Network2030 VC between a Network2030 User and a Network2030 Application

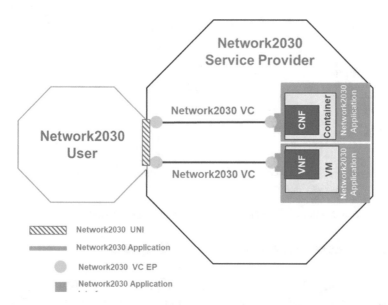

Fig. 4.27 Network2030 Application Interface for VNF and CNF

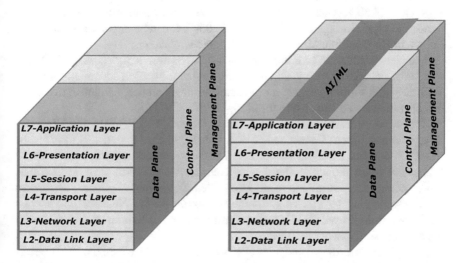

Fig. 4.28 Network2030 Application Interface Protocol Stack with and without AI/ML Functionalities

service offering is a multimedia service, then Network2030 ENNI is a Layer-7 interface.

Some of the attributes of the Network2030 ENNI are described in [31] as well. They are expected to vary from one service to another. They are expected to be defined and standardized in parallel to the evolution of Network2030.

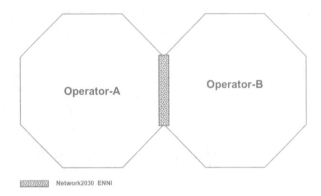

Fig. 4.29 Two Operators interfacing each other via Network2030 ENNI

4.3.3 Connections and Connection Endpoints

A Network2030 VC is an association of two or more Network2030 VC Endpoints. A Network2030 VC EP is a construct at an interface that maps a subset of the protocol data units (PDUs) that pass over the interface.

Network2030 services ride over Network2030 VCs among users and applications as depicted in Fig. 4.30. The Network2030 VC can be in point-to-point, multipoint, or three configurations among users and applications.

The Network2030 VC EP is an endpoint of a Network2030 VC when the VC is within the boundaries of one administrative domain. Interface identifier, availability, bandwidth profile, parameters of security functionalities, administrative state, and operational state are among the attributes of the VC EP.

The VC is an association of two or more Network2030 VC EPs. The VC could be an Ethernet Virtual Connection (EVC), Label-Switched Path (LSP), IP VPN, or SD-WAN connection. Identifiers of the VC EPs associated with this Network2030 VC, connection type, service-level agreement (SLA), redundancy, connection start time, connection duration, connection period, billing options, Maximum Transmission Unit (MTU) which is the maximum size of Network2030 service PDUs[1] transmitted over the VC, and administrative and operational states are among the attributes of Network2030 VC.

The VC EP is a logical endpoint of a Network2030 VC where the VC is terminated at an interface to which a particular set of Network2030 service PDUs that traverse the interface is mapped. As an example, the particular set could be identified by attributes such as application identifier, source and/or destination IP address, C-Tag VLAN ID, etc., depending on the service PDU.

---

[1] PDU exchanged at Network2030 interfaces (i.e., Network2030 UNI, ENNI, and Application Interface) is called a Network2030 service PDU.

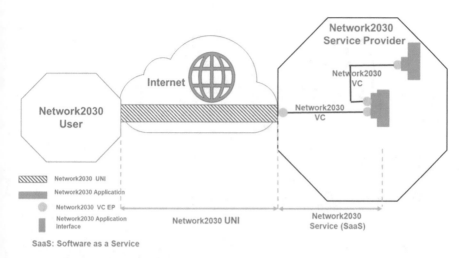

Fig. 4.30 An example of Network2030 Services

Service PDUs transported over a VC in both ingress and egress direction are tracked at interfaces to ensure alignment with service-level objectives (SLOs) and identify possible service problems.

The VC may cross multiple Operator domains as depicted in Fig. 4.31. Each domain will carry a segment of the VC. The segment in each Operator domain and its endpoints are called Operator Network2030 VC and Network2030 VC EPs, respectively.

Another example of a Network2030 service architecture is depicted in Fig. 4.32 where two Operators provide connectivity and applications, while one Operator provides just connectivity. The cloud applications form a service function chaining (SFC) via Operator Network2030 VCs connecting Netwokr2030 Application Interfaces of VNFs.

In this figure, Operator-A could be a private network Operator acting as the Service Provider in addition to providing connectivity as the Connectivity Operator, Operator-B could be a Public Cloud Operator, and Operator-C could be a Space Operator.

4.4   Management of Future Networks and Services

Section 4.3 described a service architecture that can be used in modeling and building future services explained in Chapter 3. This section will describe how to manage Network2030 networks and services.

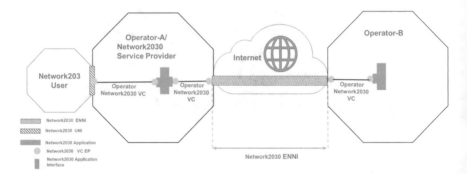

Fig. 4.31 Segments of a Network2030 VC supported by two Operators

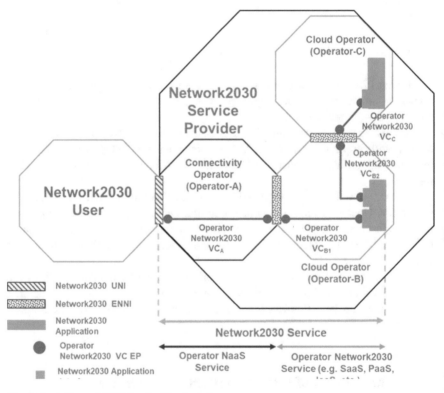

Fig. 4.32 A Network2030 Service Configuration

4.4.1 Management Functions

Some of the key emerging applications and services in Chapters 2 and 3 that are
expected to have substantial impact on the management of future networks are [1]:

- In-time and on-time services such as manufacturing automation, remote surgery, and haptic communications.
- Tightly coordinated services such as self-driven cars.
- High-throughput multimedia services such as those supporting holographic applications.

These services have the following characteristics:

- High Precision: This requirement asks for meeting stringent SLOs. It may force a decentralized management architecture to react events timely.
- No Graceful Degradation: Traditional network services degrade gracefully when service levels deteriorate. For example, when latency and/or jitter increase gradually or slightly, the quality of experience will be negatively affected and decrease by some degree. For example, reduction in the color-depth resolution of a video or reduction in its Mean Opinion Scores may result in slight service degradation. Despite of this slight deterioration, the service as a whole and applications relying on it may be usable. In contrast, Network 2030 services may not degrade gracefully; instead, even a slight deterioration may rapidly lead to a complete breakdown of the quality of experience which renders associated applications unusable.
- For example, in haptic communication services, extended latency might result in a loss of the illusion of haptic control needed to operate remote machinery confidently.
- This requirement asks for not only a highly available service design but also a decentralized management architecture.
- Mission Criticality: Some of Network2030 services are for mission-critical services for which occasional failure is not an option. For example, loss of control when operating remote machinery may result in risks for public safety.
- This requirement also asks for a highly available service design. The five of nine availability for most of Network2030 services is inadequate.
- High Accuracy Measurements: High-precision service requirement imposes the need to measure SLOs with high accuracy and in real time. For example, delay and jitter may need to be measured with microseconds or fraction of milliseconds granularity.

- This requirement impacts performance management architecture, compute, and storage requirements.

With the Network2030 service architectural constructs (i.e., interfaces, connection, and connection endpoints) that are described in the previous section, current and emerging services with key requirements listed above are expected to be modeled and implemented by identifying service-specific attributes and managing them accordingly. The services will be implemented over both underlay and overlay networks. The management of services requires management of networks and applications.

The following management functions need to be performed end to end for future services [32]:

1. **Order Fulfillment and Service Control** that support the orchestration of provisioning-related activities for the fulfillment of a customer (i.e., user) order or a service control request, including the tracking and reporting of the provisioning progress. In Network2030, we expect these functions are to be performed dynamically with a performance that is acceptable to users. These functions can be grouped as:

   (a) **Order Fulfillment Orchestration** that involves in decomposing a customer order into one or multiple service provisioning activities and orchestrating of all customer order-related fulfillment activities.
   (b) **Service Configuration Orchestration** which is responsible for the design, assignment, and activation activities for the end-to-end service and/or some or all service components.
   (c) **Service Control Orchestration** that permits the service to be dynamically changed within specific bounds described in policies that are established in advance or created on the fly with the Intent-Based Networking (see Chap. 14) approach.
   (d) **Service Delivery Orchestration** which is responsible for the service delivery via network and application implementation delegation of each service component to their respective delivery system or mechanism.
   (e) **Service Activation Testing Orchestration** that coordinates all service activation testing activities, for parts and/or the complete end-to-end service. The testing can be performed by Service Provider as well as by User.

2. **End-to-End Service Testing Orchestration** which is automating all test functions such as Service Activation Testing and In-Service Testing, and verification of services, seamlessly, across multiple Operators.

   The end-to-end service testing orchestration may require orchestration and control of the different systems capable of conducting tests and reporting on services that may be implemented within the infrastructure, the element control managers or can be deployed on demand, in the form of VNF or CNF.

   As the different locations and network elements involved in the fulfillment of end-to-end services may not all be available at the same time, the Service Testing Orchestration flexibility allows for real-time staggered testing, from simple unit level connectivity tests to end-to-end comprehensive Service Activation Testing.

   Customer (i.e., User) acceptance is received from the Customer. The Customer may view their particular services test results, or under special agreement with their Service Provider, be able to perform a set of predefined service acceptance tests.

3. **Service Problem Management** which is alarm surveillance, including the detection of errors and faults related to service, either end-to-end or per service component, and fixing failures automatically. In Network2030, we expect self-managed networks and services [33] to be implemented using artificial intelligence and machine learning techniques as described in Chap. 15

Customers are able to track the service impact of failures and status of trouble resolution.

4. **Service Quality Management** includes the collection of service performance information (e.g., delay, loss, availability, etc.) in support of key quality indicators across all Operators depicted in Fig. 4.23 who participate in delivering the service. This also includes gathering of feedback from the Customer, including Customer-provided performance measurements. Service quality is analyzed by comparing the service performance metrics with the service quality objectives described in the service-level agreement (SLA) between Customer and Service Provider. The results of the service quality analysis are provided to the Customer as well as information about known events that may impact the overall service quality (e.g., maintenance events, congestion, relevant known problems, demand peaks, etc.)

   The real-time measurements of SLA parameters and routing traffic accordingly based on customer-defined service policies are being exercised today by SD-WAN services. In Network2030, frequency of these measurements and their granularities could be enhanced.

   Service Quality Management capabilities also include capacity analysis in support of traffic engineering, traffic management, and service quality improvement.

   As new applications and verticals with new business models requiring high precision appear in future networks and services, verification of delivered service levels will become important to billing and charging.

5. **Billing and Usage Measurement** capabilities enable Operators to gather and provide usage measurements, traffic measurements, and service-related usage events (e.g., changes in service bandwidth, etc.) describing the usage of service components and associated resources. Exception reports may be generated to describe where service components and resources have been used beyond the usage commitments as defined in the SLA.

6. **Security Management** provides for the protection of management and control mechanisms, controlled access to the network and applications, and controlled access to service-related traffic that flows across the network and applications within and across Operators. Such security management capabilities support the authentication of users and applications and provide access control to the variety of capabilities on APIs supporting management and control based on the roles assigned to each authorized user. The security management capabilities include encryption and key management to ensure that only authenticated users are allowed to successfully access the management and control entities and functions, and preventing unauthorized modification/deletion of data. The security management takes responsive steps, such as applying filtering controls on specified traffic flows, when a specific threat and attack for networks is identified.

   The security management also provides audit trails for communications or ensure communications do not cross certain geographical boundaries.

As business services and cloud applications use Internet for connectivity and Service Providers supporting application development platforms for third parties, security risk increases. Secure Access Service Edge (SASE) and Zero Trust Security (ZTS) techniques for secure access to applications based on identity, context, and policy adherence are used. Some of these security capabilities are [34]:

(a) Ensuring sensitive data is accessible only under the right context and selectively route application traffic for additional security.
(b) Using DNS, preventing access to malicious domains, monitoring network traffic for connections to known phishing domains, and preventing possible data exfiltration.
(c) In-line traffic monitoring that provides real-time insights and policy control over web connections through an integrated forward proxy.
(d) Applying smart usage policies such as context aware and fine-tuned across groups, categories, and usage scenarios to protect organizations from unnecessary usage risks and ensuring that data usage is optimized in line with contextual factors.
(e) Securely adopting applications within the service edge by performing a continuous assessment of the developers that published the applications, the permissions they requested, and implementation details such as the transport security selected and the libraries and software development kits (SDKs) embedded.
(f) Ensuring that cloud-hosted applications are available only to those users and devices that are dynamically determined to meet the access requirements.
(g) Monitoring Wi-Fi wireless and cellular network traffic for unsanctioned, suspicious, or malicious behavior.
(h) Advanced malware protection against device, applications, and network risks.
(i) Identifying "normal" patterns of behavior within the service edge, flagging anomalies, and stopping risky behaviors before sensitive assets are put at risk.

Details of Security and Privacy capabilities are addressed in Chapter 13.

7. **Analytics** capabilities are for supporting the fusion and analysis of information among management and control functionality across management domains in order to assemble a relevant and complete operational picture of the end-to-end services, service components, and the supporting network and application infrastructure—both physical and virtual. Analytics ensures that information is visible, accessible, and understandable when needed and where needed to accelerate decision-making. For example, the analytics may utilize service fulfillment, control, and usage information to predict and trend service growth for the Connectivity and Cloud Operators. Chap. 15 discusses analytics application at the Edge.

8. **Policy-Based Management** is prescribing the management behavior by a set of rules under which the orchestration, management, and control logic operate. Service policies may be encoded in such rules in order to describe and design the dynamic behavior of services.

   A coordinated service relies on the orchestration of distributed capabilities across potentially Operators to enable end-to-end management. The policy-based management capabilities provide rules-based coordination and automation of management processes across administrative domains supporting effective configuration, assurance, and control of services and their supporting resources.

   Service design policies may enable the design and creation of end-to-end automated services. Service objectives may be implemented as sets of policies with event-triggered conditions and associated actions, as well as intent-based policies. Such policies would adjust the behavior of services and service resources including bandwidth, traffic priority, and traffic admission controls, allowing services to adapt rapidly to dynamic conditions in order to satisfy critical, ever-changing needs and priorities.

   The policy-based management is expected to use Intent-based Networking (IBN) and AI/ML techniques that are described in Chaps. 14 and 15.

9. **Customer Management** involves Service Provider (SP) interaction with potential Customers to determine serviceability of a Product Offering and if the underlying infrastructure is both capable and available to support the desired service for the Customer. Furthermore, the customer management may involve in dedicating physical and/or logical resources including service management resources to the Customer. This could be a part of network slicing services offered to a customer

   In future networks, we expect that Network Slices and services on a Network Slice are to be requested dynamically from Service Provider via Customer Network Management.

10. **Partner Management** involves in Service Provider interaction with Partners for service feasibility and service provisioning, and service control after the service is initiated. For certain services such as those related to Internet of Things (IoT), it is likely to have service run-time interactions between SP and Partner. Note that the Partner could be a Connectivity Operator or a Cloud Operator for the SP.

4.4.2 Management Architecture

Future Networks and Services need a management architecture that can support the functionalities described above without manual interventions. The end-to-end automation is the key component of this management architecture in a single Operator domain and multiple Operator domains. Achieving this objective may be somewhat easier by federated private networks compared to independent public networks.

Given there is no Service Provider responsible for the end-to-end management of a service over Internet, the user and/or ISP is expected to be responsible for the end-to-end service life cycle management. This can be accomplished if all the processes associated with service life cycle management are automated even if the service is supported by multiple Operators.

A high-level management architecture is depicted in Figs. 4.33 and 4.34, where each Operator providing a segment of Network2030 assigns an Orchestrator and Operation Support Systems (OSS)/Billing Support Systems (BSS) to manage all the resources and associated services in its domain and interoperate with Orchestrators and OSS/BSS of other Operators involved in the same service. The user is allowed to interact with the Orchestrator and OSS/BSS of his/her Internet Provider (ISP) as in the Lifecycle Service Orchestration (LSO) architecture in Fig. 4.34.

Figure 4.33 assumes automated end-to-end management of services riding over integrated resources of terrestrial and space infrastructure that make Network2030, via federated OSS/BSS and Orchestrators. Each domain consists of management components to manage virtualized and non-virtualized resources of sub-domains (e.g., Infrastructure Control Management, Network Function Virtualization Orchestrator (NFVO), Virtual Network Function Manager (VNFM)) and management components to manage nodes in each subdomain (e.g., Element Management System (EMS), Virtual Infrastructure Management).

In the federated OSS/BSS and Orchestrators, the interaction among OSS/BSS and Orchestrators takes place over standards interfaces. Figure 4.34 depicts the standards interfaces between a user and a SP, management entities of a SP and a Partner within their own domains, and between OSS/BSS and Orchestrators of SP and Partner, where Partner is another Operator providing a segment of the end-to-end service provided by the SP to the user. These standards interfaces have been defined in [32]. Their brief descriptions are:

- CANTATA is the interface that provides a Customer Application Coordinator with capabilities to support the operations interactions such as ordering, billing, and trouble management via trouble ticketing with the Service Provider (SP)'s

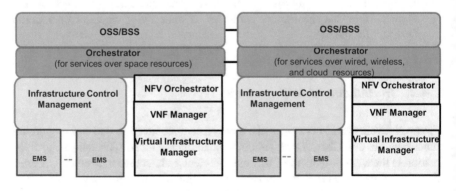

Fig. 4.33 End-to-end Orchestration of Future Networks and Services

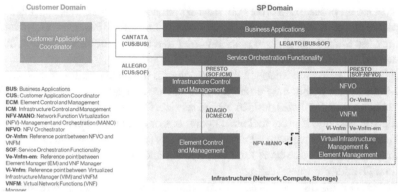

(a) LSO Architecture for a SP with both Virtualized and non-Virtualized Components

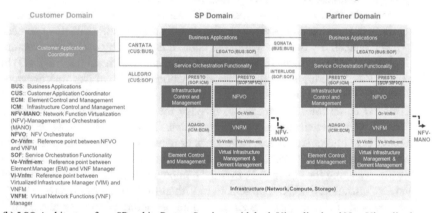

(b) LSO Architecture for a SP and its Partner Services, with both Virtualized and Non-Virtualized Components

Fig. 4.34 Lifecycle service orchestration [32]

Business Applications for a portion of the SP service capabilities related to the customer's products and services.

- ALLEGRO is the interface that allows Customer Application Coordinator supervision and control of dynamic service behavior of the LSO service capabilities under its purview through interactions with the Service Orchestration Functionality (SOF).
- LEGATO is the interface between the Business Applications and the SOF allowing management and operations interactions for supporting services. For example, the Business Applications may, based on a Customer order, use Legato to request the instantiation of a service.
- SONATA is the interface supporting the management and operations interactions such as ordering, billing, and trouble management between two network Operators. For example, the SP Business Applications may use Sonata interface

to place an order to a Partner provider for an access service that is needed as a part of an end-to-end Connectivity Service. Similarly, the SP Business Application may use Sonata interface to place an order to a Partner for an application that is needed for a Cloud Service.

- INTERLUDE is the interface that provides for the coordination of a portion of LSO services within the Partner domain that are managed by a SP's SOF within the bounds and policies defined for the service. Over the INTERLUDE, the SOF may request initiation of technical operations or dynamic control behavior associated with a service with a Partner domain. INTERLUDE interface may also be used to share service-level fault and performance information with the partner domain and/or request testing.
- PRESTO is the interface needed to manage the network infrastructure, including network and topology view-related management functions.
- For example, the SOF will use Presto (SOF:ICM) to request ICM to create connectivity or functionality associated with specific service components of an end-to-end Connectivity Service within the domain managed by each Infrastructure Control and Management (ICM). Similarly, SOF can use Presto (SOF:NFVO) to request ICM to configure virtual network functions (VNFs) or Network Services (NSs) of a Cloud Service.
- ADAGIO is the interface needed to manage the network resources, including element view-related management functions. For example, ICM can use ADAGIO to implement cross-connections or network functions on specific elements via the Element Control and Management (ECM) functionality responsible for managing the element. For virtual components, in the NFV-Management and Orchestration (MANO) architecture, Virtual Infrastructure Manager (VIM) is responsible for managing the virtualized infrastructure of an NFV-based solution, keeping an inventory of the allocation of virtual resources to physical resources. This allows for the orchestration of allocation, upgrade, release, and reclamation of Network Functions Virtualization Infrastructure (NFVI) resources and the optimization of their use. The VIM also supports the management of VNF forwarding graphs by organizing virtual links, networks, subnets, and ports.

In order to achieve true end-to-end automation for Network2030, functionalities of these interfaces need to be expanded greatly. Furthermore, Application Programming Interfaces (APIs) of the standards interfaces among Operators and between user and Operators are expected to play key roles in achieving the automation. They also need to be expanded substantially to support new services such as self-driven car and holographic services with management functionalities mentioned in Sect. 4.4.1.

The architectures described in Figs. 4.33 and 4.34 are expected to support services with best-effort SLOs such as Internet Access as well as services with very stringent SLOs such as self-driven cars. The management of services requiring very low delay and nonzero loss is expected to use Infrastructure layer (e.g., SDN Controller, VNFM) and/or EMS layer mostly while supporting the end-to-end coordination with the Orchestrator.

As we mentioned in Section 4.4.1, high-precision applications and services have been driving locating resources to close to customer locations and decentralized management architecture. This is highly visible in Mobile Edge Computing (MEC) Services [35].

The ETSI MEC management architectures with and without ETSI MANO architecture are depicted in Fig. 4.35. The MEC Orchestrator (MEO) in Fig. 4.35a is replaced by a MEC Application Orchestrator (MEAO) in Fig. 4.35b that relies on the NFV Orchestrator (NFVO) for resource orchestration. The reference points between the architectural functional blocks are defined as:

- Mm1: Reference point between multi-access edge orchestrator (MEO) and OSS to trigger instantiation and termination of MEC applications.
- Mm2: Reference point between OSS and MEC platform manager for MEC platform configuration, fault, and performance management.
- Mm3: Reference point between MEO and MEC platform manager for the management of the application lifecycle, application rules and requirements, and keeping track of available MEC services.
- Mm4: Reference point between MEO and Virtualization Infrastructure Manager (VIM) to manage virtualized resources of MEC host, including keeping track of available resource capacity and to manage application images.
- Mm5: Reference point between MEC platform manager and MEC platform to perform platform configuration, configuration of application rules and requirements, application lifecycle support procedures, management of application relocation, etc.
- Mm6: Reference point between MEC platform manager and VIM to manage virtualized resources.
- Mm7: Reference point between VIM and virtualization infrastructure to manage the virtualization infrastructure.
- Mm8: Reference point between user application lifecycle management proxy and OSS to handle device applications requests for running applications in MEC system.
- Mm9: Reference point between user application lifecycle management proxy and multi-access edge orchestrator of MEC system to manage MEC applications.
- Mx1: Reference point between OSS and the customer facing service portal to request MEC system to run applications in the MEC system.
- Mx2: Reference point between user application lifecycle management proxy and the device application to request MEC system to run an application in the MEC system, or to move an application in or out of the MEC system.
- Mm3*: Reference point between MEAO and MEPM-V (MEC Platform Manager-NFV) which is based on the Mm3 reference point, to cater for the split between MEPM-V and VNFM performing MEC applications Life Cycle Management (LCM).
- Mv1: Reference point between MEAO and NFVO which is related to Os-Ma-nfvo reference point.

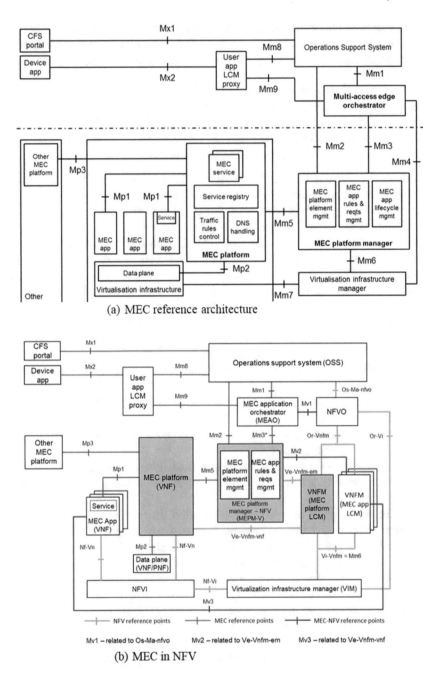

(a) MEC reference architecture

(b) MEC in NFV

Fig. 4.35 MEC system architecture and its integration with NFV architecture [35]

- Mv2: Reference point between VNF Manager (VNFM) that performs the LCM of the MEC application VNFs and the MEPM-V, allowing LCM-related notifications to be exchanged between these entities. It is related to the Ve-Vnfm-em reference point.
- Mv3: Reference point between the VNFM and the MEC application VNF instance, allowing the exchange of messages such as those related to MEC application LCM or initial deployment-specific configuration. It is related to the Ve-Vnfm-vnf reference point.

The author made an attempt to modify the LSO architecture to combine MEF LSO architecture and ETS MEC architectures [36] as depicted in Fig. 4.36. This should simplify the MEC management architecture and allow reuse of LSO APIs for the MEC services orchestration.

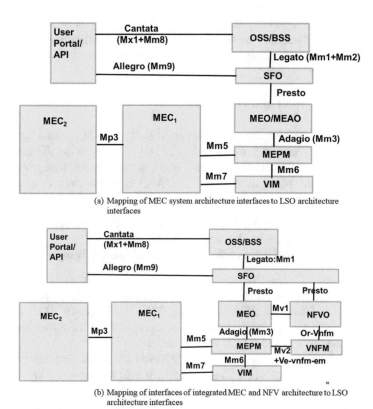

(a) Mapping of MEC system architecture interfaces to LSO architecture interfaces

(b) Mapping of interfaces of integrated MEC and NFV architecture to LSO architecture interfaces

Fig. 4.36 Management architecture that combines ETSI MEC and MEF LSO architectures [36]

4.4.3 Automation and APIs

Key components of the end-to-end automation are the APIs of standards interfaces/ reference points described in Sect. 4.4.2 to support current and future networks and services.

There are substantial efforts in standards organizations such as MEF [37], ETSI NFV[38], ETSI MEC [39], TMF [40], 3GPP [41], and IETF [42, 43]; and open-source organizations such as ONAP [44] and LF Akraino [45] to develop APIs for product qualification, quoting, and ordering; and service provisioning, on-demand service modifications, inventory, VNF and CNF management, application management,[2] and platform and element management.

Most of these APIs use REST architectural principles, JSON data format for requests and responses, and well-defined objects representing components of networks and services.

These APIs need to be expanded to tie resource views (i.e., element views) and service views and support capabilities of future networks and services.

In the following sections, we will describe example APIs for service quoting and ordering, service provisioning, and on-demand modifications.

4.4.3.1 Service Order and Provisioning

Service qualification, quoting, and ordering APIs for the LSO Sonata interface as shown in Fig. 4.34 have been developed by MEF [37]. Service qualification, quoting, and ordering APIs for Cantata interface and service provisioning APIs for Legato interface are being developed. These APIs have two main components as depicted in Fig. 4.37:

- Envelope which is independent of services and consists of information exchanges and notifications between a Buyer and a Seller of services.
- Payload which is service-specific information exchanges between a Buyer and a Seller.

The Envelope API is influenced by other standards APIs such as REST/Open APIs, RESTConf/Yang or NETCONF/Yang, TOSCA Templates, TMF, and ONF TAPI. On the other hand, the Payload carries data specific to MEF services such as Access E-Line, IP Services, SD-WAN Services, etc.

The API development for a specific LSO interface starts with use cases and requirements for that interface. The use cases for Sonata interface are shown in Figs. 4.38, 4.39, and 4.40 (Tables 4.3, 4.4, and 4.5).

APIs and related draft specifications for address validation, product qualification, quoting, ordering, inventory, and notifications for MEF Aretha release are given in [46, 48, 50–54]. The lists of API files are:

---

[2] Software Application that is not VNF or CNF

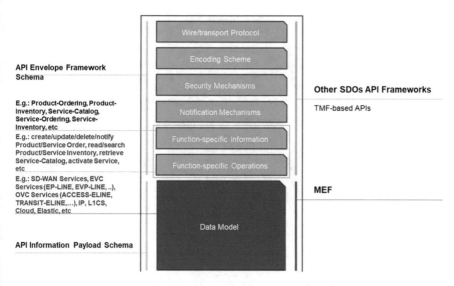

Fig. 4.37   MEF API framework [46]

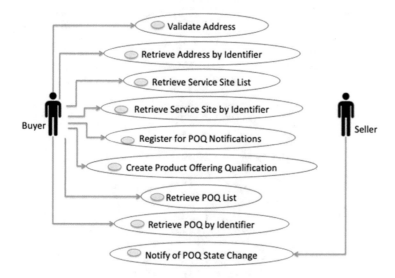

Fig. 4.38   Product offering qualification (POQ) use cases [47]

- geographicAddressManagement.api.yaml – v 5.0.0-RC2
- productOrderManagement.api.yaml – v 5.0.0-RC2
- productOrderNotification.api.yaml – v 5.0.0-RC2
- productOfferingQualificationManagement.api.yaml – v 5.0.0-RC2
- productOfferingQualificationNotification.api.yaml – v 5.0.0-RC2
- quoteManagement.api.yaml – v 5.0.0-RC2
- api/inventory/productInventoryManagement.api.yaml

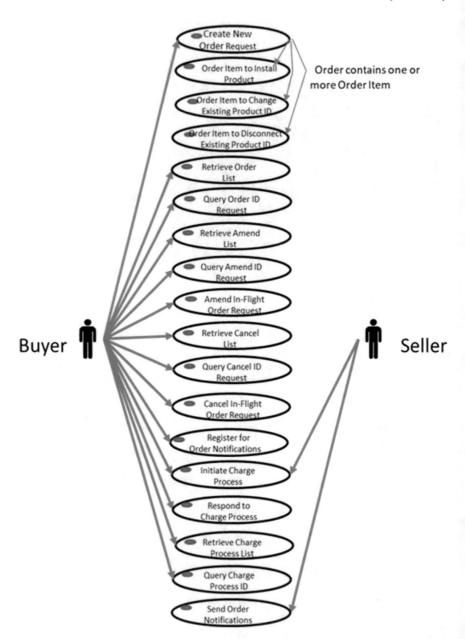

Fig. 4.39 Product order use cases [48]

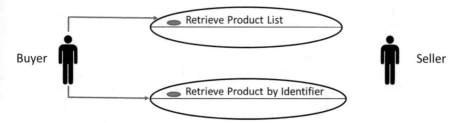

Fig. 4.40 Product inventory retrieval use cases [49]

Table 4.3 POQ use cases [47]

Use case #	Use case name	Use case description
1	Validate Address	The Buyer sends Address information known to the Buyer to the Seller. The Seller responds with a list of Addresses known to the Seller that likely match the Address information sent by the Buyer. For each Address returned, the Seller should also provide an Address Identifier, which uniquely identifies this Address within the Seller
2	Retrieve Address by Identifier	The Buyer requests the full details of a single Address based on an Address identifier that was previously provided by the Seller
3	Retrieve Service Site List	The Buyer requests that the Seller provides a list of Service Sites known to the Seller based on a set of Service Site/Address filter criteria. For each Service Site returned, the Seller also provides a Service Site Identifier, which uniquely identifies this Service Site within the Seller
4	Retrieve Service Site by Identifier	The Buyer requests the full details for a single Service Site based on a Service Site identifier that was previously provided by the Seller
5	Register for POQ Notifications	A request initiated by the Buyer to instruct the Seller to send notifications of POQ state changes (see Sect. 9) in the event the Seller uses the Deferred Response pattern to respond to a Create Product Offering Qualification request
6	Create Product Offering Qualification	A request initiated by the Buyer to determine whether the Seller can feasibly deliver a particular Product (or Products) to a specific set of geographic locations (if applicable). The Seller also provides estimated time intervals to complete these deliveries
7	Retrieve POQ List	The Buyer requests a summarized list of POQs (in any state; see Sect. 9.1) from the Seller based on a set of POQ filter criteria. For each POQ returned, the Seller also provides a POQ Identifier that uniquely identifies this POQ within the Seller
8	Retrieve POQ by Identifier	The Buyer requests the full details of a single Product Offering Qualification based on a POQ identifier.
9	Notify of POQ State Change	The Seller sends the following types of notifications to the Buyer who has subscribed to these notifications • POQ creation. • POQ state change.

Table 4.4 MEF product order use case summary [48]

Use Case #	Use case name	Use case description
1	Create New Order Request	A request initiated by the Buyer to order a new product or service component(s). A New Order Request contains at least one Order Item (Use Case # 1-a, 1-b, or 1-c) as shown below. A New Order Request may contain more than one Order Item, and Order Items do not need to have a relationship between them but the must all be covered by the same Project ID or Agreement ID
1-a	Order Item to Install Product	Order Item installs a new product
1-b	Order Item to Change Existing Product ID	Order Item changes an existing Product ID
1-c	Order Item to Disconnect Existing Product ID	Order Item disconnects an existing Product ID
2	Retrieve Order List	A request initiated by the Buyer to request a list of Orders that match the requested filter criteria
3	Query Order ID Request	A request initiated by the Buyer to query the details associated with a specific Order specified by the Order ID
4	Retrieve Amend List	A request initiated by the Buyer to request a list of Amend Requests that match the requested filter criteria
5	Query Amend Request ID Request	A request initiated by the Buyer to query the details associated with a specific Amend Request specified by the Amend Request ID
6	Amend In-Flight Order Request	A request initiated by the Buyer to modify/amend an In-Flight Order
7	Retrieve Cancel List	A request initiated by the Buyer to request a list of Cancel Requests that match the requested filter criteria
8	Query Cancel Request ID Request	A request initiated by the Buyer to query the details associated with a specific Cancel Request specified by the Cancel Request ID
9	Cancel In-Flight Order Request	A request initiated by the Buyer to cancel an In-Flight Order
10	Initiate Charge Process	Process to communicate charges from the Seller to Buyer
11	Respond to Charge Process	Process to communicate if the Buyer accepts or rejects the charges
12	Retrieve Charge Process List	A request initiated by the Buyer to request a list of Charge Processes that match the requested filter criteria
13	Query Charge Process ID	A request initiated by the Buyer to query the details associated with a specific Charge Process specified by the Charge Process ID
14	Register for Order Notifications	The Buyer requests to subscribe to notifications
15	Send Order Notifications	A notification initiated by the Seller to the Buyer providing subsequent status information on Orders

Table 4.5 Use case table [49]

Use case #	Use case name	Use case description
1	Retrieve Product List	The Buyer requests a list of Products from the Seller based on filter criteria
2	Retrieve Product by Identifier	The Buyer requests the details associated with a single Product based on a Product Identifier

- api/order/productOrderManagement.api.yaml
- api/order/productOrderNotification.api.yaml
- api/quote/quoteManagement.api.yaml
- api/quote/quoteNotification.api.yaml
- api/serviceability/address/geographicAddressManagement.api.yaml
- api/serviceability/offeringQualification/productOfferingQualificationManagement.api.yaml
- api/serviceability/offeringQualification/productOfferingQualificationNotification.api.yaml
- api/serviceability/site/geographicSiteManagement.api.yaml
- doc/cantata-sonata/carrierEthernet/epl/Carrier_Ethernet_Bandwidth_Profile.html
- doc/cantata-sonata/carrierEthernet/epl/Carrier_Ethernet_Class_of_Service.html
- doc/cantata-sonata/carrierEthernet/epl/Carrier_Ethernet_Color_Identifier.html
- doc/cantata-sonata/carrierEthernet/epl/Carrier_Ethernet_Egress_Maps.html
- doc/cantata-sonata/carrierEthernet/epl/Carrier_Ethernet_End_Point_Maps.html
- doc/cantata-sonata/carrierEthernet/epl/Carrier_Ethernet_External_Interfaces.html
- doc/cantata-sonata/carrierEthernet/epl/Carrier_Ethernet_L2CP.html
- doc/cantata-sonata/carrierEthernet/epl/Carrier_Ethernet_Link_Aggregation.html
- doc/cantata-sonata/carrierEthernet/epl/Carrier_Ethernet_Operator_UNI.html
- doc/cantata-sonata/carrierEthernet/epl/Carrier_Ethernet_Service_Level_Specification.html
- doc/cantata-sonata/carrierEthernet/epl/Carrier_Ethernet_Subscriber_UNI.html
- doc/cantata-sonata/carrierEthernet/epl/Ethernet_Private_Line_EVC.html
- doc/cantata-sonata/carrierEthernet/epl/Utility_Classes_and_Types.html

The Legato Service Catalog, Service Order, Service Inventory, and Service Notification APIs in essence allow the Business Applications (BUS) to request Service Orchestration Functionality (SOF) to configure and activate one or more services as part of an order fulfillment process.

Business Applications (BUS) in Fig. 4.34 request Service Orchestration Functionality (SOF) over the Legato interface to configure and activate one or more services as part of an order fulfillment process. The Legato Service Catalog, Service Order, Service Inventory, and Service Notification APIs support these interactions between Business Applications and Service Orchestration Functionality (SOF).

The following steps describe the high-level flow:

- As part of the ordering flow, the BUS system receives the product order (through Cantata or Sonata) which triggers the fulfillment processes in the BUS system.
- The BUS system first queries the Service Catalog to retrieve the ServiceSpecifications supported by the SOF:

  – Each specific instance of a ServiceSpecification (retrieved from the Service Catalog) minimally contains a reference to target service schema. A service schema describes the set of properties that characterize that service and are exchanged over Legato.
  – The BUS may register for notifications on specific ServiceSpecifications.

- During the service configuration and activation phase, the BUS system uses the Service Order API to instantiate the Service utilizing the ServiceSpecifications (retrieved from the *Service Catalog*).

  – The BUS achieves this by creating a ServiceOrder which contains a one or more ServiceOrderItems.
  – Each ServiceOrderItem carries some ServiceConfiguration data and the type of operation (add/delete/modify) to be performed by SOF.
  – The SOF utilizes Service schema referenced in the ServiceSpecification to validate the ServiceConfiguration data passed in by the BUS.
  – The ServiceOrder/ServiceOrderItem is processed by the SOF as per the state transition rules described in Service Order State Transitions.
  – The BUS may register for notifications on specific ServiceOrders/ServiceOrderItems.

  In such cases, the SOF also reports the ServiceOrder/ServiceOrderItem state changes as per the Service Order State Transitions.

  – The SOF performs the actions (add/delete/modify) specified in a ServiceOrderItem on the specified target Service instance in the Service Inventory as per the state transition rules described in Service State Transitions.
  – The BUS may register for notifications on Service instances.

  In such cases, the SOF also reports the Service instance state changes as per the Service State Transitions.

- The BUS system uses the same Service Order API to create new service instances as well as update **existing** service instance's properties, trigger state transitions, and delete existing service instance.

Legato draft APIs and associated draft MEF specifications are defined in [55–59]. Available draft APIs are:

- carrierEthernetCommon.yaml
- carrierEthernetEnni.yaml
- carrierEthernetEvc.yaml
- carrierEthernetOperatorUni.yaml
- o-carrierEthernetOvc.yaml

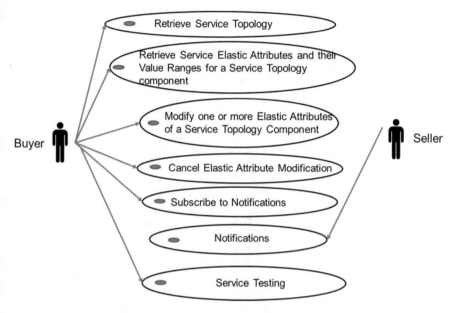

Fig. 4.41 Dynamic Service Modification Use Cases [14]

- carrierEthernetSls.yaml
- carrierEthernetSubscriberUni.yaml
- carrierEthernetVirtualUni.yaml

4.4.3.2   Dynamic Service Modification

The main objective for dynamic service modifications is to avoid service ordering process for any service changes to reduce the time interval for the modification. Dynamic modifications of Carrier Ethernet E-Line Service attributes have been addressed in [60]. The use cases associated with dynamic service modification of Connectivity and Cloud Services [14, 61] are depicted in Fig. 4.41 and Table 4.6.

The dynamic service attribute modification process is summarized as follows:

1. From LSO Allegro interface, customer requests service attribute changes within attribute bounds either Immediately or Scheduled:

   (a) Immediately.

   - With no end time for the new values of attributes.
   - With end time for the new values of attributes.

   (b) Scheduled.

   - With no end time for the new values of attributes.
   - With an end time for the new values of attributes.
   - Recurring with an end time for the new values of attributes.

Table 4.6 Dynamic modification use cases [14]

Use case #	Use case name	Use case description
1	Retrieve Service Topology	The Buyer requests retrieving the Service Topology from the Seller (e.g., UNIs, UNI location, ENNI, ENNI location, connectivity between UNIs and ENNI)
2	Retrieve Service Elastic Attributes	The Buyer with appropriate requests the list of one or more elastic attributes and their value ranges of one or more Service Topology Components
3	Modify an elastic attribute of Service Topology Component	The Buyer requests dynamic modification of an elastic attribute value of a Service Topology Component
4	Cancel Scheduled Elastic Attribute Modification	The Buyer cancels an elastic attribute modification request scheduled at a time and day in the future
5	Service Testing	The Buyer either performs or requests testing a service attribute or a service topology component: 1. Buyer requests the Seller to perform the testing. 2. Buyer requests access to test resources of the service.
6	Notifications	The Seller sends notifications to the Buyer related to the testing of an attribute or a topology component
7	Subscribe for Notifications	A request initiated by the Buyer to instruct the Seller to send Notifications

2. Time intervals for on-demand modification of an attribute Immediately can be defined in the contract between a Service Provider and a customer $(T_{sp\text{-}cust})$,[3] and between a Service Provider and a Partner $(T_{sp\text{-}part})$.[4] $T_{sp\text{-}part}$ is expected to be smaller than $T_{sp\text{-}cust}$. For example, if $T_{sp\text{-}cust}$ is 15 min, $T_{sp\text{-}part}$ could be 10 min.
3. The time interval for fulfillment between Service Provider and customer, $t_{sp\text{-}cust}$, can be recorded. In the customer contract, there can be a penalty associated with the requests that are not fulfilled within $T_{sp\text{-}cust}$.
4. The time interval for fulfillment between Service Provider and Partner, $t_{sp\text{-}part}$, can be recorded. There can be a penalty associated with the requests that are not fulfilled within $T_{sp\text{-}part}$.
5. The customer may request from customer portal a monthly history report consisting of $t_{sp\text{-}cust}$ and $t_{sp\text{-}part}$.

4.5 Conclusion

In this chapter, we have described underlay and overlay networks and services, and likely future advances in these networks and services. We provided an overall architecture for future services and their management. At the end, we described APIs that are the key components of automated future networks and services.

---

[3] $T_{sp\text{-}cust}$ may not be the same for all on-demand attributes.

[4] $T_{sp\text{-}part}$ may not be the same for all on-demand attributes.

The following chapters are aimed to provide further details of these architectures and supporting technologies.

References

1. ITU-T technical report 2020, FGNET2030 architecture framework (2020)
2. Akraino Edge Stack APIs, https://www.lfedge.org/wp content/uploads/2020/06/Akraino_ Whitepaper.pdf
3. RFC 8293, A framework for multicast in network virtualization over layer 3 (2018)
4. https://vivadifferences.com/10-difference-between-underlay-and-overlay-networks/
5. Cisco, Converged transport architecture: improving scale and efficiency in service provider backbone networks, https://www.cisco.com/c/en/us/products/collateral/routers/carrier-routing-system/white_paper_c11-728242.html
6. Juniper Networks, Evolving backbone networks with an MPLS supercore (2015), https://www.juniper.net/assets/mx/es/local/pdf/whitepapers/2000392-en.pdf
7. D. Wang et al., Toward universal optical performance monitoring for intelligent optical fiber communication networks. IEEE Commun. Mag. **58**(9), 54–58 (2020)
8. P. Lechowicz, Greenfiled gradual migration planning toward spectrally-spatially flexible optical networks. IEEE Commun. Mag. **57**(10), 14–19 (2019)
9. X. Chen et al., Building autonomous elastic optical networks with deep reinforcement learning. IEEE Commun. Mag. **57**(10), 20–26 (2019)
10. S. Huang et al., 5G-orineted optical underlay network slicing technology and challenges. IEEE Commun. Mag. **58**(2), 13–19 (2020)
11. M. Toy, MEFw76, Service control business requirements and use cases (2021)
12. MEF 22.3.1, Amendment to MEF 22.3: transport services for mobile networks (2020)
13. 3GPP TR 21.905 V17.0.0 (2020-07) technical report 3rd Generation Partnership Project; Technical Specification Group Services and System Aspects; vocabulary for 3gpp specifications (Release 17)
14. M. Toy, MEF76, Business requirements and use cases for access E-line service control (2019)
15. 3GPP TS 23.501 version 16.5.0 [Release 16] ETSI TS 123 501 V16.6.0 (2020-10), 5G; system architecture for the 5G System (5GS) (3GPP TS 23.501 version 16.6.0 Release 16)
16. 3GPP TS 23.251 V6.6.0 (2006-03) technical specification 3rd Generation Partnership Project; Technical Specification Group Services and System Aspects; network sharing; architecture and functional description (Release 6)
17. Mobile World Congress Daily, GSMA intelligence, The 5G evolution: 3GPP release 16-17 (2018)
18. https://www.statista.com/statistics/802690/worldwide-connected-devices-by-access-technology/
19. M.W. Akhtar et al., The shift to 6G communications: vision and requirements, Electronics (1) (2020)
20. D. Stiliadis, Network virtualization: overlay and underlay design (2013), https://www.nuage-networks.net/blog/network-virtualization-overlay-and-underlay-design/
21. MEF 10.4, Subscriber Ethernet service attributes (2018)
22. MEF 26.2, External Network Network Interfaces (ENNI) and operator service attributes (2016)
23. M. Toy, High availability layers and failure recovery timers for virtualized systems and services. Procedia Computer Science **114**, 126–131 (2017)
24. MEF, MEF70 SD-WAN service attributes and services (2019)
25. MEF 6.2, EVC Ethernet services definitions phase 3 (2014)
26. MEF 61.1, IP service attributes (2019)
27. MEF 63, Subscriber layer 1 service attributes technical specification (2018)

28. M. Toy, Cloud services architectures. Procedia Comput. Sci. **61**, 213–220 (2015)
29. M. Toy, Cloud services architecture, MEF68 draft specification (2020).
30. ETSI GR NFV-EVE 016 V1.1.1 (2020-09) Network Functions Virtualisation (NFV); Evolution and ecosystem; report on connection-based virtual services
31. M. Toy, OCC 1.0 reference architecture (2014), https://wiki.mef.net/display/OCC/OCC+Spec ifications?preview=%2F63185562%2F63342484%2FOCC+1.0+Reference+Architecture.pdf
32. MEF 55.1, Lifecycle service orchestration (LSO): reference architecture and framework (2021)
33. M. Toy, Self-managed networks with fault management hierarchy. Procedia Comput. Sci. **36**, 373–380 (2014)
34. M.Z. Chowdhury et al., 6G wireless communication systems: applications, requirements, technologies, challenges, and research directions. IEEE Open J. Commun. Soc. **1**, 957–975 (2020)
35. E. Strinati, C. Strinati, et al., 6G: the next frontier: from holographic messaging to artificial intelligence using subterahertz and visible light communication. IEEE Veh. Technol. Mag. **14**(3), 42–50 (2019)
36. M. Toy, MEC services architecture and lifecycle orchestration (2020), https://wiki.akraino.org/display/AK/PCEI+Files
37. ETSI GS MEC 003 V2.2.1 (2020-12), Multi-access Edge Computing (MEC); framework and reference architecture
38. https://github.com/MEF-GIT/MEF-LSO-Sonata-SDK/
39. ETSI GS NFV-SOL 013 V3.4.3 (2021-02) Network Functions Virtualisation (NFV) release 3; protocols and data models; specification of common aspects for RESTful NFV MANO APIs
40. ETSI GS MEC 033 V2.0.3 (2020-12) Multi-access Edge Computing (MEC); IoT API
41. https://www.tmforum.org/open-apis/
42. ETSI TS 129 222 V15.0.0 (2018-07) Common API framework for 3GPP northbound APIs (3GPP TS 29.222 version 15.0.0 Release 15)
43. RFC 6316 sockets application program interface (API) for Multihoming Shim (2011)
44. RFC 7807 problem details for HTTP APIs (2016)
45. https://docs.onap.org/en/latest/guides/onap-developer/apiref/index.html
46. M. Bencheck, MEF 57.2 working draft 0.1 order management requirements and use cases (2020), https://wiki.mef.net/display/LSO/2020Q4-LC-Contributions?preview=%2F11800569 4%2F121966022%2FL75009_001_MEF+W57%232+CfC_Bencheck.docx
47. MEF 79, Address, service site, and product offering qualification management requirements and use cases (2019), https://wiki.mef.net/pages/viewpage.action?pageId=33129675
48. MEF W115, LSO Sonata Quote Management API—developer guide (2020)
49. MEF 81, Product inventory management requirements and use cases (2019), https://wiki.mef.net/pages/viewpage.action?pageId=33129675
50. Sonata API Developer Guide, https://github.com/MEF-GIT/MEF-LSO-Sonata-SDK-extended/blob/develop_IIS_Quote/doc/quote/v5/MEF_W115.pdf
51. MEF W121, LSO Sonata Address Management API—developer guide (2020)
52. MEF W122, LSO Sonata Site Management API—developer guide
53. Aretha release, https://github.com/MEF-GIT/MEF-LSO-Sonata-SDK/commit/9fa76d15e2 8b541c700d15088bc3f47584cea61a
54. https://github.com/MEF-GIT/MEF-LSO-Sonata-SDK/find/working-draft
55. LSO Legato Service Provisioning
56. API (MEF W99): OAS3 API/schema definitions as YAML file, developer guide as GFM file
57. LSO Legato Service Provisioning Specification—SD-WAN (MEF W100): JSON schema definitions as YAML files, requirements document as GFM file
58. LSO Legato Service Provisioning Specification—carrier Ethernet (MEF W101): JSON schema definitions as YAML files, requirements document as GFM file
59. LSO Legato Service Provisioning Specification—L1 (MEF W103): JSON schema definitions as YAML files, requirements document as GFM file
60. M. Toy, Mapping ETSI MEC architecture to MEF LSO architecture (2020), https://wiki.akraino.org/display/AK/PCEI+Files

61. MEF W117 draft 0.071 SASE service attributes and service framework (2021)
62. ETSI GS NFV-MAN 001 V1.1.1 (2014-12), NFV; management and orchestration
63. MEF 57.1/J-SPEC-001.1, Ethernet ordering technical specification business requirements and use cases (2018), https://wiki.mef.net/pages/viewpage.action?pageId=33129675
64. https://github.com/MEF-GIT/MEF-LSO-Legato-SDK-extended/tree/working-draft/spec/legato/carrierEthernet
65. https://www.tutorialspoint.com/gsm/gsm_architecture.htm
66. K.B. Letaief et al., The roadmap to 6G—AI empowered wireless networks, in *IEEE ComSoC JSAC Machine Learning Series*, 2019
67. M. Toy, Elastic metro Ethernet and elastic cloud services. Procedia Comput. Sci. **185**, 19–27 (2021)

Chapter 5
Access and Edge Network Architecture and Management

Jane Shen and Jeff Brower

5.1 Introduction

The need for network edge computing is constant and relentless. Since the 1970s when the first network edges had no storage or computing resources, CDNs were deployed to speed up web pages, telecom operators increased bandwidth to allow video playback, gateways and on-premise data centers appeared—all pushing computing closer to users. Today, the rise of AI is driving another massive increase in network edge computing.

With the advent of 5G, coupled with new computing technologies, the network edge is evolving a new paradigm where network and compute/storage combine to offer advanced services such as ultralow latency, enhanced mobile broadband, better control for users of their privacy and data, and more energy-efficient computing. User applications spanning multiple vertical domains stand to benefit, from connected vehicles to intelligent fleet management, from multiplayer mobile gaming to AR/VR real-time rendering, and from industrial IoT to manufacturing. With enhancements in computing and memory technologies, it becomes possible to support these applications by devices located at the edge of future networks and/or customer premises.

These advances are being driven by the following trends:

- Densification of the edge through placing micro data center capabilities.
- Innovation in future use cases, e.g., industrial automation, security, and proactive monitoring, robotic surgery.

J. Shen (✉)
VP of Technology Strategy, Mavenir Systems, San Jose, CA, USA

J. Brower
CEO, Signalogic, Inc., Dallas, TX, USA
e-mail: jbrower@signalogic.com

© The Author(s), under exclusive license to Springer Nature
Switzerland AG 2021
M. Toy (ed.), *Future Networks, Services and Management*,
https://doi.org/10.1007/978-3-030-81961-3_5

- Economics of network by optimizing backhaul and transport capacity through localization of content, e.g., augmented reality/virtual reality (AR/VR) content, HD, ultra HD media content.
- Economics of network through multi-access edge computing (MEC) federation, collaboration, and infrastructure sharing.

Existing access and edge network operation is already capable of localized traffic steering, e.g., local internet breakout or local content mixing in entertainment. The aforementioned trends further extend such concepts in network engineering, with more innovation in technology and service domains expected.

A rapid increase in MEC deployment, localization of user plan, and data plane processing near ultradense access networks will require innovative approaches to designing future networks. These approaches need to be service oriented, adaptive to change in operating conditions including environment, secure, and capable of supporting multiple technologies at access and edge layers. Future networks need to be structured to provide easy integration with networks of multi-domains and collaboration between operators and users.

The following are a few elementary capabilities that future edge network will support:

- Use cases emerging from service designs in the areas of in-time and on-time services. Therefore, access and edge networks must provide guaranteed performance to support latency requirements associated with in-time and on-time services.
- Access networks need to operate in uncertain environments (e.g., wireless access prone to weather disturbance) and balance rapid and dynamic change in capacity utilization. They also must deal with natural and artificial noise that may affect appropriate service delivery. Therefore, access networks need a design that caters to complex operating scenarios so they can provide desired QoS by efficiently adapting to changes.
- Access and edge networks will support multi-access and user plane data routing to the most optimal access technology based on service and user profiles, as users may have subscribed to multiple access technologies, or they may be using telecom service for an essential or critical service.
- Access and edge networks are prone to security and privacy breaches. Therefore, specific security and data privacy considerations associated with emerging use cases will be supported in future networks.

5.1.1 Edge Definition

The term "edge" does not have a widely accepted definition. Depending on who you talk to, the definition may vary. In this chapter, the term "network edge" refers to communication and computing infrastructure in locations such as central offices, cell towers, stadiums, first responder sites, and others. These exist as either Telco premises or points of presence (PoPs). Usually owned and operated by telecom operators, a network edge is always tightly coupled with a communication network, as shown in Fig. 5.1, future network access and edge architecture.

5.1.2 Access and Edge Components

Access and edge network components can be categorized as shown in Fig. 5.2, access and edge network components:

Future edge network devices may be classified as:

- Human use devices
- Machine-operated devices
- Sensors

These devices must work intelligently in association with mobile or fixed-line networks and may also need to implement peer-to-peer communication. Device properties and characteristics that form their role in the network access layer become important in considering future network innovation.

Devices can access the connected world through fixed access or through the radio network. The radio network may be based on any technology; the first level of edge computing may happen just after termination of radio traffic. Normally, the termination is a radio access network (RAN) unit based on 5G technology or any similar unit or future technology. Further network traffic to next logical computing stages is provided by a network known as fronthaul. Typically, this is handled by a DU (distributed unit) or any similar future network unit. It is at this point enhanced edge computing capabilities can be deployed.

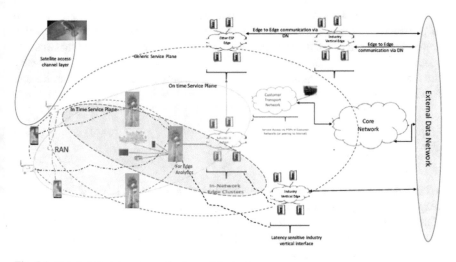

Fig. 5.1 Future network access and edge architecture

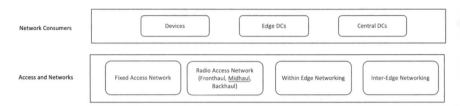

Fig. 5.2 Access and edge network components

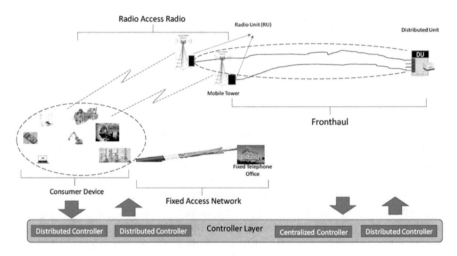

Fig. 5.3 Consumer devices and fronthaul

Use cases and communication service delivery platforms for quality-sensitive services (QoS) will require a truly integrated last mile in which customer services must access network agnostic. Therefore, access and edge network solutions need to provide similar performance irrespective of their underlying technology. Edge computing at each compute node provides the comprehensive outcome that ensures customers are able to consume services over diverse access networks. This is why the diagram in Fig. 5.3, consumer devices and fronthaul, includes both radio and fixed access scenarios.

5.1.3 Network Edge Roles

With rapid growth of network edge-related commerce, more roles will come into play, such as edge infrastructure owner, edge operator, and edge service provider. These roles are not restricted to telecom operators, although operators are always involved, due to their ownership of underlying components in the communication network(s) to which the network edge is connected.

5.1.4   MEC, 5G Edge, and Beyond

MEC (multi-access edge computing) is the ideal platform for network edge infrastructure and services. MEC tightly couples general edge computing and mobile networks, bringing public clouds closer to users. MEC allows users to access applications via the best route, be it mobile, Wi-Fi, or other supported access method. This was not possible with earlier edge computing attempts (examples include fog computing or cloudlet). Thus, MEC is the centerpiece of new network edge services.

The term MEC was first introduced by ETSI MEC SIG, to refer to mobile edge computing. This reflected the focus when ETSI MEC SIG WG was established in 2014: provide lower latency, context, and location awareness, and higher bandwidth. With the advance of edge computing, it became clear that "mobile" is only one of several accesses to the edge, although remaining the most important. Recently, ETSI MEC SIG has refined MEC terminology to include multi-access edge computing [1]. Currently, the full definition of MEC is "A system which provides an IT service environment and cloud-computing capabilities at the edge of an access network which contains one or more type of access technology, and in close proximity to its users." Multi-access includes radio network access, fixed network access, Wi-Fi access, and more [2].

5G is a game changer for network edge computing. For the first time, Telco operators are exposing key components of their core networks, essentially creating "smart pipes" instead of the "dumb pipe" model that has persisted for so many years. As one example, CUPS (Control and User Plane Separation) pushes computing closer to users. It can reduce the round-trip time (RTT) between mobile devices and edge computing nodes to several milliseconds (in URLLC with edge at the RAN). This is a significant RTT reduction from 80 ms + in 4G. NEF (Network Exposure Functions) [3] further enable applications to interact with the network in real time for traffic steering and QoS control. Multiple non-3GPP access methods support in 3GPP Release 16 brings standardization into multi-access aggregation.

MEC and 5G are closely related. However, since its inception, MEC does not mandate 5G, while 5G plays an essential role in MEC. A few early versions of ETSI MEC specifications were drafted with 4G as the access and core network. There are operators that started MEC in their 4G network and later migrated to 5G. MEC can work with all seven (7) 3GPP 4G/5G deployment model options. New 5G capabilities such as ultralow-latency, MBB, and eMTC are key enablers of MEC services and applications. 5G needs MEC to deliver its promises. Applications leveraging MEC edge services highlight the economic and business value of 5G networks.

5G MEC deployment locations are shown in Fig. 5.4, Typical 5G MEC deployment. These have the following characteristics and requirements:

- Type 1: Access edge location: limited space and power.
- Type 2: CO and other aggregation edge—restrictions in power and network wiring, limited space.
- Type 3: Regional DC (Data Center) edge—standard Telco DC.
- Type 4: Open space requires pre-integrated all-in-one MEC solutions.

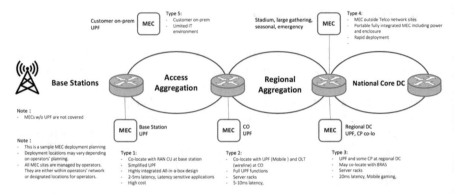

Fig. 5.4 Typical 5G MEC deployment [4]

- Type 5: Customer edge varies, usually providing a limited IT environment.

This broad scope adds complexity to hardware infrastructure. To accommodate underlying hardware varieties, we need to provide unified, consistent APIs open to upper layer applications, allowing platform implementations to vary.

MEC platforms also need to carefully address MEC-unique requirements:

- Accelerated hardware support, open for future additions.
- Multitenancy support.
- Edge-cloud collaboration.
- Inter-edge networking support.

There are many opinions of what a MEC platform looks like. Traditional platform vendors believe a unified MEC platform for all edge applications would help reduce MEC management complexity. Vertical industry application vendors would prefer a customized platform to target specific end user applications. Hardware vendors are competing to promote performance, reduced power consumption, reliability, and accelerator support. Some hyperscalers have their own customized hardware platforms, some even with their own virtualization technologies. Clearly, MEC platforms mean different things to different people.

As the edge is where we can expect innovative, new applications to emerge, there are always concerns that available platforms may not support new application requirements. For instance, with the heated competition of new AI chip rollouts, there could be applications that need AI chips and their connection fabrics for memory and data transport not yet supported in existing platforms. Operators must balance their natural desire to host easy-to-manage and consistent platforms with the need to keep up with fast-moving edge computing application requirements.

Because of this rapid pace in edge technology development, open is the buzz word in MEC platform yet open doesn't mean easy. How to make open-ness work in practice is challenging. Typically, operators do not have big engineering teams for platform development, relying instead on the vendor ecosystem. Even operational maintenance depends heavily on platform vendors. New platform requirements often arise from new applications with evolving platform needs, which require operators to integrate with existing platforms. Differences in platforms from various vendors bring extra operational maintenance work for operators. Successfully managing the diversity of vendors, new requirements, and reducing development cost is crucial for operators.

5.2   Edge Applications and Services

Application developers and users see the promise of new edge services. Applications that are latency sensitive, data intensive, and/or demand privacy and trustworthiness are at the top of the list. They span multiple vertical domains, from connected vehicles to intelligent fleet management, from multiplayer mobile gaming to AR/VR real-time high-resolution rendering, and from industrial IoT to discrete manufacturing digital transformation. Such applications are the driving force in edge service innovations. The synergistic reciprocal relationship between applications and edge services is moving both technologies forward.

Edge computing will be the opportunity for SPs (service providers) to realize new revenue generation. SPs need to combine computing and connectivity thoughtfully through edge innovations. The advantage of edge-based services comes from either proximity to end user locations, or shorter processing chains for applications. Low latency, data privacy, and security are edge service hallmarks, allowing applications to leverage edge services to improve performance and functionality. In this sense, edge services are considered by applications as premium services. Moreover, edge services can enable new application creations based on computing and connectivity innovations. These new edge services were not possible before.

The edge introduces business complexity. Edge is where many stakeholders collaborating to offer services. Understanding roles and players in the edge helps to bring in perspectives of edge service providers and supporters.

5.2.1   Service Types

5.2.1.1   Latency-Sensitive Edge

Low latency is one of the biggest attractions of an edge service. Improved latency derives from proximity to end user devices connected to the edge service deployment location. In a 5G network, lower latency can further be achieved by interacting with network capabilities such as application function influencing or network exposure functions, etc. For example, lower latency can be achieved through real-time QoS control.

There are two (2) types of low latencies: human-to-machine (H2M) and machine-to-machine (M2M). The former emphasizes applications where initial or intermittent lags would negatively impact users' technology experience. The latter emphasizes applications where sustained throughput is a defining characteristic because machines are able to process data much faster than humans.

5.2.1.2 Data Intensive

This includes use cases where data volume, cost, or bandwidth issues make it impractical to transfer over the network directly to the cloud or from the Telco and edge computing. Examples include smart cities, smart factories, smart homes/buildings, high-definition content distribution, high-performance computing, restricted connectivity, virtual reality, and oil and gas digitization.

- Bandwidth.
- Distributed massive data processing.

5.3 Architecture

Because the edge sits between users and clouds, multiple players must collaborate, including IT (Information Technology), CT (Communication Technology), and OT (Operation Technology). An architecture considering all of these requires both technical feasibility and operational flexibility.

Users are connected to the edge via an access network. Edge-to-edge or edge-to-cloud connectivity is provided by a transport network. There are three (3) types of networks involved in edge architecture: access network, edge interconnect network, and in-edge network. Requirements of these networks are different and so are the objectives and challenges.

In 5G, the network edge extends to within the RAN (radio access network), in order to offer ultralow latency services or network optimization such as x-hauls (see Sect. 5.3.2.2, Operational Architecture) to support determination of the shortest path from user devices to an edge computing node.

5.3.1 Requirements to Support Edge Services

The requirement of future network to support new edge services has been studied and investigated over years. With progressive 5G deployments across many countries, such services and their network requirements are increasingly getting focused.

- Future networks need to support use cases emerging from service designs in area of in-time and on-time services. Therefore, access network and edge network need to be designed that they can provide guaranteed performance to support latency requirements associated with in-time and on-time services.

- Access network needs to operate in very uncertain environment (e.g., it is prone to disturbance in weather if access is wireless); it needs to balance very rapid and dynamic change in capacity utilization, etc. It is also affected by natural and artificial noise that may affect appropriate service delivery. Therefore, access network needs a design that caters to complex operating scenarios so that it can provide desired QoS by adapting to changes in most efficient way.
- Access and edge network need to support multi-access technology and user plane data routing to the most optimal access technology based on service and user profile because user may have subscribed to multiple access technologies or user may be using telecom service for some essential or critical service.
- Access and edge network is area that is prone to security and privacy breaches. Therefore, specific security and data privacy considerations associated with emerging use cases need to be supported in future network.

5.3.2   General Framework

5.3.2.1   Functional Architecture

The main purpose of edge is to serve applications with better performance and data security, leveraging the close proximity of edge nodes to end user devices. In a typical edge functional diagram shown in Fig. 5.5, edge functional architecture, key functional blocks consist of edge access, edge infrastructure, an application enabler, and application itself. A wide range of interfaces, including physical hardware, functional software in various forms, data flow, and APIs will need to be specified and designed to meet the functional requirements. The edge border gateway and application enabler layers are key to abstract and expose edge access and edge infrastructure capabilities to end user applications.

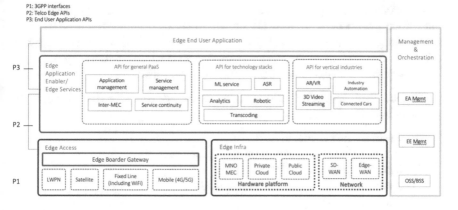

Fig. 5.5 Edge functional architecture [4]

5.3.2.2   Edge Border Gateway

An Edge Border Gateway faces the Telco network. Its mission is to allow operators to expose network information per 3GPP standard to support edge applications, such as AF traffic influencing, UPF reselection, QoS, etc. It provides industry standard APIs which are consumed by an application enabler. An Edge Border Gateway can be viewed as an "API gateway" into the core Telco network.

APIs exposed by an Edge Border Gateway require fundamental Telco network knowledge to understand and consume correctly. For example, typical RNIS (Radio Network Information Specification) APIs may specify a data model containing S1 bearer information. It's reasonable to expect mobile gateway developers to have this level of Telco network knowledge, but not mobile device application developers.

Abstracting 3GPP interaction yields significant benefits:

1. Provides a bridge between Telco network and edge applications, hiding Telco network function level interface complexity.
2. Allows Telco operator expose network capabilities to edge service developers.
3. Allows easy upgrade for future 3GPP standard evolution.
4. Provides a buffer zone to the Telco core network for better security control.
5. Allows operator to better control service differentiation.

It's expected that an Edge Border Gateway is owned and maintained by operators. It can function as a customizable non-3GPP network function for operators to offer various network enhancements.

5.3.2.3   Application Enabler

An application enabler [5] sits between the Edge Border Gateway and edge application developers. Its mission is to provide developer-friendly APIs, allowing app developers to consume and manage application-specific Telco network capabilities without extensive Telco network knowledge.

It is expected that application enablers are owned and maintained by public/private cloud providers, who must (a) support edge applications and services, lifecycle management on edge nodes, (b) connect edge nodes to cloud data centers, and (c) allow edge applications to run temporarily disconnected from the cloud.

An application enabler may include three (3) categories:

1. General PaaS layer APIs are application management, service management, including:

   (a) Resource management.
   (b) Application service management, e.g., registration.
   (c) Monitor, reporting, and notification.
   (d) Authorization, certificates, authentication.
   (e) Package manager.

2. Technology functional stacks, e.g., IoT, ML, and analytics. A Telco edge application stack is one of these. Functional stacks may include:

   (a) Message bus/broker.
   (b) Event bus.
   (c) Device management.
   (d) Data analytics service.
   (e) ML inference or learning service.

3. Vertical domain edge stacks, for example:

   (a) Gaming.
   (b) AR/VR.
   (c) Video streaming.
   (d) Connected cars.

   There could be multiple vendors providing one or a mix of the above functions, on one or more platforms, which may be different.

5.3.2.4 Edge Function Layer Ownership and Operational Models

Depending on individual Telco operator MEC strategies, edge stack ownership and business models will vary. Figure 5.6, Edge Layer ownership and operational models, shows four (4) models of stack ownership and operational responsibilities. Each rounded box represents one single ownership and associated operational responsibility. An operator may adopt a mix of models in order to achieve its business goal.

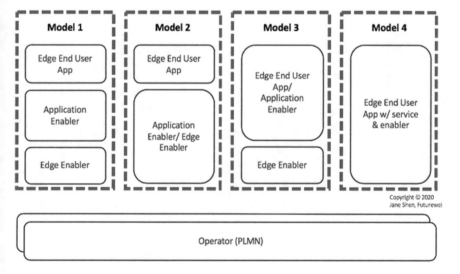

Fig. 5.6 Edge layer ownership and operational models [4]

Model 1

Model 1 follows a clear, layered ownership and operational responsibility matching our edge layer diagram in Fig. 5.6, edge layer ownership, and operational models. Here typically the edge enabler is owned by a Telco operator, application enablers are owned by one or more edge service providers, and end user edge applications are owned by application vendors.

APIs between layers help hide implementation details. A consistent, versioned API definition set can increase API adoption rate.

An example of this ownership model might look like this:

- In a smart city scenario, an operator will provide an edge enabler.
- A smart city platform vendor provides one or more application enablers supporting various smart city applications, e.g., smart meter, intelligent surveillance, smart traffic management.
- End user smart city application vendors deploy their respective applications on the smart city platform.

Model 2

In Model 2, an edge and application enabler combo can be provided by operators or trusted edge service providers, e.g., hyperscalers or neutral hosts. This is depicted in Fig. 5.7, network access and edge operational architecture. For example, Operator B is offering a solution similar in concept to Model 2, with a software stack including IaaS, PaaS, and edge management. The solution also provides a client facing service portal for easy DevOps and may be implemented in various hardware form factors.

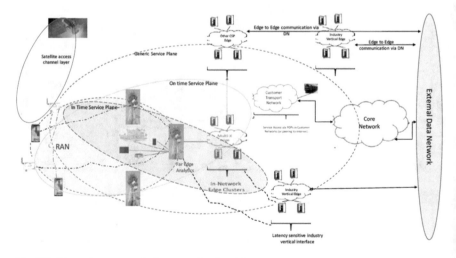

Fig. 5.7 Network access and edge operational architecture

For hyperscalers, this enabler combo can include edge extensions with unified cloud/edge application development/deployment platform. In this model, edge service providers include an edge enabler. This means they have the expertise to directly access the network and also an agreement with operators to do so.

An example of this model is a smart factory. An operator deploys a vertically integrated edge and application enabler with application management support, containing factory-specific AI-based maintenance (predictive analytics) and AR remote assistance service. Factory applications can be built on the enabler platform and perform factory equipment/environment anomaly detection, robotic production line inspection.

Model 3

In Model 3, the edge enabler is owned and operated by the operator. Application vendors have their own vertical stacks to directly interface with Edge Enabler northbound APIs. This saves the application the hassle of keeping track of operator network changes. The operator gains better separation between its network and application layers, which improves network security. In addition, the edge enabler gives the operator a buffer zone in which to provide additional and enhanced services.

In Model 3, typical application entities are large X2C service providers, for example, video streaming service providers who typically have developed and optimized the application platform for their applications. The only piece missing when they move to the Telco edge is an edge enabler—exactly what a Telco operator can offer. By providing consistent edge enabler APIs to 2C service providers, operators can open new revenue streams.

Model 4

In Model 4, the edge enabler, application enabler, and application are all owned and operated by one entity, which can be an operator or a major application vendor. In the case of an operator, they may provide a vertically integrated service such as a port-management edge service. As an example of a major application vendor, a global gaming vendor may reach agreement with an operator to directly access the operator's network at the 3GPP interface level and operate edge services on its own.

Operational Architecture

Future network access and edge operational architecture must be intelligently structured to support extreme operating conditions, e.g., in-time and on-time service delivery, security and privacy, energy efficiency, dynamic service configuration, ubiquitous coverage, technology independence, and rapid self-healing.

Future network access and edge operational architecture must adopt the following principles:

• Service-oriented virtual network-driven design to provide on-demand service plane for each service type, e.g., a dedicated plane for in-time service.
• Multi-technology access networks to provide unified features across heterogeneous technology. This allows services to be seamlessly ported between one device or network or technology to another.
• Distributed edge computing close to radio nodes for lower latency and rapid decision-making.
• Controllers to provide service, network, and infrastructure management in distributed and centralized formats, allowing localized configuration, management, and operation instructions to be implemented in real time, while centralized decisions are made at the end-to-end network controller agent layer.

Future network access and edge operational architecture are divided into the following parts:

• Device-centric network. Device-centric networks provide device-to-device direct links for local communication, e.g., device-to-device file transfer. Multiple devices can come together to form a specific purpose-based network to implement specific use cases, for example, local community discussions during lockdown. Setup and management can remain at macro-cell level or can be handed over to micro cells, whereas data flow is directly device-to-device.
• Radio access network (RAN). RAN is similar to prevailing radio units of mobile/wireless network:

  – Macro Cell—Macro cells are for wide-area coverage ranging in few miles.
  – Micro Cell—Micro cells are for very short distance and can be further segmented into personal cell, FEMTO cell, PICO cell, or any other form.

• Contextual/on-demand cell. Service providers may provision a cell for some contextual service that may be on-demand. For example, a virtual cell provisioned to support government administrative activity in high security service context during lockdown.
• Access radio termination. Access radio termination happens at the radio unit level and provides a backhaul link and is therefore a major point of interface in access networks. Remote edge compute capabilities may be established near radio units; therefore, access radio termination points provide a demarcation area for traffic related decision-making.
• Fronthaul, midhaul, and backhaul. Traditional backhaul is divided into these three segments. The idea is to provide different compute and decision-making capabilities at different points in backhaul networks. Fronthaul edge computing will have less latency but will be more sensitive to storage and algorithm complexity, whereas any edge computing capability in backhaul may end up adding more latency but help in more centralized operation and hence storage and

algorithm complexity can be built up. Midhaul may be considered as a trade-off between fronthaul and backhaul.

- X-haul termination. X-haul termination refers to termination points for fronthaul, midhaul, or backhaul used to create demarcation points and specify network segment as one of the three.
- Edge computing and analytics. Edge computing and analytics is a critical aspect of future networks, providing localization of data management, decision-making, and traffic offloading, therefore reducing latency as well as dependency on core networks.
- Far edge computing and analytics (Near Radio Unit). NRU capability refers to antenna and other associated radio equipment, hosted at locations most remote from all other network equipment.

5.3.2.5   Edge Network

An edge network usually refers to either the network within an edge node (intra-edge) or the network connecting edge nodes (inter-edge), as shown in Fig. 5.8, edge network model.

The intra-edge network addresses connectivity among edge infrastructure components including ingress/egress and application network. An in-edge network is typically not multilayered since individual edge nodes are typically small and constructed with a few servers or a server rack. However, an edge node may have multiple external network interfaces which require routing information exchange. For

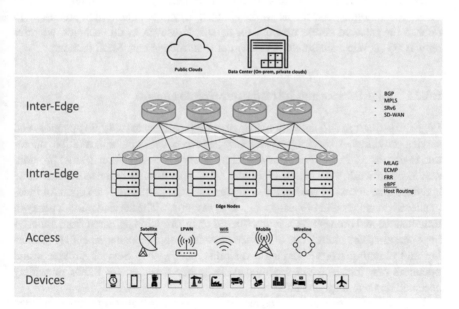

Fig. 5.8 Edge network model

example, for a typical 5G MEC edge node, ingress/egress traffic may come from various sources: RAN, Core, Enterprise DC, Cloud, etc. Making content aware switching and routing is crucial to efficiency and high performance. The network should support switching and routing on routing information carried within edge node functional components (e.g., UPF) as well as the edge node infrastructure components (such as VMs or containers). ECMP (equal-cost multi-path) routing is typically used for VM or physical server host routing. In MEC edge node scenario, ideally an integrated one-layer routing (e.g., in form of a MEC gateway) is deployed to provide all routing functions.

Inter-edge network addresses connectivity among edge nodes or between edge nodes and other data centers (public cloud, private cloud, or DCs). Depending on ownerships of edge nodes, an inter-edge network may span several operators' networks. Latency and fast convergence time are the two major requirements of an inter-edge network. Supporting 5G slicing is often required for MEC nodes, which makes the overlay network a must-have feature of inter-edge network. Inter-edge network can use either MPLS network or the SD-WAN technologies depending on the edge node latency requirement and cost considerations. In recent years, SRv6 technology further enhances SDN capabilities by allowing fast service provisioning, fast VPN connection establishment across network segments, simplified protocol stacks, and simplified system integration.

Connectivity between an end user device and an edge node is usually addressed by the access network. Two major access networks are mobile network and fixed-line network. There are also satellite access networks, LPWN (low-power wireless network) or LPWAN (low power wide-area network). These networks are positioned between devices and edge nodes. Since they each have employed unique access technologies, and the networks adopted vary significantly, it's preferable to refer to the relevant access network for details. However, in this chapter, we refer only to 5G mobile access networks in order to introduce new MEC features.

5.3.2.6   Edge Service and Infrastructure Management

Edge infrastructure and network management typically span multiple operators and service providers. A service provider may have to collaborate with multiple operators to deliver a service. For example, a low-latency service (under 10 ms) provider may need to work with MBB, FBB, and MEC operators in order to cover mobile and fixed-line access as well as the computing service platform. Each operator independently manages and orchestrates its own resources. The service provider may in turn manage and orchestrate resources based on information gathered from underlying operators. The main focus of service provider resource management is monitoring and planning from a service availability perspective. Service providers and operators can interface via common APIs, either specified by SDOs or widely accepted de facto implementations.

Edge infrastructure and service management are distributed by nature. An edge infrastructure has a combination of the following characteristics:

- Distributed computation and/or storage.
- Heterogeneous hardware resources.
- Support a plethora of access methods.
- Power/space constraint.

A useful analogy of this management architecture is a computer operating system (OS). A computer has hardware resources including CPU, memory, and non-volatile storage, as well as basic peripherals such as keyboard and display, and open, well-defined interfaces such as NIC, USB, and PCIe bus. Hardware components make themselves available to the OS by complying with a defined interface. Application software running on the computer can then access hardware components regardless of make, model, and version.

Like the computer OS, edge management architecture must provide open, well-defined interfaces to allow independently owned and managed resources to plug in.

In Sect. 5.1.4, MEC, 5G Edge, and beyond, we discussed a variety of MEC deployment locations. Clearly, managing and orchestrating among widely dispersed and functionally different MEC locations present a challenge.

These orchestration-level APIs are intended to provide unique portal and management interfaces to end customers.

When Telco and public clouds cooperate at the edge, integration happens on not only the function side but also the management side. End customers—whether enterprise or vertical integrators—want to see a unique management interface, which means the customer can turn to the appropriate operating team as soon as possible when they need support.

Good orchestration can accelerate service onboarding, automate full life-cycle management, enhance customer experience, transform seamlessly from VNF to CNF, and simplify interoperation between Telco and public clouds. Especially for MEC applications, customers need on-premise service to meet their low-latency expectations. Low-latency requirements are inherent not only on the function side but also on the management side. Some enterprise customers even demand to have a self-controlled portal that integrates Telco and public cloud orchestration functions.

For most Telco operators, the generic VNF Manger and NFV orchestrator components are standards based, MANO compliant architecture. Some open-source orchestration projects such as ONAP and OpenNESS aim for adoption by numerous operators. However, the MANO layer is still very specific to different Telco operators and is integrated with northbound OSS/BSS systems. It is even related to organizational hierarchy and geography aspects and operator team technical backgrounds. It is still a long way for different Telcos to build unique orchestration architectures for public clouds and third parties to integrate.

Another challenge comes from the different architectures of Telco CT and IT infrastructure. It remains difficult to manage VNF for Telco core functions and container-based applications. As Telco core functions evolve to cloud-native functions, based on differences of IT and CT and operator regulatory and uptime requirements, it will be challenging to use a single orchestration platform to manage both sides. Unifying Telco operators' network and IT environments and connecting them to private enterprise clouds, edge clouds, and public clouds is ongoing work.

MEC promises to reduce latency and cost of customer service. Its most important advantages are related to less physical distance to customer locations. However, Telco and public clouds have different hierarchies, which means orchestration distances are not uniform. Such differences in operating and management granularity may bring uneven customer experience. Unifying Telco and public cloud edge orchestration will be a big advantage of MEC and lead to end-to-end solutions for customers.

An enabler layer containing both Telco and public cloud orchestration APIs is a solution that can potentially integrate management modules of both sides, providing unified, flexible, and rapid operating capabilities that will enhance customer experience.

5.3.3 Edge-Edge Federations and Edge-Cloud Collaboration

Growth in service innovation provides both revenue opportunities and technical challenges for service providers (SPs). Challenges come in many forms; some are related to management of complex ecosystem of service platforms, seamless integration with non-telecom capabilities, managing balance between cost and benefits, and remaining innovative from a new feature perspective.

One of the sharpest arrows in the quiver of SP and Edge/Access Operators (E/AO) is federating capability: collaborating to share infrastructure. By sharing infrastructure, SPs and Edge/Access Operators (E/AO) share cost and risk, increasing telecom industry sustainability.

Future communication services are evolving toward platforms, free from complex, monolithic, vendor-proprietary core network systems historically built by each SP and E/AO. By pushing more capability toward the network edge, MEC allows most analytical functions to be logically hosted on platforms, with service configurations routed through MEC-based capability.

By adopting federation and service platform collaboration, MEC layer capabilities will reduce cost for SP and E/AOs who must keep pace with service domain innovation. MEC federation and collaboration provides a unique value proposition in multi-industry services such as industry vertical solutions and industrial automation.

There are multiple paths for SP and E/AOs to take advantage of MEC federation and collaboration. As one example, GSMA has put forward its "Operator Platform Concept" in a white paper published in January 2020 [6].

5.3.3.1 GSMA Operator Platform

Operator platform is a set of functional modules that enables an operator to place the solutions or applications of enterprises in close proximity to their customers, as shown in Fig. 5.9, operator platform overview. SPs and E/AOs can monetize and exploit service capabilities such as edge cloud computing capabilities, IP communications, or slicing in a scalable way and in a federated manner with other SPs and E/AOs.

5.3.3.2 ETSI MEC Federated Edge Specifications

The GSMA "operator platform" proposed concept can be further extended and converted into a completely open and collaborative MEC platform, per ETSI MEC federated edge specifications. In this case, service capabilities are hosted on MEC platforms and offered as independent services to any E/AO or enterprise. MEC capabilities can be extended from MEC-to-MEC integration of MEC capability providers via point-to-point communication or wide-area network cloud. This is shown in Fig. 5.10, E/AO provider MEC platform and collaboration.

Service capability built and offered as a MEC platform by an E/AO or industry vertical solution provider can be extended to any other service providers or E/AO or enterprise consumer. There can be "pure play" independent service capability providers specializing in build service capability to offer them to any service provider or service consumer. GSMA identifies role of aggregators in its operator platform where an aggregator can create a composite service, i.e., a service involving multiple capabilities or aggregate services offered by multiple players. In broader terms, aggregators can be further generalized in form of traders or brokers that collaborate with multiple service providers in multiple geolocations. These traders or brokers

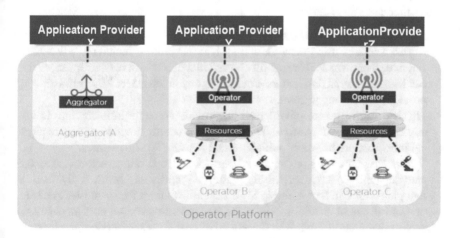

Fig. 5.9 Operator platform overview

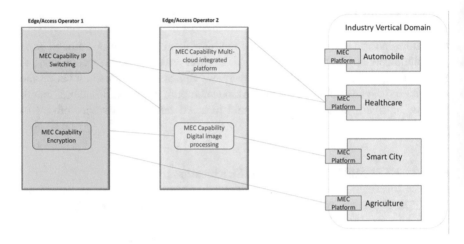

Fig. 5.10 E/AO provider MEC platform and collaboration

negotiate on commercial and service performance parameters to provide optimum offer to respective service provider or Edge/Access Service Provider or enterprise consumer. Here, these are called as "MEC capability broker and aggregator." MEC capability broker and aggregator can create end-to-end service by joining MEC capabilities from various MEC capability providers in a form of MEC capability chain and offer this chain as complete service to service provider or Edge/Access Service Provider or Enterprise Customer [7].

In this way, federated environment of MEC capabilities creates an ecosystem of additional business models, shared risk, and optimized operations for E/AO.

The diagram in Fig. 5.11 provides an end-to-end view of MEC capability collaboration.

Figure 5.11, end-to-end view of MEC capability collaboration, shows the following components and actors:

- E/AO-E/AOs are Access and Edge network operators providing respective Edge/Access services using network and system capabilities including MEC. E/AO can have their own MEC capability hosted in their respective datacenters, or they can host in external datacenters provided by any independent MEC infrastructure provider or MEC aggregator.
- Physical infrastructure providers, who provide common infrastructure to host any virtual system or platform. Physical infrastructure providers can share physical infrastructure between multiple virtual platform/system owners.
- Virtual MEC platform providers, who provide virtual infrastructure and associated capabilities, e.g., OS/Hosting Platform/Security and MEC. Individual virtual platforms can host multiple MEC capabilities from the same E/AO or MEC or from different E/AO or MECs. By implementing shared cost mechanisms, such collaboration optimizes the effective cost for each service provider and Edge/Access Service Provider (E/AO).

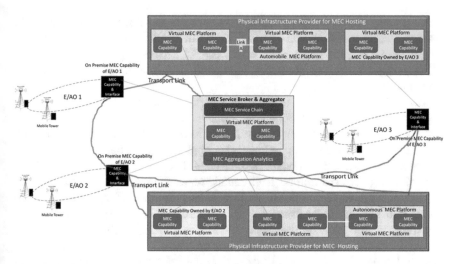

Fig. 5.11 End-to-end view of MEC capability collaboration

- MEC capability provider. Anyone offering well-executed MEC value can market MEC capability to service providers and E/AO or industry vertical solution providers. E/AO and industry vertical solution providers can also host their own MEC capabilities and offer to other service providers, industry vertical solution provider, or other enterprise customers. For example, an end-to-end encryption service can be hosted as MEC and offered to any service provider, enterprise, or industry player.
- MEC service broker and aggregator. MEC service brokers and aggregators play important roles in negotiation and establishing complete service delivery by combining MEC capabilities offered by different providers. Service providers and E/AO may have direct commercial and technical service delivery relationships with other MEC capability providers or else they may do business through MEC service brokers or aggregators. The latter become more relevant in the value chain when each MEC capability provider provides atomic MEC services not sufficient for complete delivery of end customer services.
- Transport link providers. SP and E/AOs can have their own respective transport and connectivity services between different MEC platforms, or they may utilize other long-distance connectivity providers.

5.3.4   Edge-Cloud Interfacing

We have described edge layer ownership and operational responsibilities, with underlying implementations decided by layer owners. Operators or other entities may hope to architect a universal platform for all edge enabler, application enabler, and edge applications. This is a natural tendency to reduce development cost and make product management easier. But in reality, a one-size-fits-all platform will have challenges in operational responsibilities. An operator's edge business strategy may require multiple models described in Sect. 5.1.3 to be adopted. Hence a flexible underlying platform implementation to support functional modules would be the proper approach.

Let's take a look at a Telco mobile edge site system anatomy from an edge stack point of view. In Fig. 5.12, edge stack options, each edge node has two resource groups: one for Telco mobile network functions (box labeled Mobile Network Services) and one for edge computing (Edge Applications and Services). Edge computing is the newcomer in a typical mobile edge site. Most likely, it is an additional rack of servers. Traffic outputs from mobile network functions flow directly into edge computing servers at the IP routing level. From the traffic content point of view, there is no difference from when EC servers are placed miles away. Oftentimes that traffic is only partial of the total traffic M is processing. Techniques like a local breakout can be applied to do the split. There are also other ways such as a hardwire split. Telco functions usually include but are not limited to user plane processing functions such as S/PGW-U for 4G or UPF for 5G [8], as opposed to edge computing servers, which usually terminate traffic for processing. This is depicted at left in the diagram in Fig. 5.12, edge stack options.

In most current implementations, mobile network services and edge computing are adopting two different infrastructure and platform technologies: (a) NFV architecture and (b) an IT flavored architecture typically layered as IaaS, PaaS, and SaaS/ Application stacks. Both mobile network services and edge computing have their own infrastructure and system-level management. Mobile edge management is part of the overall mobile network OSS. In the public cloud case, edge computing extension management is part of global cloud management.

This diagram is a simplified view on player groups in Telco 5G Edge. It is meant to highlight the differences between two major player groups (Telco and IT). Each group has its own ecosystem which plays various roles in Telco 5G Edge. The Edge is where these two groups meet and collaborate. Collaboration interfaces may vary across operators and solutions.

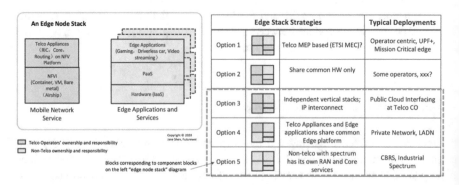

Fig. 5.12 Edge stack options

At first look, it seems natural that mobile network services and edge computing reside in physically separated racks managed by separate Telco and public cloud teams. This avoids concerns related to regulatory requirements, service-level agreements (SLAs), etc. Essentially, it is a co-location arrangement, as depicted in the third option in the right-hand column of the table in Fig. 5.12, edge stack options.

There are certainly other options in that column. Different colors represent different ownership and operation responsibility. In option 1, the Telco not only provides mobile connectivity but also an edge computing platform for applications.

Option 2 is one step further toward convergence between Telco and public clouds, with a common shared infrastructure layer. Sharing can include only the hardware layer or also lower platform layers such as VMs and containers. The main benefit of option 2 is a unified infrastructure layer extending from the Telco core network to edge nodes. As mentioned earlier, infrastructure that meets typical edge application requirements might look quite different from the NFV VIM layer. This implies Telcos either have moved or will move to a cloud-native (non-NFV) Telco edge or a layer above NFV VIM to create a suitable environment for typical edge applications (the former would likely be the case). Operators with in-house infrastructure expertise might be interested in this option. They most likely are in the process of building a cloud-native Telco network and extending the infrastructure layer to the edge seems logical.

Options 4 and 5 probably will not happen in Telco edge data centers. These are the scenarios where Telco RAN or Core user planes are considered more or less to be access options. A typical case would be private enterprise networks, which usually require various access methods. Option 4 deploys Telco edge core functions in enterprise premises. Option 5 only has shared Telco RAN as an option; not all operators are ready to take responsibility for a non-Telco site. Although some operators have announced they are working on an option 4 solution, most likely option 5 will be more widely adopted. Both options have unified IaaS and PaaS layers, with Telco RAN/Core functions deployed as special applications. There can be SLA differences between Telco appliances vs. typical edge applications. In an enterprise private network environment, it is manageable.

5.4   Key Enabling Technologies

Technology advancement is the key supporter behind every major CT (Communication Technology) generation. With many promises from edge services, here is a list of the key technological enablers: deterministic network, ultra-reliable communication, AI at edge, trustworthy edge, and block chain. Note here only technologies applicable to edge itself are listed.

With the architecture laid out, what are the key technology enablers? There are a few technology enablers essential across all future networks, e.g., virtualization, containerization, and micro services.

5.4.1 Opportunistic Multicast

Most current Internet traffic is due to unicast or multicast delivery of relatively immutable content such as video or software to large client groups. This has resulted in a high amount of redundancy in network traffic, as well as capacity bottlenecks—both in core networks and in server infrastructure serving the content. Technologies such as content delivery networks (CDNs) help to spread out the network load and reduce redundancy but are complex to manage, with inherent limits in terms of how rapidly they can react to changing network and server conditions. CDNs cannot fundamentally reduce the network overhead arising from redundant unicast streams.

In contrast, opportunistic multicast delivery as a basic service promises to automatically deliver responses to quasi-concurrent requests in a single lightweight multicast transmission over L2. Unlike traditional IP multicast, this approach has no additional setup time overhead, and it does not require per-flow state in the network. The time period (which we call the catchment interval) over which this process takes place can be flexibly configured on a per-service basis, further improving the opportunity for multicast delivery. For latency-sensitive services such as video chunk delivery short timescales are appropriate (e.g., 100 ms–1 s), whereas for delivering software updates, carrying out DB or cloud service synchronization, and other relatively delay-tolerant service much longer timescales can be used. The gains from multicast delivery can be especially dramatic for highly popular content at peak request times (e.g., new episodes of a popular series becoming available). As an optimization (that can again be enabled on per-service basis), we can combine opportunistic multicast delivery with request suppression where the origin server does not even receive the redundant requests that would be replied within a time multicast transmission, thereby reducing the server load and costs for content delivery even further.

Any deep-edge service architecture needs to provide means to opportunistic multicast delivery by allowing for efficient multicast transmission at the level of the transport network in order to reduce traffic load.

Most of the current Internet traffic is due to unicast delivery of relatively immutable content such as video or software to very large client groups. This has resulted in large amount of redundancy in network traffic, as well as creating capacity bottlenecks both in the core network and the server infrastructure serving the content. Technologies such as content delivery networks (CDNs) help to spread out the network load but are complex to manage, have inherent limits in terms of how rapidly they can react to changing network and server conditions, and cannot fundamentally reduce the network overhead arising from redundant unicast streams.

In contrast, opportunistic multicast delivery as a basic service is proposed to automatically deliver responses to quasi-concurrent requests in a single lightweight multicast transmission over L2. Unlike traditional IP multicast, this approach has no additional setup time overhead, and it does not require per-flow state in the network. The time period (which we call the catchment interval) over which this process takes place can be flexibly configured on a per-service basis, further improving the opportunity for multicast delivery. For latency-sensitive services such as video chunk delivery short timescales are appropriate (e.g.,100 ms–1 s), whereas for delivering software updates, carrying out DB or cloud service synchronization, and other relatively delay-tolerant service much longer timescales can be used. The gains from multicast delivery can be especially dramatic for highly popular content at peak request times (e.g., new episodes of a popular series becoming available). As an optimization (that can again be enabled on per-service basis), we can combine opportunistic multicast delivery with request suppression where the origin server does not even receive the redundant requests that would be replied within a time multicast transmission, thereby reducing the server load and costs for content delivery even further.

5.4.2   Resource Fairness

Recent interest in novel transport protocols such as QUIC has shown that the traditional end-to-end resource management model the Internet is based on is often suboptimal for modern services, with rapidly changing routing patterns between several virtual service endpoints rendering classical TCP congestion control inefficient. Opportunistic multicast decouples the resource management of the access link (which will be handled by whichever protocol the client uses to access the involved service, typically TCP or QUIC) from resource management of the transport network. This creates first of all the opportunity to support fairness between different resource management mechanisms (with UDP and TCP being the extreme classical example) and also to optimize the network more aggressively than enabled by traditional endpoint-centric solutions.

Any deep-edge service architecture needs to provide means for fair transport resource management at an end-to-end as well as edge-to-edge level.

5.4.3   Flow Setup

One of the key latency bottlenecks in the current Internet is caused by the high flow setup latency, especially when transport (or higher) layer security is involved. Furthermore, many applications still rely (for reliability reasons and to simplify development) on nonpersistent connections that get rebuilt for every individual request for content items, even if served by the same origin server. In contrast, our proposal enables (but does not require) splitting of the connection at the network ingress point. Since this is usually very close latency-wise to the end user, optimizing the residual latency in the core translates to substantial latency reduction at the edge, even if the client-to-edge connection establishment is not modified. Such approaches have been successfully used in the wireless community to deal with extreme latencies (e.g., as found in satellite communications), and our approach enables deploying them transparently at the network edge as well.

Any deep-edge service architecture needs to separate setup of long-term end-to-end as well as edge-to-edge flows from short-term end-to-end transactions to reduce setup latency of the latter.

5.4.4 Deterministic Networking

As the end-to-end latency and latency requirements for edge services are expected to be in the order of milliseconds, or even sub-millisecond in extreme cases, low-latency services must be provided through access to local edge computing and storage resources.

Depending on the specific application requirements, there should be a need to implement deterministic networking and/or time-sensitive networking (TSN) profiles.

This places requirements on the incorporation of specific queuing algorithms/disciplines, such as priority and frame pre-emption queuing, synchronized port gating, and persistent and semi-persistent scheduling superimposed over random access or request/grant access procedures. In some cases, delay variation requirements may be met through the use of buffering, but it will often be the case that precise playout times for the user data will be required. There may be other requirements for precise time synchronization of network elements, for example, in highly accurate localization.

To meet such requirements, the deep-edge service architecture needs to be able to support the use of precision timing protocols, enabling time synchronization to nanosecond accuracy.

5.4.5 Ultra-Reliable Communications

No-loss network optimizes the network from network delay, packet loss, and throughput aspects. Correspondingly, RDMA, PFC (priority-based flow control) and ECN (explicit congestion notification), DCQCN (data center quantized congestion notification) are the state-of-art techniques to achieve low delay, no packet loss, and high throughput.

Some of edge services require a packet delivery "guarantee" (typically specified as 99.9999%, or "5 nines" reliability) over networks limited by noise, interference, or congestion. Example applications include emergency services, remote equipment operation, augmented reality, industrial automation (Industry 4.0) with remote control/operation of equipment and machinery, and vehicle-to-vehicle and vehicle-to-infrastructure communication for (semi-)autonomous driving.

For such mission-critical applications, enhanced forward error correction and coding schemes should be applied, where these may need to take into account short control message lengths. This may require some joint L1/L2 mechanisms.

Reliability should be augmented by mechanisms such as packet/frame replication, forwarding over diverse paths, and duplicate elimination. In many cases, the requirements for ultra-reliable communications will intersect with those for low latency. Thus, new joint encoding schemes and frame replication and duplicate elimination mechanisms *must* be latency sensitive.

5.4.6 Content Addressable Networks

Content-addressable networks can efficiently switch and route traffic using non-IP information. This can still be implemented as an overlay on top of the IP network. However, applications may be agnostic to underlying networks and utilize content aware information. This is very useful since edge is a multi-domain hub with various protocols and addressing schemes. The common delimiter is application content. Content is the new addressable information across all domains at the edge.

5.5 Conclusion

In this chapter, we described solutions to challenges posed by edge computing in future networks. The relationship between edge computing and 5G and MEC and interfaces to Telco core networks and hyperscaler public clouds are explained. The edge architecture, including the concepts of an edge border gateway and federation (both edge-edge and edge-cloud), and key enabling technologies are described.

References

1. ETSI MEC whitepaper, MEC in 5G networks
2. https://forge.etsi.org/
3. 3GPP TS 29.522 V16.3.0, 5G systems, network exposure function northbound APIs
4. LF edge whitepaper, Cloud interfacing at the telco 5G edge
5. 3GPP TR23.758 V17.0.0, Study on application architecture for enabling edge applications
6. GSMA operator platform concept, https://www.gsma.com/futurenetworks/resources/operator-platform-concept-whitepaper/
7. 3GPP TR 23.748, Study on enhancement of support for edge computing in 5G core network (5GC)
8. 3GPP TS 23.501, System architecture for the 5G system (5GS)

Chapter 6
Data Center Architecture, Operation, and Optimization

Kaiyang Liu, Aqun Zhao, and Jianping Pan

6.1 Introduction

The explosive growth of workloads driven by data-intensive applications, e.g., web search, social networks, and e-commerce, has led mankind into the era of big data [1]. According to the IDC report, the volume of data is doubling every 2 years and thus will reach a staggering 175 ZB by 2025 [2]. Data centers have emerged as an irreplaceable and crucial infrastructure to power this ever-growing trend.

As the foundation of cloud computing, data centers can provide powerful parallel computing and distributed storage capabilities to manage, manipulate, and analyze massive amounts of data. A special network, i.e., data center network (DCN), is designed to interconnect a large number of computing and storage nodes. In comparison with traditional networks, e.g., local area networks and wide area networks, the design of DCN has its unique challenges and requirements [3], which are summarized as follows:

Hyperscale: Currently, over 500 hyperscale data centers are distributed across the globe. We are witnessing the exponential growth of scale in modern data centers. For example, Range International Information Group located in Langfang, China, which is one of the largest data centers in the world, occupies 6.3 million square feet of space. A hyperscale data center hosts over a million servers spreading across

K. Liu (✉)
Department of Computer Science, University of Victoria, Victoria, BC, Canada

School of Computer Science and Engineering, Central South University, Changsha, China
e-mail: liukaiyang@uvic.ca

A. Zhao
School of Computer and Information Technology, Beijing Jiaotong University, Beijing, China

J. Pan
Department of Computer Science, University of Victoria, Victoria, BC, Canada

© The Author(s), under exclusive license to Springer Nature
Switzerland AG 2021
M. Toy (ed.), *Future Networks, Services and Management*,
https://doi.org/10.1007/978-3-030-81961-3_6

hundreds of thousands of racks [4]. Data centers at such a large scale put forward severe challenges on system design in terms of interconnectivity, flexibility, robustness, efficiency, and overheads.

Huge Energy Consumption: In 2018, global data centers consumed about 205 tWh of electricity, or 1% of global electricity consumed in that year [5]. It has been predicted that the electricity usage of data centers will increase about 15-fold by 2030 [6]. The huge energy consumption prompts data centers to improve the energy efficiency of the hardware and system cooling. However, according to the New York Times report [7], most data centers consume vast amounts of energy in an incongruously wasteful manner. Typically, service providers operate their facilities at maximum capacity to handle the possible bursty service requests. As a result, data centers can waste 90% or more of the total consumed electricity.

Complex Traffic Characteristics: Modern data centers have been applied to a wide variety of scenarios, e.g., Email, video content distribution, and social networking. Furthermore, data centers are also employed to run large-scale data-intensive tasks, e.g., indexing Web pages and big data analytics [8]. Driven by diversified services and applications, data center traffic shows complex characteristics, i.e., high fluctuation with the long-tail distribution. In fact, most of the flows are short flows, but most of the bytes are from long flows [9]. Short flows are processed before optimization decision takes place. Furthermore, data centers suffer from fragmentation with intensive short flows. It is a challenge to handle traffic optimization tasks in hyperscale data centers.

Tight Service-Level Agreement: The service-level agreement (SLA) plays the most crucial part in a data center lease, spelling out the performance requirements of services that data centers promise to provide in exact terms. It has been increasingly common to include mission-critical data center services in SLAs such as power availability, interconnectivity, security, response time, and delivery service levels. Considering the inevitable network failures, congestion, or even human errors, constant monitoring, agile failure recovery, and congestion control schemes are necessary to provide tight SLAs.

To solve these significant technical challenges above, DCNs have been widely investigated in terms of network topology [3], routing [10], load balancing [11], green networking [12], optical networking [13], and network virtualization [14]. This book chapter presents a systematic view of DCNs from both the architectural and operational principle aspects. We start with a discussion on the state-of-the-art DCN topologies (Sect. 6.2). Then, we examine various operation and optimization solutions in DCNs (Sect. 6.3). Thereafter, we discuss the outlook of future DCNs and their applications (Sect. 6.4). The main goal of this book chapter is to highlight the salient features of existing solutions which can be utilized as guidelines in constructing future DCN architectures and operational principles.

6.2   Data Center Network Topologies

Currently, the research on the DCN topologies can be divided into two types: switch centric schemes and server centric schemes. Some of the topologies originate from the interconnection networks in supercomputing for both categories. Furthermore, some researchers have introduced optical switching technology into the DCN and proposed some full optical and optical/electronic hybrid topologies. Others have introduced wireless technology into the DCN and proposed some wireless DCN topologies. Besides, in the real world, most service providers have built their production data centers with some specific topologies.

6.2.1   Switch-Centric Data Center Network Topologies

In the switch-centric DCN topologies, the network traffic is all routed and forwarded by the switches or routers. These topologies include Fat-tree [15], VL2 [16], Diamond [17], Aspen Trees [18], F10 [19], F^2 Tree [20], Scafida [21], Small-World [22], and Jellyfish [23]. In this section, some representative schemes are selected for introduction.

Fat-tree: Fat-tree is proposed by Al-Fares et al., which drew on the experience of Charles Clos et al. in the field of telephone networks 50 years ago [15]. A general Fat-tree model is a k-port n-tree topology [24]. In the data center literature, a special instance of it with $n = 3$ is usually adopted. In this Fat-tree topology, each k-port switch in the edge level is connected to $\frac{k}{2}$ servers. The remaining $\frac{k}{2}$ ports are connected to $\frac{k}{2}$ switches at the aggregation level. The $\frac{k}{2}$ aggregation level switches, $\frac{k}{2}$ edge level switches, and the connected servers form a basic cell of a Fat-tree, called a pod. At the core level, there are $\left(\frac{k}{2}\right)^2$ k-port switches, each connecting to each of the k pods. Figure 6.1 shows the Fat-tree topology with 4-port switches. The maximum number of servers in a Fat-tree with k-port switches is $\frac{k^3}{4}$.

Fat-tree has many advantages. Firstly, it eliminates the throughput limitation of the upper links of the tree structure and provides multiple parallel paths for communication between servers. Secondly, its horizontal expansion reduces the cost of building a DCN. Finally, this topology is compatible with Ethernet structure and IP-configured servers used in existing networks. However, the scalability of Fat-tree is limited by the number of switch ports. Another drawback is that it is not fault-tolerant enough and is very sensitive to edge switch failures. Finally, the number of switches needed to build Fat-tree is large, which increases the complexity of wiring and configuration.

VL2: VL2 is proposed by Greenberg et al. using a Clos Network topology in [16]. VL2 is also a multi-rooted tree. When deployed in DCNs, VL2 usually consists of three levels of switches: the Top of Rack (ToR) switches directly connected to servers, the aggregation switches connected to the ToR switches, and the intermediate switches connected to the aggregation switches. The number of the switches is determined by the number of ports on the intermediate switches and aggregation switches. If each of these switches has k ports, there will be k aggregation switches and $\frac{k}{2}$ intermediate switches. There is exactly one link between each intermediate switch and each aggregation switch. The remaining $\frac{k}{2}$ ports on each aggregation switch are connected to $\frac{k}{2}$ different ToR switches. Each of the ToR switches is connected to two different aggregation switches, and the remaining ports on the ToR switches are connected to servers. There are $\frac{k^2}{4}$ ToR switches because $\frac{k}{2}$ ToR switches are connected to each pair of aggregation switches. While intermediate switches and aggregation switches must have the same number of ports, the number of ports on a ToR switch is not limited. If k_{ToR} ports on each ToR switch are connected to servers, there will be $\frac{k^2}{4} \cdot k_{\text{ToR}}$ servers in the network. Figure 6.2 shows the VL2 topology with $k = 4$, $k_{\text{ToR}} = 2$.

The difference of VL2 with Fat-tree is that the topology between intermediate switches and aggregation switches forms a complete bipartite graph, and each ToR switch is connected to two aggregation switches. VL2 reduces the number of cables by leveraging higher speed switch-to-switch links, e.g., 10 Gbps for switch-to-switch links and 1 Gbps for server-to-switch links.

6.2.2 Server-Centric Data Center Network Topologies

In the server-centric DCN designs, the network topologies are constructed by recursion. The servers are not only computing devices but also routing nodes and will actively participate in packet forwarding and load balancing. These topologies avoid the bottleneck in the core switches through recursive design, and there are multiple disjoint paths between servers. Typical topologies include DCell [25], BCube [26],

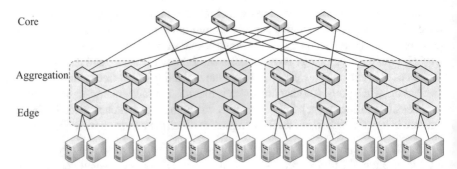

Fig. 6.1 The Fat-tree topology with $k = 4$ ports

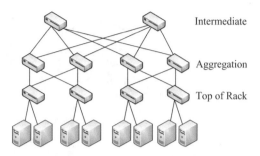

Fig. 6.2 The VL2 topology with $k = 4$ ports and $k_{ToR} = 2$

FiConn [27], DPillar [28], MCube [29], MDCube [30], PCube [31], Snowflake [32], HFN [33], and so on.

DCell: DCell topology was proposed by Guo et al. [25] in 2008 which uses lower-level structures as the basic cells to construct higher-level structures. $DCell_0$ is the lowest cell in DCell, which is composed of one n-port switch and n servers.[1] The switch is used to connect all of the n servers in the $DCell_0$. Let $DCell_k$ be a level-k DCell. Firstly, $DCell_1$ is built from a few $DCell_0$s. Each $DCell_1$ has $(n + 1)$ $DCell_0$s, and each server of every $DCell_0$ in a $DCell_1$ is connected to a server in another $DCell_0$, respectively. Therefore, the $DCell_0$s are connected, with only one link between every pair of $DCell_0$s. A similar method is applied to build a $DCell_k$ from a few $DCell_{k-1}$s. In a $DCell_k$, each server will finally have $k + 1$ links: the first link or the level-0 link connected to a switch when forming a $DCell_0$, and level-i link connected to a server in the same $DCell_i$ but a different $DCell_{i-1}$. Suppose that each $DCell_{k-1}$ has t_{k-1} servers, then a $DCell_k$ will consist of t_k $DCell_{k-1}$s, and thus $t_{k-1} \cdot t_k$ servers. Obviously, we have $t_k = t_{k-1} \cdot (t_{k-1} + 1)$. Figure 6.3 shows a $DCell_1$ with $n = 4$.

DCell satisfies the basic requirement of DCNs such as scalability, fault tolerance, and increased network capacity. The main idea behind DCell not only depends on switches but also takes the advantage of the network interface card (NIC) deployed within servers to design the topology. The number of servers in a DCell increases double-exponentially with the number of server NIC ports. A level-3 DCell can support 3,263,442 servers with 4-port servers and 6-port switches. DCell also overcomes the constraint of a single point of failure as in tree-based topologies.

BCube: To overcome the issue of traffic congestion bottleneck and NIC installation, Guo et al. proposed a new hypercube-based topology known as BCube for shipping-container-based modular data centers [26]. It is also considered a module version of DCell. The most basic element of a BCube, which is named as $BCube_0$, is also the same as a $DCell_0$: n servers connected to one n-port switch. The main difference between BCube and DCell lies in how they scale up. BCube makes use of more switches when building higher-level structures. While building a $BCube_1$, n extra switches are used, connecting to exactly one server in each $BCube_0$. Consequently, a $BCube_1$ contains n $BCube_0$s and n extra switches, which means if the switches in the $BCube_0$s are considered, there are $2n$ switches in a $BCube_1$. In general, a $BCube_k$ is built from n $BCube_{k-1}$s and n^k extra n-port switches. These

---

[1] Unlike Fat-tree and VL2, n is used to represent the number of ports in DCell and BCube.

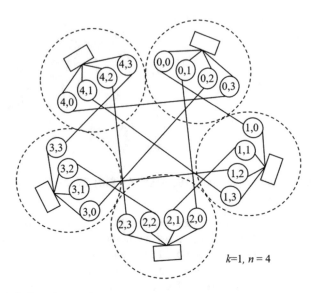

Fig. 6.3 The DCell topology with $n = 4$ ports

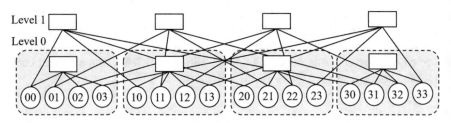

Fig. 6.4 The BCube topology with $n = 4$ ports

extra switches are connected to only one server in each BCube$_{k-1}$. In a level-k BCube, each level requires n^k n-port switches. BCube makes use of more switches when building higher-level structures, while DCell uses only level-0 n-port switches. However, both BCube and DCell require servers to have $k + 1$ NICs. The implication is that servers will be involved in switching more packets in DCell than in BCube. Figure 6.4 shows a BCube$_1$ with $n = 4$.

Just like DCell, the number of levels in a BCube depends on the number of ports on the servers. The number of servers in BCube grows exponentially with the levels, much slower than DCell. For example, when $n = 6$, $k = 3$, a fully constructed BCube can contain 1296 servers. Considering that BCube is designed for container-based data centers, such scalability is sufficient.

6.2.3 Data Center Network Topologies Originated from Interconnection Networks

Some of the DCN topologies originate from the interconnection network topologies, which have been applied to connect the multiple processors in the supercomputing domain. The typical examples include Clos [34], FBFLY [35], *Symbiotic* [36], Hyper-BCube [37], and so on, some of which are switch-centric topologies and the others are server-centric ones.

FBFLY: FBFLY [35] was proposed by Abts et al., which originates from the flattened butterfly, a cost-efficient topology for high-radix switches [38]. The FBFLY k-ary n-flat topology takes advantage of recent high-radix switches to create a scalable but low-diameter network. FBFLY is a multidimensional directed network, similar to a k-ary n-cube torus. Each high-radix switch with more than 64 ports interconnects servers and other switches to form a generalized multidimensional hypercube. A k-ary n-flat FBFLY is derived from a k-ary n-fly conventional butterfly. The number of supported servers is $N = k^n$ in both networks. The number of switches is $n \cdot k^{n-1}$ with port number $2k$ in the conventional butterfly and is $\frac{N}{k} = k^{n-1}$ with port number $n(k - 1) + 1$ in FBFLY. The dimension of FBFLY is $n - 1$. Figure 6.5 shows an 8-ary 2-flat FBFLY topology with 15-port switches. Here, c is the abbreviation for concentration, which means the number of servers and $c = 8$. Although it is similar to a generalized hypercube, FBFLY is more scalable and can save energy by modestly increasing the level of oversubscription. The size of FBFLY can scale from the original size of $8^4 = 4096$ to $c \cdot k^{n-1} = 6144$. The level of oversubscription is moderately raised from 1:1 to 3:2.

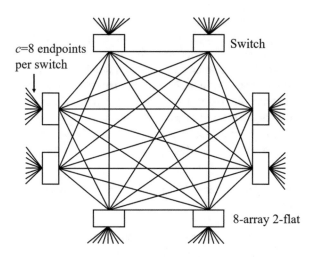

Fig. 6.5 The 8-ary 2-flat FBFLY topology

Symbiotic: Symbiotic [36] was proposed by Hussam et al. to build an easier platform for distributed applications in modular data centers. Communication between servers applies a k-ary 3-cube topology, which is also known as a direct-connect 3D torus and is employed by several supercomputers on the TOP500 list. This topology is formed by having each server directly connected to six other servers, and it does not use any switches or routers. In each dimension, k servers logically form a ring. Symbiotic can support up to k^3 servers. Each server is assigned an address, which takes the form of an (x, y, z) coordinate that indicates its relative offset from an arbitrary origin server in the 3D torus. They refer to the address of the server as the server coordinate, and, once assigned, it is fixed for the lifetime of the server.

6.2.4 Optical Data Center Network Topologies

The electronic data center architectures have several constraints and limitations. Therefore, with the ever-increasing bandwidth demand in data centers and the constantly decreasing price of optical switching devices, optical switching is envisioned as a promising solution for DCNs. Besides offering high bandwidth, optical networks have significant flexibility in reconfiguring the topology during operation. Such a feature is important considering the unbalanced and ever-changing traffic patterns in DCNs. Optical DCN can be classified into two categories, i.e., optical/electronic hybrid schemes such as c-Through [39] and Helios [40], and fully optical schemes such as optical switching architecture (OSA) [41].

c-Through: c-Through [39] is a hybrid network architecture that makes use of both electrical packet switching networks and optical circuit switching networks. Therefore, it is made of two parts: a tree-based electrical network part which maintains the connectivity between each pair of ToR switches and a reconfigurable optical network part which offers high bandwidth interconnection between some ToR electrical switches. Due to the relatively high-cost optical network and the high bandwidth of optical links, it is unnecessary and not cost-effective to maintain an optical link between each pair of ToR switches. Instead, c-Through connects each ToR switch to exactly one other ToR switch at a time. Consequently, the high-capacity optical links are offered to pairs of ToR switches transiently according to the traffic demand. The estimation of traffic between ToR switches and reconfiguration of the optical network is implemented by the control plane of the network. Figure 6.6 shows a c-Through network.

To configure the optical network part of c-Through, the traffic between ToR switches should be estimated. c-Through estimates the rack-to-rack traffic demands by observing the occupancy of the TCP socket buffer. Since only one optical link is offered to each ToR switch, the topology should be configured so that the most amount of estimated traffic can be satisfied. In [39], the problem is solved using the max-weight perfect matching algorithm [42]. The topology of the optical network is configured accordingly.

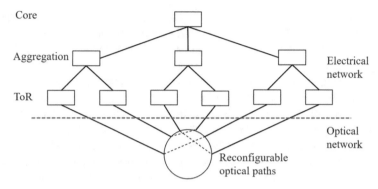

Fig. 6.6 The c-Through network topology [39]

Helios: Helios [40] is another hybrid network with both electrical and optical switches, which are organized as two-level multi-rooted ToR switches and core switches. Core switches consist of both electrical switches and optical switches to make full use of the two complementary techniques. Unlike c-Through, Helios uses the electrical packet switching network to distribute the bursty traffic, while the optical circuit switching part offers baseline bandwidth to the slow-changing traffic. On each of the pod switches, the uplinks are equipped with an optical transceiver. Half of the uplinks are connected to the electrical switches, while the other half are connected to the optical switch through an optical multiplexer. The multiplexer combines the links connected to it to be a "superlink" and enables flexible bandwidth assignment on this superlink.

Helios estimates bandwidth demand using the max-min fairness algorithm [43] among the traffic flows measured on pod switches, which allocates fair bandwidth for TCP flows. Similar to c-Through, Helios computes optical path configuration based on max-weighted matching. Unlike c-Through, both the electrical switches and the optical switches are dynamically configured based on the computed configuration. Helios decides where to forward traffic, the electrical network or the optical network.

OSA: Unlike optical/electronic hybrid schemes, OSA [41] explores the feasibility of a pure optical switching network, which means that it abandons the electrical core switches and use only optical switches to construct the switching core. The ToR switches are still electrical, converting electrical and optical signals between servers and the switching core. OSA allows multiple connections to the switching core on each ToR switch. However, the connection pattern is determined flexibly according to the traffic demand. Since the network does not ensure a direct optical link between each pair of racks with traffic demand, the controlling system constructs the topology to make it a connected graph, and ToR switches are responsible to relay the traffic between other ToR switches.

OSA estimates the traffic demand with the same method as Helios. However in OSA, there can be multiple optical links offered to each rack, and the problem can no longer be formulated as the max-weight matching problem. It is a multi-commodity flow problem with degree constraints, which is NP-hard. In OSA, the problem is simplified as a max-weight b-matching problem and approximately solved by applying multiple max-weight matching.

Compared with electronic switching, optical switching has potentially higher transmission speed, more flexible topology, and lower cooling cost, so it is an important research direction of DCNs. However, the existing optical DCN, such as OSA, still faces some problems. For example, the current design of OSA is for container data centers (i.e., small data centers that can be quickly deployed for edge computing applications in the IoT world), and its scale is limited. To design and build large-scale DCNs is very challenging from the perspective of both architecture and management. Also, the ToR switch used in OSA is the switch that supports optical transmission and electric transmission at the same time, which makes it difficult to be compatible with pure electric switch in traditional data centers.

6.2.5 Wireless Data Center Network Topologies

Wireless/wired hybrid topologies: Wireless technology has the flexibility to adjust the topologies without rewiring. Therefore, Ramachandran et al. introduced wireless technology into DCNs in 2008. Subsequently, Kandula et al. designed Flyways [44, 45], by adding wireless links between the ToR switches to alleviate the rack congestion problem to minimize the maximum transmission time. However, it is difficult for the wireless network to meet all the requirements of DCNs by itself, including scalability, high capacity, and fault tolerance. For example, the capacity of wireless links is often limited due to interference and high transmission load. Therefore, Cui et al. introduced wireless transmission to alleviate the congestion of the hotspot servers, which took wireless communication as a supplement to wired transmission, and proposed a heterogeneous Ethernet/wireless architecture, which is called WDCN [46]. In order not to introduce too many antennas and interfere with each other, Cui et al. regarded each rack as a wireless transmission unit (WTU). This design makes the rack not block the line of sight transmission.

The wireless link scheduling mechanism proposed by Cui et al. includes two parts: collecting traffic demand and link scheduling. A specific server in a WTU is designated as the unit head of the WTU. The unit head is responsible for collecting local traffic information and executing the scheduling algorithm. Each unit head is equipped with a control antenna, and all unit heads broadcast their traffic load in push mode through a common 2.4/5 GHz channel. Therefore, all units can get global traffic load distribution and can schedule the wireless link independently. After collecting the traffic demand information, the head server needs to allocate channels for wireless transmission. Cui et al. proposed a heuristic allocation method [46] to achieve this goal.

The application of wireless technology makes the DCN topologies no longer fixed and saves the complex wiring work, so it has a certain application prospect in the DCN environment. The introduction of Flyways and WDCN alleviates the bandwidth problem of hot servers, and achieved certain results, and made the traffic demand and wireless link scheduling become the focus of research. However, under the premise of providing enough bandwidth, the transmission distance of wireless technology is limited, which limits its deployment in large-scale data centers. Besides, WDCN uses the broadcast method to collect traffic demand, which makes it face the problems of clock synchronization and high communication overhead. Moreover, the measurement results show that the data center traffic is constantly changing, and this makes the location of the hot servers uncertain, which poses a greater challenge to topology adjustment.

All wireless DCN topologies: Based on the 60 GHz wireless communication technology, Shin et al. proposed a DCN with all wireless architecture [45, 47]. They aggregated the switch fabric to the server nodes and expected to arrange the server nodes to be closely connected, low stretch, and support failure recovery. To achieve this requirement, the network card of each server was replaced by Y-switch [47]. The servers are also arranged in a cylindrical rack, so that the communication channels between and within the racks can be easily established, and these connections together form a closely linked network topology.

The topology is modeled as a mesh of Cayley graphs [48]. When viewed from the top, connections within a story of the rack form a 20-node, degree-k Cayley graph, where k depends on the signal's radiation angle (Fig. 6.7). This densely connected graph provides numerous redundant paths from one server to multiple servers in the same rack and ensures strong connectivity. The transceivers on the exterior of the rack stitch together Cayley sub-graphs in different racks. There is great flexibility in how a data center can be constructed out of these racks, but they pick the simplest possible topology by placing the racks in rows and columns for ease of maintenance. Figure 6.7 illustrates an example of the two-dimensional connectivity of four racks in 2 by 2 grids: small black dots represent the transceivers and the lines indicate the connectivity. A Cayley graph sits in the center of each rack: lines coming out of the Cayley graphs are connections through the Y-switches. Relatively long lines connecting the transceivers on the exterior of the racks show the wireless inter-rack connections. Further, since the wireless signal spreads in a cone shape, a transceiver is able to reach other servers in different stories in the same or different racks.

There are still many problems in the scalability and performance of all wireless DCNs. First of all, the competition of the MAC layer greatly affects the performance of the system. Secondly, the performance of the wireless network is greatly affected by the number of network hops. Finally, the performance of multi-hop restricts the scalability of the Cayley data center. However, the advantages and continuous development of wireless technology make it possible to build small- and medium-sized data centers for specific applications.

Fig. 6.7 The Cayley data center topology [47]

6.2.6 Production Data Center Network Topologies

Nowadays, production data centers have become indispensable for the service providers, such as Google, Facebook, Microsoft, Amazon, and Apple. Large IT companies built several production data centers to support their business. Others rented out to provide services to medium-sized and small-sized enterprises that cannot afford their own data centers.

Google's Jupiter: Singh et al. introduced Google's five generations of DCNs based on Clos topology in the last 10 years [49]. The newest-generation Jupiter is a 40G datacenter-scale fabric equipped with dense 40G capable merchant silicon. Centauri switch is employed as a ToR switch, which includes four switch chips. Four Centauris composed a Middle Block (MB) for use in the aggregation block. The logical topology of an MB is a two-stage network. Each ToR chip connects to 8 MBs with 2 × 10G links to form an aggregation block. Six Centauris are used to build a spine block. There are 256 spine blocks and 64 aggregation blocks in Jupiter.

Facebook's next-generation data center fabric: After the "four-post" architecture [51], Facebook proposed its next-generation data center fabric [50]. As is shown in Fig. 6.8, the basic building block of this data center fabric is the server pod, which is composed of 4 fabric switches and 48 rack switches. The network scale can be extended by increasing the number of server pods. To implement building-wide

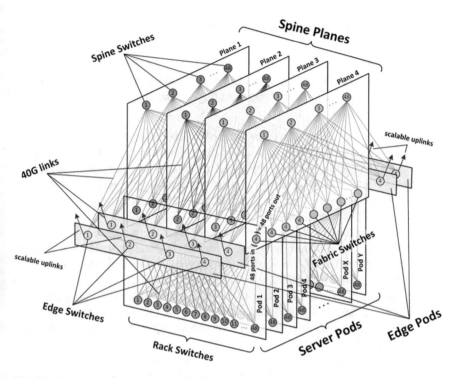

Fig. 6.8   The Facebook data center fabric topology [50]

connectivity, 4 independent planes have been created, each scalable up to 48 spine switches within a plane. Each fabric switch in each server pod connects to each spine switch within its local plane.

Both Google's Jupiter and Facebook's next-generation data center fabrics are based on switch-centric DCN topologies, so they have similar characteristics which are those of switch-centric topologies.

6.3   Operations and Optimizations in Data Center Networks

In a dynamic environment characterized by demand uncertainties, high energy consumption, low average resource utilization, and ever-changing technologies, how to optimize the performance and efficiency of data center operations is a continuous challenge. This section presents a series of operation and optimization schemes that are implemented to enable secure, on-demand, and highly automated services in data centers.

6.3.1 Data Forwarding and Routing

On the basis of a given DCN topology, data forwarding and routing schemes determine how data packets are delivered from a sender to a receiver. Due to the unique features of DCNs, data forwarding and routing are not the same as other networks, e.g., Internet or LAN.

1. Hyperscale data centers have hundreds of thousands switches that interconnect millions of servers. Operating at such a large scale exerts great pressure on forwarding and routing schemes. Unlike the Internet, the sender already knows the topology of DCNs. So the sender also knows the number of available paths to the receiver. How to use the regularities of the DCN topology to scale up forwarding and routing is a critical challenge.
2. People deploy data centers to deliver communication- and computation-intensive services, such as web searching, video content distribution, and big data analytics on a large scale. All these applications are delay sensitive. Major cloud providers, e.g., Amazon and Google, have reported that a slight increase in the service latency may cause observable fewer user accesses and thus a considerable revenue loss [52]. This means forwarding and routing schemes should be highly efficient with less introduced overheads when handling intensive data requests.
3. Servers in data centers may experience downtime frequently [53]. All these failures should be transparent to the client. Therefore, agility has become increasingly important in modern DCNs as there is a requirement that any servers can provide services to any kinds of applications.[2] Facing network crash or server failure, data forwarding and routing schemes should be able to route traffic using alternate paths without interrupting the running application.

Currently, data forwarding and routing schemes have been extensively investigated for DCNs [15, 16, 25, 26, 36, 43, 54–56], which can be classified into different categories based on different criteria. A summary of important data forwarding and routing schemes is presented in Table 6.1, which shows the comparison of these schemes based on five criteria which are briefly described as follows:

Topology: Is the scheme designed for a particular data center topology or can be applied to a generic network?

Implementation: Is the scheme performed in a distributed or centralized manner?

Structure: Which component is involved in packet forwarding with the proposed routing scheme? Typically, forwarding occurs at switches. But in several designs, servers perform forwarding between their equipped network interface cards (NICs).

---

[2] If a specific number of servers are fixed for specific applications without agility, data centers may operate at low resource utilization with growing and variable demands.

Table 6.1 Summary of data forwarding and routing schemes in DCNs

Scheme	Topology	Implementation	Structure	Traffic	Stack
M. Al-Fares [15]	Fat-tree	Centralized	Switch-centric	Unicast	L3
DCell [25]	Custom	Distributed	Server-centric	Unicast	L3
BCube [26]	Hypercube	Distributed	Server-centric	Unicast	L3
VL2 [16]	Clos	Distributed	Switch-centric	Unicast	L2.5
PortLand [54]	Fat-tree	Centralized	Switch-centric	Multicast	L2
Hedera [43]	Fat-tree	Centralized	Switch-centric	Multicast	L3
Symbiotic [36]	CamCube	Distributed	Switch-centric	Multicast	L3
XPath [55]	Generic	Centralized	Switch-centric	Unicast	L3
FatPaths [56]	Custom	Distributed	Switch-centric	Unicast	L2
PBARA [57, 58]	Generic	Distributed	Both	Unicast	L3

Traffic: Can the scheme support unicast or multicast?

Stack: Is the scheme performed at Layer-2 (Ethernet) or Layer 2.5 (shim layer) or Layer-3 (network) of the TCP/IP stack?

Then, we introduce the data forwarding and routing schemes in detail.

Fat-tree Architecture by M. Al-Fares et al. [15]: This work presented Fat-tree topology, which uses interconnected identical commodity Ethernet switches for full end-to-end bisection bandwidth. A Fat-tree contains k pods and $\left(\dfrac{k}{2}\right)^2$ core switches. Each pod is assigned with $\dfrac{k}{2}$ edge switches and $\dfrac{k}{2}$ aggregation switches. Each edge switch is connected with $\dfrac{k}{2}$ servers. A novel addressing scheme is designed to specify the position of a switch or a server in the Fat-tree. The core and edge/aggregation switches are addressed in the form of 10. *n. j. i*, $i, j \in \left[1, \dfrac{k}{2}\right]$, and 10.pod.switch.1, *pod, switch* $\in [0, k-1]$, respectively. The $\dfrac{k}{2}$ servers connected to an edge switch are addressed as 10.pod.switch.server, *server* $\in \left[2, \dfrac{k}{2}+1\right]$. In Fat-tree, all inter-pod traffic is transmitted through the core switches. For load balancing, the proposed data forwarding and routing scheme evenly allocates the traffic among core switches. To achieve this goal, a centralized controller is deployed to maintain two-level forwarding tables at all routers. In this way, the data forwarding is simplified to a static two-level table lookup process, with no need to execute the scheme when a packet arrives. However, how to handle failure is not provided in this work. Another limitation is that the wiring cost is high due to the Fat-tree topology.

DCell [25]: As the first recursive DCN topology, DCell addresses its server with a $(k + 1)$-tuple $\{a_k, a_{k-1}, \ldots, a_1, a_0\}$, where a_k indicates the server belongs to which DCell$_k$, and a_0 denotes the number of servers in DCell$_0$. Let t_k denotes the number of servers in DCell$_k$. Each server is assigned with a unique ID $uid_k = a_0 + \sum_{j=1}^{k} a_j \times t_{j-1}$.

The server in DCell$_k$ can be recursively defined as $[a_k, uid_k]$. Then, DCellRouting, a divide-and-conquer-based routing scheme, is proposed with the consideration of server failures. We assume that the sender and receiver belong to different DCell$_{k-1}$ but are in the same DCell$_k$. The link $\{n_1, n_2\}$ between two DCell$_{k-1}$ can be obtained. The sub-paths are: from the sender to n_1, from n_1 to n_2, and from n_2 to the receiver. We repeat the above process until all sub-paths are direct links. The shortcoming is that DCell cannot guarantee the shortest path can be obtained.

BCube [26]: As a recursive topology, the server in BCube is similarly identified with $\{a_k, \ldots, a_1, a_0\}$. We have $a_i \in [0, n-1]$ as BCube$_0$ contains n servers and BCube$_k$ contains n BCube$_k$. The server address is defined as baddr $= \sum_{i=0}^{k} a_i \times n^i$. Similarly, the switch is identified as $\{l, s_{k-1}, \ldots, s_0\}$, where l denotes the level of switch, $s_j \in [0, n-1], j \in [0, k-1], 0 \leq l \leq k$. With this design, servers connected to the same switch only have a single-digit difference in the addresses. For data forwarding, the relay node will calculate the difference in the addresses and change one digit in each step. Theoretical analysis shows that a data packet needs to be transmitted with $k+1$ switches and k servers at most in routing. Compared with DCell and Fat-tree, BCube provides better network throughput and fault tolerance performance. However, the scalability of BCube is limited. For example, with $k = 3$, $n = 8$, only 4096 servers are supported.

VL2 [16]: VL2 is a flexible and scalable architecture based on the Clos topology, which provides uniform high capacity among servers in the data center. For flexible addressing, VL2 assigns IP addresses based on the actual service requirements. To achieve this goal, two types of IP addresses are designed, i.e., location-specific IP addresses (LA) for network devices (e.g., switches and interfaces) and application-specific IP addresses (AA) for applications. LA is used for routing, which varies if the physical location of the server changes or VMs migrate. In contrast, AA is assigned to an application, which remains unchanged. Each AA is mapped to an LA. A centralized directory system is deployed to maintain the mapping information. VL2 designs a new 2.5 shim layer into the classic network protocol stack to answer the queries from users. The shim layer replaces AA with LA when sending packets and replaces LA back to AA when receiving packets. The shortcoming of VL2 is that it does not provide the bandwidth guarantee for real-time applications.

PortLand [54] and Hedera [43]: PortLand is a scalable, fault-tolerant layer 2 forwarding and routing protocol for data center networks. Observing that data centers are with known baseline topologies and growth models, PortLand adopts a centralized fabric manager to store the network configuration. A 48-bit hierarchical Pseudo MAC (PMAC) addressing scheme is proposed for efficient routing and forwarding. Each host (physical or virtual) is assigned with a unique PMAC, encoding the physical location and the actual MAC (AMAC) address. For routing, when an edge switch receives a data packet, it sends the Address Resolution Protocol (ARP) request for IP to MAC mapping. The request is sent to the fabric manager for its PMAC. After receiving the PMAC address, the edge switch creates an ARP reply and sends it to the source node for further routing. As an extension work, Hedera implementation augments the PortLand routing and fault tolerance protocols. Hedera tries to maximize the utilization of bisection bandwidth with the least scheduling overhead in DCNs. Flow information from constituent switches is collected, which is used to compute nonconflicting paths. This instructs switches to reroute traffic accordingly. The shortcoming is that Portland and Hedera only maintain a single fabric manager, which is at risk of malicious attacks.

Symbiotic [36]:   Symbiotic routing allows applications to implement their routing services in the CamCube topology. CamCube [59] directly connects a server with several other servers without any switches or routers, forming a 3D torus topology which uses content addressable network (CAN). Symbiotic routing provides a key-based routing service, where the key space is a 3D wrapped coordinate space. Each server is assigned an (x, y, z) coordinate which determines the location of the server. With the key-based routing, all packets are directed toward the receiver. Each application is assigned with a unique ID on each server. When a data packet arrives at the server, the kernel determines which application the packet belongs to and then queues the packet for execution. The limitation of symbiotic routing is that it requires a CamCube topology; thus, it may not be applicable to other widely used data center network topologies.

XPath [55]:   By using existing commodity switches, XPath explicitly controls the routing path at the flow level for well-structured network topologies, e.g., Fat-tree, BCube, and VL2. To address the scalability and deployment challenges of routing path control schemes, e.g., source routing [60], MPLS [61], and OpenFlow [62], XPath preinstalls all desired paths between any two nodes into IP TCAM tables of commodity switches. As the number of all possible paths can be extremely large, a two-step compression algorithm is proposed, i.e., path set aggregation and path ID assignment for prefix aggregation, compressing all paths to a practical number of routing entries for commodity switches. The limitation of XPath is that it needs to contact the centralized controller for every new flow in the DCN to obtain the corresponding path ID.

FatPaths [56]:   FatPaths is a simple, generic, and robust routing architecture for low-diameter networks with Ethernet stacks. Currently, many low-diameter topologies, e.g., Slim Fly [63], Jellyfish [23], and Xpander [64], have been designed to improve the cost-performance trade-off when compared with the most commonly used Clos topologies. For example, Slim Fly is 2× more cost- and power-efficient than Fat-tree and Clos while reducing the latency by about 25%. However, traditional routing schemes are not suitable for low-diameter topologies. As only one shortest path exists between any pair of servers, limiting data flow to the shortest path does not utilize the path diversity, which may incur network congestion. FatPaths improves the path diversity of low-diameter topologies by using both minimal and non-minimal paths. Then, the transport layer is redesigned based on new advances for Fat-tree topology [65], removing all TCP performance issues for ultimately low latency and high bandwidth. Furthermore, flowlet switching is used to prevent packet reordering in TCP, ensuring simple and effective load balancing.

PBARA [57, 58]: Port-based addressing and routing architecture (PBARA) applies a kind of port-based source-routing address for forwarding, which renders the table-lookup operation unnecessary in the switches. By leveraging the characteristics of this addressing scheme and the regularity of the DCN topologies, a simple and efficient routing mechanism is proposed, which is completely distributed among the servers, without switch involvement, control message interaction, and topology information storage. The load-balancing and fault-tolerance mechanisms have been designed based on the addressing and routing schemes. The PBARA architecture places all the control functions in servers and keeps the switches very simple. The switches can really realize the goal of no forwarding tables, state maintenance, or configuration, and can implement high-speed packet forwarding through hardware. It may lead to the innovation of the switch architecture. Therefore, it can improve the forwarding performance and reduce the cost and energy consumption of the entire DCN. The PBARA architecture is originally proposed based on the Fat-tree topology [57] and then is extended to the other DCN topologies including F10, Facebook's next-generation data center fabric and DCell [58], which shows that it can be used in a generic DCN topology.

6.3.2 Traffic Optimization

The characteristics of data traffic determine how network protocols are designed for efficient operation in DCNs. Most workloads in data centers fall into two categories: online transaction processing (OLTP) and batch processing. Typically, OLTP is characterized by different kinds of queries that need to analyze massive data sets, e.g., web search. The characteristics of OLTP traffic can be summarized as follows:

Patterns: Most of the data exchanging traffic happens within servers on the same rack. To be more specific, the probability of exchanging traffic for server pairs inside the same rack is 22× higher than that for server pairs in different racks [66]. The inter-rack traffic is incurred by the distributed query processing, e.g., MapReduce applications, where a server pushes or pulls data to many servers across the cluster. For example, the server either does not talk to any server outside its rack or talks to about 1–10% of servers outside its rack [66].

Congestion: Data center experiences network congestion frequently. For example, 86% links observe congestion lasting at least 10 s and 15% of them observe congestion lasting at least 100 s [66]. Furthermore, most congestions are short-lived. Over 90% of congestions are no longer than 2 s while the longest lasted for 382 s [66].

Flow Characteristics: Most of the flows are short flows, but most of the bytes are from long flows [67].

Batch processing is the execution of jobs, e.g., big data analytics, that can run without end user interaction. Each job consists of one or more tasks, which use a direct acyclic graph (DAG) as the execution plan if task dependencies exist. Further, each task creates one or multiple instances for parallel execution. In a recently released Alibaba trace [68, 69], the percentages of jobs with single task and tasks with single instance are over 23% and 43%, respectively. Furthermore, 11% jobs only create one instance. In addition, 99% jobs have less than 28 tasks and 95% tasks have less than 1000 instances. The complexity of batch jobs brings more pressure on the data center operation. The system requires a decent amount of expenses in the beginning, which needs to be trained to understand how to schedule the traffic of batch jobs. Furthermore, debugging these systems can be tricky.

Traffic optimization in DCNs has been explored extensively, which mainly focuses on load balancing and congestion control. The main goal of traffic optimization in DCNs is the flow scheduling to improve network utilization and other quality of service (QoS) parameters (e.g., delay, jitter, data loss, etc.), which can be categorized into link-based and server-based scheduling [11].

6.3.2.1 Link-Based Scheduling

Link-based scheduling schemes aim to balance traffic among links, which contain two crucial procedures, i.e., congestion information collection and path selection.

Congestion Information Collection: Existing scheduling schemes use explicit congestion notification (e.g., FlowBender [70] and CLOVE [71]), data sending rate (e.g., Hedera [43], CONGA [72], HULA [73], and Freeway [74]), or queue length at switch (DeTail [75] and Drill [76]) to represent the congestion information.

Path Selection: With the congestion information, the scheduling schemes determine the paths of data flows. The static ECMP [77] scheme randomly hashes the flows to one of the equal-cost paths, which is convenient to be deployed in DCNs. However, this random scheme cannot properly schedule large flows, which easily incurs network congestions. Some schemes, e.g., CONGA [72], HULA [73], DeTail [75], and Freeway [74], forward flows to the least congested paths. CONGA [72] is a distributed load balancing scheme which obtains the global congestion information among leaf switches. The TCP flows are split into flowlets to achieve fine-grained load balancing. The flowlet is assigned to the least congested path based on the congestion table maintained by each leaf switch. However, due to the limited size of switch memory, CONGA is not scalable to large topologies as only a small amount of congestion information can be maintained. To address this limitation, HULA [73] only tracks congestion for the best path to the destination in each switch. Furthermore, HULA is designed for programmable switches without requiring custom hardware. DeTail [75] presents a new cross-layer network stack to reduce the long tail of flow completion times (FCT). In the link layer, a combined input/output queue is used to sort packets based on their priority in each switch. In the network layer, local egress queue occupancies indicate network congestion. In the transport layer, the out-of-order packets are processed. In the application layer, the flow priorities are determined based on the deadlines. However, as the network stack is modified, DeTail may not be suitable for traditional hardware. Freeway [74] divides the traffic into long-lived elephant flows and latency-sensitive mice flows. According to the path utilization, the shortest paths are dynamically divided into low-latency paths for mice flows and high-throughput paths for elephant flows. The limitation of these schemes is that the precise path utilization is hard to be obtained in real time.

To relieve the requirement of precise path utilization information, some other methods, e.g., Hedera [43], Multipath TCP (MPTCP) [78], and Drill [76], forward flows to the less congested paths. Hedera [43] is a centralized scheme for passive load balancing. Hedera detects large flows in DCNs[3] and assigns them to the paths with enough resources for the maximization of path utilization. However, Hedera is not optimal as the large flows which do not reach the rate limitation may not be scheduled to proper paths. To improve throughput, MPTCP [78] establishes multiple TCP connections between servers by using different IP addresses or ports. Data flows are stripped into subflows for parallel transmission. However, MPTCP may increase the FCT with more additional TCP connections. With no need to collect global congestion information, Drill [76] uses local queue occupancies at each switch and randomized algorithms for load balancing. When a data packet arrives, DRILL randomly chooses two other available ports and compares them with the port which transmits the last packet. The port with the smallest queue size is selected. Drill is a scalable scheme for large-scale data center network topologies as (1) no extra overhead is introduced, and (2) there is no need to modify the existing hardware and protocols.

Some other schemes, e.g., DRB [79] and CLOVE [71], adopt the weighted-round-robin method to achieve load balancing. By utilizing the characteristics of Fat-tree and VL2 topologies, DRB [79] evenly distributes flows among available paths. For each packet in a data flow, the sender selects one of the core switches as the bouncing switch and transmits the data packet to the receiver through that switch. DRB selects the bouncing switch by digit-reversing the IDs of core switches, ensuring that no two successive data packets pass the same path. CLOVE [71] uses ECN and in-band network telemetry (INT) [80] to detect the congestion information and calculate the weight of paths. Then, flowlets are assigned to different paths by rotating the source ports in a weighted round-robin manner.

With the blooming of machine learning techniques in solving complex online optimization problems, recent research efforts proposed learning-based schemes for automatic traffic optimization. CODA [81] utilized an unsupervised clustering scheme to identify the flow information with no need for application modification. AuTO [9] is a two-level deep reinforcement learning (DRL) framework to solve the scalability problem of traffic optimization in DCNs. Due to the non-negligible computation delay, current DRL systems for production data centers (with more than 10^5 servers) cannot handle flow-level traffic optimization as short flows are gone before the decisions come back. AuTO mimics the peripheral and central nervous

---

[3] The large flows are detected if their rates are larger than 10% of the link capacity.

systems in animals. Peripheral systems are deployed at end-hosts to make traffic optimization decisions locally for short flows. A central system is also deployed to aggregate the global traffic information and make individual traffic optimization decisions for long flows, which are more tolerant of the computation delay. Iroko [82] analyzes the requirements and limitations of applying reinforcement learning in DCNs and then designs an emulator which supports different network topologies for data-driven congestion control.

6.3.2.2   Server-Based Scheduling

Server-based scheduling schemes are designed to balance loads between servers for the improvement of system throughput, resource utilization, and energy efficiency. From the viewpoint of computation, server-based scheduling can be classified into Layer-4- and Layer-7-based schemes. Layer-4 load balancing schemes work at the transport layer, e.g., Transmission Control Protocol (TCP) and User Datagram Protocol (UDP). Without considering application information, the IP address and port information are used to determine the destination of the traffic. Ananta [83] is a Layer-4 load balancer that contains a consensus-based reliable controller and several software multiplexers (Muxes) for a decentralized scale-out data plane. The Mux splits all incoming traffic and realizes encapsulation functionality in software. Google Maglev [84] provides distributed load balancing that uses consistent hashing to distribute packets across the corresponding services. However, software load balancers suffer high latency and low capacity, making them less than ideal for request-intensive and latency-sensitive applications. For further performance improvement, Duet [85] embedded the load balancing functionality into switches and achieved low latency, high availability, and scalability at no extra cost. SilkRoad [86] implements a fast load balancer in a merchant switching ASIC, which can scale to ten million connections simultaneously by using hashing to maintain per-connection state management.

In contrast, Layer-7 load balancers operate at the highest application layer, which are aware of application information to make more complex and informed load balancing decisions. Traditional Layer-7 load balancers are either dedicated hardware middleboxes [87] or can run on virtual machines (VMs) [88]. The key problem of Layer-7 load balancers is that when a load balancing instance fails, the TCP flow state for the client-server connections is lost, which breaks the data flows. To solve this problem, Yoda [89] keeps per-flow TCP state information with a distributed store.

As discussed above, the load balancing can be actively achieved by scheduling computing tasks into appropriate servers. From another viewpoint of storage, traffic optimization can also be passively achieved by optimizing the storage location with data replica placement and erasure code schemes in DCNs. By creating full data copies at storage nodes near end users, data replication can reduce the data service latency with good fault tolerance performance. An intuitive heuristic is hash—hash data and replicas to data centers so as to optimize for load balancing, which has been widely adopted in the distributed storage systems today, such as HDFS [90]. Nevertheless, this simple heuristic is far from ideal as it overlooks the skewness of data requests. Yu et al. [91] designed a hypergraph-based framework for associated data placement (ADP), achieving low data access overheads and load balancing among geo-distributed data centers. However, the centralized ADP is not effective enough in terms of the running time and computation overhead, making it slow to react to the real-time changes in workloads. Facebook Akkio [92] is a data migration scheme, which adapts to the changing data access patterns. To improve the scalability of the solution for petabytes of data, Akkio groups the related data with similar access locality into a migration unit. DataBot [93] is a reinforcement learning-based scheme which adaptively learns optimal data placement policies, reducing the latency of data flows with no future assumption about the data requests. The limitation of data replication is that it suffers from high bandwidth and storage costs with the growing number of replicas.

With erasure codes, each data item is coded into K data chunks and R parity chunks. The original data item can be recovered via the decoding process from any K out of $K + R$ chunks. Compared with replication, erasure codes can lower the bandwidth and storage costs by an order of magnitude while with the same or better level of data reliability. EC-Cache [94] provides a load-balanced, low-latency caching cluster that uses online erasure coding to overcome the limitations of data replication. Hu et al. [95] designed a novel load balancing scheme in coded storage systems. When the original storage node of the requested data chunk becomes a hotspot, degraded reads[4] are proactively and intelligently launched to relieve the burden of the hotspot. Due to the non-negligible decoding overhead, erasure codes may not be suitable for data-intensive applications.

6.4 Future Data Center Networks and Applications

According to the prediction of Oracle Cloud, 80% of all enterprises plan to move their workloads to the cloud data centers [96]. The amount of stored and processed data continues to increase, from 5G and Internet of Things (IoT) devices to emerging technologies, e.g., artificial intelligence, augmented reality, and virtual reality. These new technologies are dramatically reshaping the data center in order to meet the rising demands. However, Forbes reported that only 29% engineers said their data centers can meet the current needs [97]. Here, we list several future data center trends that let network infrastructure meet the ultimate challenges of the upcoming days.

Low Latency: From Milliseconds to Microseconds and Nanoseconds. Currently, a significant part of the communication traffic is within DCNs. Network latency can affect the performance of delay-sensitive applications, e.g., web search, social networks, and key-value stores in a significant manner. The latencies for current

---

[4]The action of parity chunk retrieval for decoding is defined as degraded read.

DCNs are in the order of milliseconds to hundreds of microseconds, which use (1) the mainstream Hadoop and HDD/SSD as the storage solution, (2) TCP as the communication protocol, and (3) statistical multiplexing as the communication link sharing mechanism. The latencies are planned to be reduced by an order of magnitude to microseconds or even nanoseconds with the evolving of future data center architectures. The storage access latencies have been reduced to tens of microseconds by all using NVMe SSD or even tens of nanoseconds with the emerging storage class memory (SCM). Through network virtualization, the data center is virtualized to a distributed resource pool for scalable all-IP networks, improving the resource utilization. Furthermore, TCP introduces many extra overheads. This means the processors need to spend a lot of time in managing network transfers for data-intensive applications, reducing the overall performance. In contrast, remote direct memory access (RDMA) allows servers to exchange data in the memory without involving either one's processor, cache, or operating system. RDMA is the future of data center storage fabrics to achieve low latency.

In-Network Computing. The newly emerged programmable network devices (e.g., switches, network accelerators, and middleboxes) and the continually increasing traffic motivate the design of in-network computing. In future DCNs, the computing will not start and end at the servers but will be extended into the network fabric. The aggregation functions needed by the data-intensive applications, e.g., big data, graph processing, and stream processing, have the features that make it suitable to be executed in programmable network devices. The total amount of data can be reduced by arithmetic (add) or logical function (minima/maxima detection) that can be parallelized. By offloading computing tasks onto the programmable network devices, we can (1) reduce network traffic and relieve network congestion, (2) serve user requests on the fly with low service latency, and (3) reduce the energy consumption of running servers. How to enable in-network computing inside commodity data centers with complex network topologies and multipath communication is a challenge. The end-to-end principle which has motivated most of the networking paradigms of the past years is challenged when in-network computing devices are inserted on the ingress-egress path.

Data Center Automation. With the explosive growth of traffic and the rapid expansion of businesses in data centers, manual monitoring, configuration, troubleshooting, and remediation are inefficient and may put businesses at risk. Data center automation means the process of network management, e.g., configuration, scheduling, monitoring, maintenance, and application delivery, can be executed without human administration, which increases the operational agility and efficiency. Massive history traces have been accumulated during the operation of data centers. Machine learning, which gives computers the ability to learn from history, is a promising solution to realize data center automation. The purposes of the learning-based data center automation are to (1) provide insights into network devices and servers for automatic configurations, (2) realize adaptive data forwarding and routing according to network changes, (3) automate all scheduling and monitoring tasks, and (4) enforce data center to operate in agreement with standards and policies.

High Reliability and Availability in the Edge. Providing highly available and reliable services has always been an essential part of maintaining customer satisfaction and preventing potential revenue losses. However, downtime is the enemy of all data centers. According to the Global Data Center Survey report in the year 2018, 31% of data center operators reported they experienced a downtime incident or severe service degradation [98]. The time to full recovery for most outages was 1–4 h, with over a third reporting a recovery time of 5 h or longer. The downtime in data centers is costly. It has been reported that about $285 million have been lost yearly due to failures [99]. According to the global reliability survey in 2018, 80% of businesses required a minimum uptime of 99.99% [100]. To achieve high reliability and availability, equipment redundancy is widely utilized in the data center industry. Compared with redundancy in hyperscale data centers, providing services at the edge of the network is attracting increasing attention. In the not-too-distant future, edge data centers are likely to explode as people continue to offload their computing and storage tasks from end devices to centralized facilities. With data being captured from so many different sources, edge data centers are going to become as common as streetlights to ensure high reliability and availability. The hyperscale data centers may work together with edge computing to meet the computing, storage, and latency requirements, which creates both opportunities and threats to the design of the existing system.

References

1. Y. Mansouri, A.N. Toosi, R. Buyya, Data storage management in cloud environments: taxonomy, survey, and future directions. ACM Comput. Surv. **50**(6), 1–51 (2018)
2. D. Reinsel, J. Gantz, J. Rydning, The digitization of the world from edge to core, IDC white paper (2018), [Online]: https://www.seagate.com/em/en/our-story/data-age-2025/
3. W. Xia, P. Zhao, Y. Wen, H. Xie, A survey on data center networking (DCN): infrastructure and operations. IEEE Commun. Surv. Tuts. **19**(1), 640–656 (2017)
4. T. Ye, T.T. Lee, M. Ge, W. Hu, Modular AWG-based interconnection for large-scale data center networks. IEEE Trans. Cloud Comput. **6**(3), 785–799 (2018)
5. N. Jones, How to stop data centres from gobbling up the world's electricity. Nature **561**(7722), 163–167 (2018)
6. A.S.G. Andrae, T. Edler, On global electricity usage of communication technology: trends to 2030. Challenges **6**(1), 117–157 (2015)
7. Power, pollution and the internet (2012), [Online]: https://www.nytimes.com/2012/09/23/technology/data-centers-waste-vast-amounts-of-energy-belying-industry-image.html
8. T. Benson, A. Akella, D.A. Maltz, Network traffic characteristics of data centers in the wild, in *ACM IMC*, 2010, pp. 267–280
9. L. Chen, J. Lingys, K. Chen, F. Liu, AuTO: scaling deep reinforcement learning for datacenter-scale automatic traffic optimization, in *ACM SIGCOMM* (2018), pp. 191–205
10. K. Chen, C. Hu, X. Zhang, K. Zheng, Y. Chen, A.V. Vasilakos, Survey on routing in data centers: insights and future directions. IEEE Netw. **25**(4), 6–10 (2011)
11. J. Zhang, F.R. Yu, S. Wang, T. Huang, Z. Liu, Y. Liu, Load balancing in data center networks: a survey. IEEE Commun. Surv. Tuts. **20**(3), 2324–2352 (2018)

12. E. Baccour, S. Foufou, R. Hamila, A. Erbad, Green data center networks: a holistic survey and design guidelines, in IEEE IWCMC, 2019, pp. 1108–1114
13. C. Kachris, I. Tomkos, A survey on optical interconnects for data centers. IEEE Commun. Surv. Tuts. **14**(4), 1021–1036 (2012)
14. M.F. Bari, R. Boutaba, R. Esteves, L.Z. Granville, M. Podlesny, M.G. Rabbani, Q. Zhang, M.F. Zhani, Data center network virtualization: a survey. IEEE Commun. Surv. Tuts. **15**(2), 909–928 (2013)
15. M. Al-Fares, A. Loukissas, A. Vahdat, A scalable, commodity data center network architecture, in *ACM SIGCOMM*, 2008, pp. 63–74
16. A. Greenberg, J.R. Hamilton, N. Jain, S. Kandula, C. Kim, P. Lahiri, S. Sengupta, VL2: a scalable and flexible data center network, in *ACM SIGCOMM*, 2009, pp. 51–62
17. Y. Sun, J. Chen, Q. Liu, W. Fang, Diamond: an improved fat-tree architecture for large-scale data centers. J. Commun. **9**(1), 91–98 (2014)
18. M. Walraed-Sullivan, A. Vahdat, K. Marzullo, Aspen trees: balancing data center fault tolerance, scalability and cost, in *ACM CoNEXT*, 2013, pp. 85–96
19. V. Liu, D. Halperin, A. Krishnamurthy, T. Anderson, F10: a fault-tolerant engineered network, in *USENIX NSDI*, 2013, pp. 399–412
20. G. Chen, Y. Zhao, D. Pei, D. Li, Rewiring 2 links is enough: accelerating failure recovery in production data center networks, in *IEEE ICDCS*, 2015, pp. 569–578
21. L. Gyarmati, T.A. Trinh, Scafida: a scale-free network inspired data center architecture. ACM SIGCOMM Comput. Commun. Rev. **40**(5), 4–12 (2010)
22. J.Y. Shin, B. Wong, E.G. Sirer, Small-world datacenters, in *ACM SOCC*, 2011, pp. 1–13
23. A. Singla, C.Y. Hong, L. Popa, P.B. Godfrey, Jellyfish: networking data centers randomly, in *USENIX NSDI*, 2012, pp. 225–238
24. X.Y. Lin, Y.C. Chung, T.Y. Huang, A multiple LID routing scheme for fat-tree-based InfiniBand networks, in *IEEE IPDPS*, 2004, p. 11
25. C. Guo, H. Wu, K. Tan, L. Shi, Y. Zhang, S. Lu, DCell: a scalable and fault-tolerant network structure for data centers, in *ACM SIGCOMM*, 2008, pp. 75–86
26. C. Guo, G. Lu, D. Li, H. Wu, X. Zhang, Y. Shi, C. Tian, Y. Zhang, S. Lu, BCube: a high performance, server-centric network architecture for modular data centers, in *ACM SIGCOMM*, 2009, pp. 63–74
27. D. Li, C. Guo, H. Wu, K. Tan, Y. Zhang, S. Lu, FiConn: using backup port for server interconnection in data centers, in *IEEE INFOCOM*, 2009, pp. 2276–2285
28. Y. Liao, D. Yin, L. Gao, DPillar: scalable dual-port server interconnection for data center networks, in *IEEE ICCCN*, 2010, pp. 1–6
29. C. Wang, C. Wang, Y. Yuan, Y. Wei, MCube: a high performance and fault-tolerant network architecture for data centers, in *IEEE ICCDA*, 2010, pp. 423–427
30. H. Wu, G. Lu, D. Li, C. Guo, Y. Zhang, MDCube: a high performance network structure for modular data center interconnection, in *ACM CoNEXT*, 2009, pp. 25–36
31. L. Huang, Q. Jia, X. Wang, S. Yang, B. Li, PCube: improving power efficiency in data center networks, in *IEEE CLOUD*, 2011, pp. 65–72
32. X. Liu, S. Yang, L. Guo, S. Wang, H. Song, Snowflake: a new-type network structure of data center. Chin. J. Comput. **34**(1), 76–85 (2011)
33. Z. Ding, D. Guo, X. Liu, X. Luo, G. Chen, A MapReduce-supported network structure for data centers. Concurr. Comput. Pract. Exp. **24**(12), 1271–1295 (2012)
34. Y. Liu, J.K. Muppala, M. Veeraraghavan, D. Lin, M. Hamdi, *Data center network topologies: current state-of-the-art* (Springer International Publishing, Cham, 2013), pp. 7–14
35. D. Abts, M.R. Marty, P.M. Wells, P. Klausler, H. Liu, Energy proportional datacenter networks. ACM SIGARCH Comput. Archit. News **38**(3), 338–347 (2010)
36. H. Abu-Libdeh, P. Costa, A. Rowstron, G. O'Shea, A. Donnelly, Symbiotic routing in future data centers, in *ACM SIGCOMM*, 2010, pp. 51–62
37. D. Lin, Y. Liu, M. Hamdi, J. Muppala, Hyper-BCube: a scalable data center network, *IEEE ICC*, 2012, pp. 2918–2923

38. J. Kim, W.J. Dally, D. Abts, Flattened butterfly: a cost-efficient topology for high-radix networks. ACM SIGARCH Comput. Archit. News **35**(2), 126–137 (2007)
39. G. Wang, D. Andersen, M. Kaminsky, K. Papagiannaki, T. Ng, M. Kozuch, M. Ryan, c-Through: part-time optics in data centers. ACM SIGCOMM Comput. Commun. Rev. **40**(4), 327–338 (2010)
40. N. Farrington, G. Porter, S. Radhakrishnan, H. Bazzaz, V. Subramanya, Y. Fainman, G. Papen, A. Vahdat, Helios: a hybrid electrical/optical switch architecture for modular data centers. ACM SIGCOMM Comput. Commun. Rev. **40**(4), 339–350 (2010)
41. K. Chen, A. Singla, A. Singh, K. Ramachandran, L. Xu, Y. Zhang, X. Wen, Y. Chen, OSA: an optical switching architecture for data center networks with unprecedented flexibility, in *USENIX NSDI*, 2012, pp. 239–252
42. J. Edmonds, Paths, trees, and flowers. Can. J. Math. **17**(3), 449–467 (1965)
43. M. Al-Fares, S. Radhakrishnan, B. Raghavan, N. Huang, A. Vahdat, Hedera: dynamic flow scheduling for data center networks, in *USENIX NSDI*, 2010, pp. 89–92
44. J.P.S. Kandula, P. Bahl, Flyways to de-congest data center networks, in *ACM HotNets*, 2009, pp. 1–6
45. D. Halperin, S. Kandula, J. Padhye, P. Bahl, D. Wetherall, Augmenting data center networks with multi-gigabit wireless links, in *ACM SIGCOMM*, 2011, pp. 38–49
46. Y. Cui, H. Wang, X. Cheng, B. Chen, Wireless data center networking. IEEE Wirel. Commun. **18**(6), 46–53 (2011)
47. J.Y. Shin, E.G. Sirer, H. Weatherspoon, D. Kirovski, *On the Feasibility of Completely Wireless Data Centers. Technical report* (Cornell University, Ithaca, 2011) [Online]: http://hdl.handle.net/1813/22846
48. A. Cayley, On the theory of groups. Am. J. Math. **11**(2), 139–157 (1889)
49. A. Singh, J. Ong, A. Agarwal, G. Anderson, A. Armistead, R. Bannon, S. Boving, G. Desai, B. Felderman, P. Germano et al., Jupiter rising: a decade of Clos topologies and centralized control in Google's datacenter network, in *ACM SIGCOMM*, 2015, pp. 183–197
50. A. Andreyev, Introducing data center fabric, the next-generation Facebook data center network (2014), [Online]: https://code.facebook.com/posts/360346274145943/
51. N. Farrington, A. Andreyev, Facebook's data center network architecture, in *IEEE OI*, 2013, pp. 49–50
52. Q. Pu, G. Ananthanarayanan, P. Bodik, S. Kandula, A. Akella, P. Bahl, I. Stoica, Low latency geo-distributed data analytics, in *ACM SIGCOMM*, 2015, pp. 421–434
53. O. Khan, R.C. Burns, J.S. Plank, W. Pierce, C. Huang, Rethinking erasure codes for cloud file systems: minimizing I/O for recovery and degraded reads, in *USENIX FAST*, 2012, pp. 20–33
54. R. Mysore, A. Pamboris, N. Farrington, N. Huang, P. Miri, S. Radhakrishnan, A. Vahdat, Portland: a scalable fault-tolerant layer 2 data center network fabric, in *ACM SIGCOMM*, 2009, pp. 39–50
55. S. Hu, K. Chen, H. Wu, W. Bai, C. Lan, H. Wang, C. Guo, Explicit path control in commodity data centers: design and applications, in *USENIX NSDI*, 2015, pp. 15–28
56. M. Besta, M. Schneider, K. Cynk, M. Konieczny, E. Henriksson, S. Di Girolamo, T. Hoefler, FatPaths: routing in supercomputers, data centers, and clouds with low-diameter networks when shortest paths fall short. *arXiv preprint arXiv:1906.10885* (2019)
57. A. Zhao, Z. Liu, J. Pan, M. Liang, A simple, cost-effective addressing and routing architecture for fat-tree based datacenter networks, in *IEEE INFOCOM Workshop on DCPerf*, 2017, pp. 36–41
58. A. Zhao, Z. Liu, J. Pan, M. Liang, A novel addressing and routing architecture for cloud-service datacenter networks. IEEE Trans. Serv. Comput. (2019). https://doi.org/10.1109/TSC.2019.2946164
59. P. Costa, A. Donnelly, G. O'Shea, A. Rowstron, CamCube: a key-based data center, *Technical Report MSR TR-2010-74*, Microsoft Research, 2010
60. C.A. Sunshine, Source routing in computer networks. ACM SIGCOMM Comput. Commun. Rev. **7**(1), 29–33 (1977)

61. E. Rosen, A. Viswanathan, R. Callon, Multiprotocol label switching architecture, RFC 3031 (2001)
62. N. McKeown, T. Anderson, H. Balakrishnan, G. Parulkar, L. Peterson, J. Rexford, S. Shenker, J. Turner, OpenFlow: enabling innovation in campus networks. ACM SIGCOMM Comput. Commun. Rev. **38**(2), 69–74 (2008)
63. M. Besta, T. Hoefler, Slim fly: a cost effective low-diameter network topology, in *IEEE SC*, 2014, pp. 348–359
64. A. Valadarsky, M. Dinitz, M. Schapira, Xpander: unveiling the secrets of high-performance datacenters, in *ACM HotNets*, 2015, pp. 1–7
65. M. Handley, C. Raiciu, A. Agache, A. Voinescu, A.W. Moore, G. Antichi, M. Wójcik, Re-architecting datacenter networks and stacks for low latency and high performance, in *ACM SIGCOMM*, 2017, pp. 29–42
66. S. Kandula, S. Sengupta, A. Greenberg, P. Patel, R. Chaiken, The nature of data center traffic: measurements & analysis, in *ACM IMC*, 2009, pp. 202–208
67. M. Alizadeh, A. Greenberg, D.A. Maltz, J. Padhye, P. Patel, B. Prabhakar, M. Sridharan, Data center TCP (DCTCP), in *ACM SIGCOMM*, 2010, pp. 63–74
68. Alibaba Cluster Trace, 2018. [Online]: https://github.com/alibaba/clusterdata
69. Q. Liu, Z. Yu, The elasticity and plasticity in semi-containerized co-locating cloud workload: a view from Alibaba trace, in *ACM SoCC*, 2018, pp. 347–360
70. A. Kabbani, B. Vamanan, J. Hasan, F. Duchene, FlowBender: flow-level adaptive routing for improved latency and throughput in datacenter networks, in *ACM CoNEXT*, 2014, pp. 149–160
71. N. Katta, A. Ghag, M. Hira, I. Keslassy, A. Bergman, C. Kim, J. Rexford, CLOVE: congestion-aware load balancing at the virtual edge, in *ACM CoNEXT*, 2017, pp. 323–335
72. M. Alizadeh, T. Edsall, S. Dharmapurikar, R. Vaidyanathan, K. Chu, A. Fingerhut, G. Varghese, CONGA: distributed congestion-aware load balancing for datacenters, in *ACM SIGCOMM*, 2014, pp. 503–514
73. N. Katta, M. Hira, C. Kim, A. Sivaraman, J. Rexford, HULA: scalable load balancing using programmable data planes, in *ACM SOSR*, 2016, pp. 1–12
74. W. Wang, Y. Sun, K. Zheng, M.A. Kaafar, D. Li, Z. Li, Freeway: adaptively isolating the elephant and mice flows on different transmission paths, in *IEEE ICNP*, 2014, pp. 362–367
75. D. Zats, T. Das, P. Mohan, D. Borthakur, R. Katz, DeTail: reducing the flow completion time tail in datacenter networks, in *ACM SIGCOMM*, 2012, pp. 139–150
76. S. Ghorbani, Z. Yang, P.B. Godfrey, Y. Ganjali, A. Firoozshahian, Drill: micro load balancing for low-latency data center networks, in *ACM SIGCOMM*, 2017, pp. 225–238
77. C.E. Hopps, Multipath issues in unicast and multicast next-hop selection, RFC 2991 (2000)
78. C. Raiciu, S. Barre, C. Pluntke, A. Greenhalgh, D. Wischik, M. Handley, Improving datacenter performance and robustness with multipath TCP, in *ACM SIGCOMM*, 2011, pp. 266–277
79. J. Cao, R. Xia, P. Yang, C. Guo, G. Lu, L. Yuan, Y. Zheng, H. Wu, Y. Xiong, D. Maltz, Per-packet load-balanced, low-latency routing for clos-based data center networks, in *ACM CoNEXT*, 2013, pp. 49–60
80. C. Kim, A. Sivaraman, N. Katta, A. Bas, A. Dixit, L.J. Wobkeret, In-band network telemetry via programmable dataplanes, in *ACM SIGCOMM*, 2015
81. H. Zhang, L. Chen, B. Yi, K. Chen, M. Chowdhury, Y. Geng, CODA: toward automatically identifying and scheduling coflows in the dark, in *ACM SIGCOMM*, 2016, pp. 160–173
82. F. Ruffy, M. Przystupa, I. Beschastnikh, Iroko: a framework to prototype reinforcement learning for data center traffic control, in *NIPS*, 2018
83. P. Patel, D. Bansal, L. Yuan, A. Murthy, A. Greenberg, D.A. Maltz, R. Kern, H. Kumar, M. Zikos, H. Wu, C. Kim, N. Karri, Ananta: cloud scale load balancing, in *ACM SIGCOMM*, 2013, pp. 207–218
84. D.E. Eisenbud, C. Yi, C. Contavalli, C. Smith, R. Kononov, E. Mann-Hielscher, A. Cilingiroglu, B. Cheyney, W. Shang, J.D. Hosein, Maglev: a fast and reliable software network load balancer, in *USENIX NSDI*, 2016, pp. 523–535

85. R. Gandhi, H.H. Liu, Y.C. Hu, G. Lu, J. Padhye, L. Yuan, M. Zhang, Duet: cloud scale load balancing with hardware and software, in *ACM SIGCOMM*, 2015, pp. 27–38
86. R. Miao, H. Zeng, C. Kim, J. Lee, M. Yu, Silkroad: making stateful layer-4 load balancing fast and cheap using switching asics, in *ACM SIGCOMM*, 2017, pp. 15–28
87. F5 load balancer, 2020, [Online]: http://www.f5.com
88. HAProxy load balancer, 2020, [Online]: http://haproxy.1wt.eu
89. R. Gandhi, Y. C. Hu, M. Zhang, Yoda: a highly available layer-7 load balancer, in *EuroSys*, 2016, p. 21
90. HDFS Architecture Guide, 2019, [Online]: https://hadoop.apache.org/
91. B. Yu, J. Pan, A framework of hypergraph-based data placement among geo-distributed data-centers. IEEE Trans. Serv. Comput. **13**(3), 395–409 (2020)
92. M. Annamalai, K. Ravichandran, H. Srinivas, I. Zinkovsky, L. Pan, T. Savor, D. Nagle, M. Stumm, Sharding the shards: managing datastore locality at scale with Akkio, in *USENIX OSDI*, 2018, pp. 445–460
93. K. Liu, J. Wang, Z. Liao, B. Yu, J. Pan, Learning-based adaptive data placement for low latency in data center networks, in *IEEE LCN*, 2018, pp. 142–149
94. K.V. Rashmi, M. Chowdhury, J. Kosaian, I. Stoica, K. Ramchandran, EC-cache: load-balanced, low-latency cluster caching with online erasure coding, in *USENIX OSDI*, 2016, pp. 401–417
95. Y. Hu, Y. Wang, B. Liu, D. Niu, C. Huang, Latency reduction and load balancing in coded storage systems, in *ACM SoCC*, 2017, pp. 365–377
96. Oracle's Top 10 Cloud Predictions, 2019, [Online]: https://questoraclecommunity.org/learn/blogs/oracles-2019-top-10-cloud-predictions/
97. The Data Center of The Future, 2020, [Online]: https://www.forbes.com/sites/insights-vertiv/2020/01/22/the-data-center-of-the-future/
98. Data Center Industry Survey Results, 2018, [Online]: https://uptimeinstitute.com/2018-data-center-industry-survey-results
99. B. Snyder, J. Ringenberg, R. Green, V. Devabhaktuni, M. Alam, Evaluation and design of highly reliable and highly utilized cloud computing systems. J Cloud Comput. **4**(1), 1–16 (2015)
100. 80% of businesses now require uptime of 99.99% from their cloud service vendors, 2018, [Online]: https://www.techrepublic.com/article/80-of-businesses-now-require-uptime-of-99-99-from-their-cloud-service-vendors/

Chapter 7
Public Cloud Architecture

Matt Lehwess

7.1 Introduction

Since 2006, Amazon Web Services (AWS), along with Microsoft Azure (Azure), and Google Cloud Platform (GCP) have been leading the charge in IT systems innovation. Cloud providers have since been completely changing how small businesses and large enterprises have been deploying applications, building IT systems, and networks too.

When using cloud providers to deploy IT systems, there are several key areas to consider: (1) the underlying hardware capabilities such as servers, performance, security, and connectivity and (2) the products, features, and services that run on top of this underlying hardware, that provide you, the user of cloud and builder of IT systems, the most frictionless way to spend more time doing whatever it is your business does, and less time in managing datacenters, hardware deployments, and building connectivity to these systems.

To be able to provide slices of their underlying hardware systems, cloud providers needed to abstract the experience of using and deploying IT systems away from the process of physically deploying the actual hardware and its required components such as servers, storage, and networking, much in the same way private cloud solutions such as VMware and OpenStack have successfully done, just at much greater scale, with millions of potential users on the same infrastructure deployment to keep in mind. To create this abstraction, cloud providers built a first level of control plane for themselves that would enable the management of the underlying hardware at a scale that would support these thousands or even millions of users, and then a second customer-facing control plane for their users, allowing a user to deploy virtual infrastructure, in a similar manner to how they would have previously

M. Lehwess (✉)
Amazon.com, Inc., San Francisco, CA, USA
e-mail: mlehwess@amazon.com

© The Author(s), under exclusive license to Springer Nature
Switzerland AG 2021
M. Toy (ed.), *Future Networks, Services and Management*,
https://doi.org/10.1007/978-3-030-81961-3_7

deployed infrastructure on-premises in their own datacenters. To enable this user-level control plane separation, cloud providers abstracted and enabled every step of the user's configuration process through application programming interfaces (APIs). It's these API-driven systems that have enabled users of cloud to deploy systems that consist of 1000s of virtual servers, globally, and in a mere matter of minutes. These tasks would have previously taken many months to achieve if they were performed within traditional datacenters and had to rely on traditional networks.

Cloud technology is one of the greatest recent innovation enablers, not just accessible to large enterprises who have traditionally had the funds to build large datacenters. Cloud technology is available to anyone who has a little bit of know-how, and a credit card, with virtual server prices charged by the hour and as low as $0.0065 per hour (e.g., the *T2.nano* instance from AWS), and offering virtual machines, or commonly called *instances*, that have networking speeds from the megabits per second up to 100s of gigabits per second.

Underneath the great innovation that is cloud, the glue that has made cloud technology possible is the network. Ethernet, Internet Protocol (IP), Transmission Control protocol (TCP), and User Datagram Protocol (UDP), all of our favorite networking protocols that have been around for decades, are all still alive and well within the cloud. The only difference between building datacenters on-premises *yesterday* and building virtual datacenters in the cloud *today*, outside of the ease of use and programmable interfaces, is layers. Networking has always been layered by design, considering the Open Systems Interconnect (OSI) model, or the TCP/IP stack, and cloud networking is no different, just with more depth and additional layers to consider, not layers of the OSI model, but rather layers of infrastructure management as well as abstraction of the user space from the infrastructure itself.

Datacenter networks in the 2000s and 2010s tended to be split horizontally into domains, usually using a technology called virtual routing forwarding (VRF) at Layer-3 of the OSI model, or Virtual LANs (VLAN)s at layer 2 of the OSI model. VRFs or VLANs would enable folks to build isolated sections of their datacenter that could contain particular applications or functions. For our purposes here, VLANs and VRFs can be considered horizontal slicing, with normally an intermediary such as a router, switch, or even a firewall, providing interconnectivity between these slices or segments of network.

When building initial cloud platforms beyond just public storage and internet facing only compute like Amazon's EC2 classic, it was soon apparent that giving customers the ability to *horizontally* slice networks was still required, both within their deployment like in times past with VRFs or VLANs, but also to segment customers from each other. To enable both the consumer of the cloud's user plane and control plane and the underlying cloud provider control plane, additional to *horizontal* slicing, now *vertical* slicing of networks would be also required, usually using mechanisms such as encapsulation or overlay networks also previously mentioned.

To enable vertical slices or overlays within a cloud providers platform, virtual private cloud was introduced in 2009 on AWS, similar functions were released on Microsoft's Azure (Vnet) and Google's GCP (also called VPC) shortly thereafter.

VPCs or Vnets now allowed customers to build a segmented area of the public cloud to house their virtual compute and control access to and from this domain, similar to the datacenter construct we saw in Fig. 7.1. Interestingly, before virtual private clouds, we would see single instances on the cloud providers platform, with public access to the public internet via a public IP associated to the virtual machine directly, just like with Fig. 7.2. After VPC, we would see whole domains that could consist of thousands of instances within a single VPC, and either connectivity to an on-premises datacenter through VPN or internet access through a centralized gateway, each of these VPCs being separate from the underlying platform and other users of the platform through both horizontal and vertical slicing as shown in Fig. 7.3.

Virtual private cloud technology provided the users of cloud providers platforms with a similar construct to what they were used to in datacenters (VRFs or VLANs), and the subsequent growth of IT systems on the cloud has been astronomical.

This chapter isn't designed to be an overview of cloud; even though we have talked about of the origins of cloud networking, there are many more in-depth books out there that dive into cloud and how to successfully deploy applications and networking within the cloud. This chapter will however talk about general cloud networking constructs and the additions we've seen in recent years to the fundamental platform that is virtual private cloud, we'll then take a step back to talk about where we see VPC or Vnet capabilities and architectures moving toward in the future as we get closer to Network 2030. Before we do that however, we'll take a look at the underlying infrastructure that cloud providers employ in the next section.

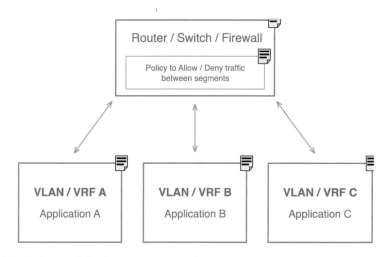

Fig. 7.1 Horizontal slicing in datacenter networks to segment applications or users

Public Cloud before VPC

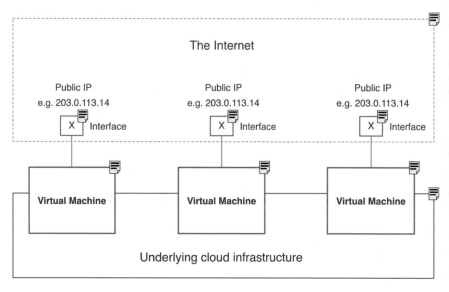

Fig. 7.2 Initial public cloud deployments of instances

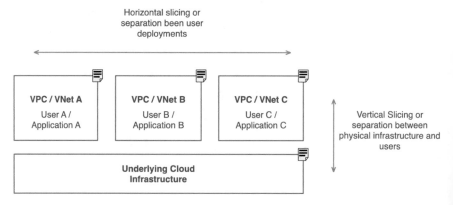

Fig. 7.3 Horizontal and vertical slicing of domains within the cloud

7.2   Platform

7.2.1   Core Public Cloud Infrastructure

To understand the responsibilities and different mechanisms employed between both cloud providers and users of the cloud, we need to understand the shared responsibility model that cloud providers follow. Figure 7.4 shows the breakdown

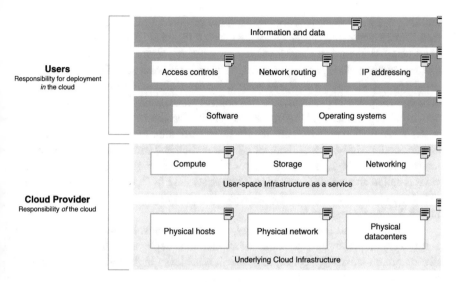

Fig. 7.4   Shared security responsibility model [1]

of where cloud provider's responsibilities for cloud infrastructure start and end, and where the user of a cloud platforms responsibilities start and end.

The shared responsibility model is normally used when referencing security responsibilities of using the cloud. The cloud provider's responsibility being the security *of* the cloud, and the user's responsibility being the security *in* the cloud through using security mechanisms provided by the cloud provider's platform.

This section uses the shared security responsibility model, removes the focus on security, and then uses the demarcation between cloud provider and user of the cloud to take a look into the responsibilities of cloud providers, the core platform overall, and where we see it evolving in the future.

The components that cloud providers use include physical servers, datacenters, groupings of datacenters, isolation of fault domains, global deployments, and edge deployments. The following defines each of these constructs.

> **Not all providers use the same naming conventions; we'll use the AWS method primarily for ease of naming and call out the differences between the next two largest providers, GCP and Azure, where needed.**

7.2.1.1   Availability Zone Infrastructure

Cloud providers still manage hardware, even with the cloud, someone needs too. Hardware, compute, storage, and networking are all deployed in datacenters which usually have redundant network connectivity and power. When you have multiple datacenters in a very close geographic area, it isn't always possible to have each building use redundant network connectivity or power that is completely different from a datacenter that might be located across the road. It is however possible to have many high-capacity and low-latency network connections between buildings that are within a close geographic proximity, and it is because of this high-scale and low-latency connectivity that makes it easier for multiple buildings to share a specific control-plane implementation. By control plane, we mean the mechanisms for which cloud providers offer APIs and the ability for users to deploy IT systems.

As these datacenters that are within close proximity have high-speed and low-latency connectivity, shared control planes, and shared redundant fiber and power feeds in and out of the region, it makes a lot of sense to group these together into what is called an availability zone (AZ), as shown in Fig. 7.5 and named as such due to the group of datacenters having a shared fate or availability (or unavailability during an AZ-level outage).

Given the shared fate of datacenters, and shared control plane within the availability zone, it makes a lot of sense for applications deployed in the cloud to utilize two or more availability zones, and a region achieves high availability by having multiple availability zones for use.

Due to the nature of availability zones, users of the cloud can rest easier knowing that applications deployed in one availability zone will not share any of the same systems that are used in another availability zone.

Availability zones are a key concept when deploying infrastructure within a geographic region, giving users of the cloud the ability to control the fate of their application deployments in relation to physical hardware or control plane failure.

> **We'll dive further into cases where the control plane of an availability zone can span multiple geographic areas when we talk about distributed cloud. Availability zones also may not constitute the same control plane separation in all cloud providers; however, the concept is similar across the three main public providers.**

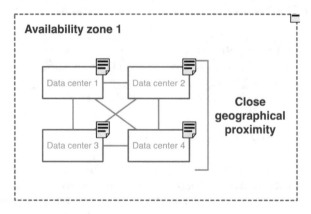

Fig. 7.5 Availability zone topology

7.2.1.2 Regional-Level Infrastructure

Availability zones, or grouped datacenters, split into isolated fault domains, all would generally exist in the same geographic region, also referred to by cloud providers as just a *region*.

An availability zone would consist of datacenters with high-speed and low-latency connectivity between them and would be considered a deployment domain for something like a virtual machine—which would exist on a single server, in a rack, in a datacenter. In this case, it makes a lot of sense for a virtual machine to be tied to an availability zone; however, there are services which behind the scenes operate across multiple availability zones and give the impression that they are *region-level* services. One such service that we'll talk about in great depth is VPC or virtual private cloud, which appears as though it spans an entire region.

A VPC spanning an entire region and no further is true for AWS and Azure; however, when using GCP, a VPC construct is global and can span multiple regions.

Figure 7.6 shows a region, which spans a large geographic area, consists of multiple availability zones, with region-level connectivity generally going through a centralized transit point to the internet or on-premises via a point of presence or meet-me room.

7.2.1.3 Global Infrastructure and Edge

There are cloud services that are considered global in nature such as each provider's content delivery network, or DNS services such as Amazon's Route 53 or Google's Cloud DNS. These global services utilize points of presence (POP)s as their point of existence. Alongside regions, POPs are smaller locations than their respective regions and normally far vaster in number. A POP would normally exist in another provider's location such as a colocation facility.

A POP would be used to deploy content delivery networks, or any cast-like services. In some cases, functionality such as a web application firewalls or serverless functions can be deployed at the edge at one of these edge locations or POPs.

For each cloud provider, as region counts slowly increase, for example, AWS going from 12 regions to 24 regions in the past 5 years, POP counts for services such as CDN or any cast have increased from in far greater numbers, for example, from 100 to 200 POPs in total; this is mainly due to the limited services offered at these points of presence vs. the services offered at a fully-fledged region which make it far easier for a cloud provider to deploy edge locations vs. regions. Edge in the sense of global cloud infrastructure is not to be confused with distributed cloud infrastructure; edge is for serving customers out on the public internet, and in these

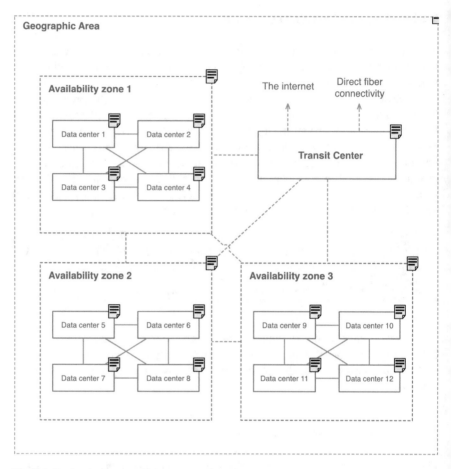

Fig. 7.6 Region-level connectivity

terms, distributed cloud infrastructure is the ability to have VPC infrastructure or compute at a location closer—and more than likely privately connected—to where you users are. The next section describes distributed cloud infrastructure.

7.2.1.4 Distributed Cloud Infrastructure

Distributed cloud is a new concept that has appeared over the last few years, yet something that has the potential to expand the reach of public providers even more than regional deployments that providers have been rolling out at an increased pace.

Distributed cloud is the concept of offering the technology that cloud providers have built in their regions, yet on hardware closer to users. There are a few versions of distributed cloud available, and we'll cover these at a VPC or user level in-depth below in our edge cloud section; however, from an infrastructure perspective, there are a few flavors:

AWS Outposts: Deployed as a rack or multiple racks at a user's location (usually a colocation facility or private datacenter), AWS Outposts takes the same hardware as AWS uses in their regions and makes it available for use just like an Amazon region, yet at the customers location. The hardware includes servers, networking, and storage. Instead of on-premises constructs such as virtual storage area networks (VSANs) or relying on customers to deploy virtualization technology, the outpost uses AWS hardware and then offers AWS services such as their Simple Storage Service (S3), Elastic Block Store (EBS), or Elastic Compute Cloud (EC2). All of these constructs use the Amazon VPC, yet now in the outpost, instead of being confined to the region. Physical security at an outpost is also increased by the use of a physical security key called the Nitro Security Key, which, when removed from a server and destroyed, renders all data on the server useless, which is very useful for when returning hardware back to the cloud provider. An example Outposts VPC architecture has been shown in Fig. 7.7.

AWS Local Zone: Similar to AWS Outposts and built upon the same technology, AWS Local Zone is an availability zone that sits outside the geographic area that a region is deployed. For example, the AWS Region new Portland, US-West-2 is extended down to Los Angeles through the Los Angeles Local Zone. Local zones are an important construct for users of cloud, as it allows cloud providers to deploy

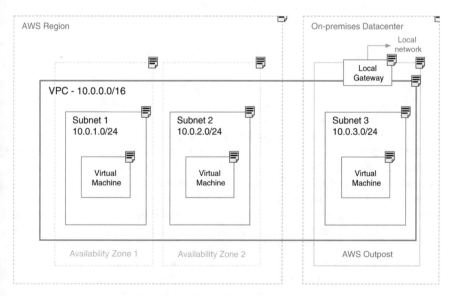

Fig. 7.7 Showing an AWS Outposts deployment, extending the VPC from the region down to the infrastructure deployed on-premises at the user's location [2]

services in a remote locality without having to invest in a fully-fledged region deployment.

AWS Wavelength: Again, using AWS Outposts technology, and deploying a similar construct to an AWS Local Zone, however this time connected natively to a provider's mobile network. AWS Wavelength is perfect for folks that want to deploy mobile applications at the edge, with low-latency connectivity to providers that offer 5G and mobile connectivity such as Verizon.

Microsoft Azure Stack: Similar to AWS Outposts, taking Azure on-premises on hardware that is deployed closer to a user's geographic area, however with AWS Outposts, customers must use AWS hardware, whereas Azure Stack relies on users purchasing hardware from a specific set of vendors that are qualified to run Azure Stack.

Google Anthos: A Kubernetes forward stack that can run on many different types of hardware vendors; Google Anthos is more of an umbrella of services that take GCP capabilities on-premises. Also similar to Azure Stack, whereby users can deploy on non-Google hardware and essentially pay by the hour for the software stack itself.

Overall, most cloud providers have some form of infrastructure or software that users can now deploy closer to their end clients. This has had a large effect on networking deployments, as the cloud components such as VPCs are now extended to on-premises and are easily bridged back to a cloud providers region. More on this below in the user-space distributed cloud section.

7.2.1.5 Future State of Core Public Cloud Infrastructure

The future of cloud provider infrastructure is quite clear; we've seen cloud providers consistently deploy additional regions in additional geographic areas to serve local users of that area over the last 10 years or so. Leading up to Network 2030, we'll continue to see this same pace of regional growth. At some point however, most geographic areas of the globe will be within 25–50 ms reach of a public cloud providers region. The biggest question is if that low of a latency between cloud and your users is low enough, probably not when it comes to low-latency applications such as video or game streaming, and this is where distributed cloud will help even more. Most cloud providers will probably reach region saturation from a geographic perspective by 2030, and from there, the focus will be expanding services in existing regions, deploying more capacity in existing regions, and distributed cloud.

Distributed cloud such as Google Anthos, Microsoft Azure Stack, and AWS Outposts allows customers to pay for an infrastructure deployment at a location that they need it, instead of waiting for a cloud provider's economies of scale and geographic deployment schedule to make sense for the cloud provider to deploy a full region at a location that the user needs. The downside of using distributed cloud technology however is that it often does not have the same breadth of services that a centralized cloud provider region does; this means that the latest and greatest services that users of cloud may want to use will only be accessible in the region. Heading toward Network 2030, we will likely see feature parity or close to it, across both distributed cloud and centralized cloud provider regions, although this will be difficult to achieve given the scale that some cloud services require to operate, which just isn't available in distributed cloud (with one or two racks of compute that services can use for distributed deployments vs. the region's almost infinite supply of hardware). For distributed cloud to also matter, we'll also need to see far easier scale up and scale down like we see at region-based deployments; however, this can only be achieved through cloud-friendly commitment contracts (pay as you go vs. the current 1-, 2-, and 3-year contracts we see with products like Outposts); this coupled with a faster pace of deployment vs. the several weeks it takes for a cloud providers to deploy a distributed cloud deployment today.

The last curiosity of distributed cloud is the pre-announcement of AWS's small form factor Outposts. Small form factor outposts are essentially distributed cloud, with Amazon VPC, in a single-rack unit, or two-rack unit-sized server. With the release of small form factor Outposts, we could see distributed cloud in even more locations such as Point of Sale or small-office deployments.

Overall, distributed cloud technology coupled with traditional regions could see the deployment of IT systems becomes even more frictionless, with the use of APIs to deploy virtual private clouds on cloud providers regional deployments and then anywhere else you are able to have a cloud provider come in and deploy hardware— or, where existing hardware that is supported by Azure Stack or Google Anthos is located.

7.2.2   User-Space Platform

The user space defines anything on top of the cloud providers' network that is configurable or within the control of the end user. There are many different services that run inside the user-space platform, but many of these share the same networking construct—the virtual private cloud. This section dives into each of the core components and their use when deploying IT systems within the cloud.

7.2.2.1   VPC Networking Components and Architecture

Virtual private cloud is a mechanism for creating a private, isolated section of the cloud, similar to VRFs or VLANs in the traditional networking world. To achieve this isolation, any packets that leave instances that are deployed on a physical host, within a specific VPC, will be encapsulated and isolated within the virtual network construct that is the VPC.

To achieve a similar experience to which engineers were used to in the on-premises world, the following components were created for curators of the user space to use; Fig. 7.8 shows the overall topology with each of these components and how they would work together.

VPC Addressing

Within the VPC, each instance or virtual machine needs to have an IP address, just like in datacenter networking. When the VPC is created, a CIDR address range is chosen, normally within the RFC 1918 range. All subnets will usually be created from this range; in some cases, an additional range can be added to the VPC. It is a common practice to use/16 ranges when creating a VPC, which nets approximately 65,000 addresses. Even though VPCs are isolated domains, there are some

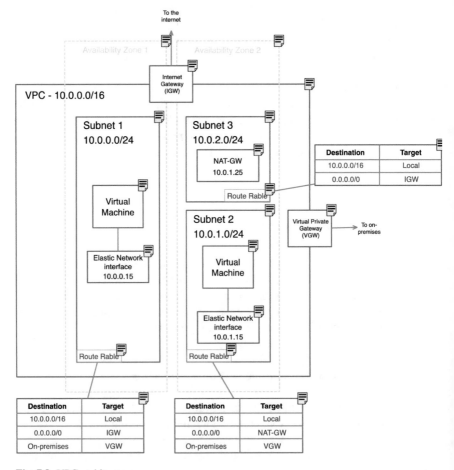

Fig. 7.8 VPC architecture

interconnectivity mechanisms such as Direct Connect or VPC Peering that will advertise the VPC range outside of the VPC; it is therefore a good idea to use an IP Address Management (IPAM) system and allocate unique ranges to each of your VPCs, as at some point you may want to connect your VPC to other VPCs, or to your on-premises, more on this below in our VPC connectivity section.

Subnets

Created within the supernet(s) assigned to the VPC. A subnet contains a subsection of the VPC range that can be used by objects within the subnet such as instances deployed in the subnet. Subnets are generally tied to an availability zone and are considered an availability zone construct, meaning, to deploy instances in multiple availability zones, a user would need to deploy those instances in different subnets, with each subnet tied to a difference availability zone.

Availability Zones

Covered previously when referencing cloud infrastructure and the platform itself, an availability zone or grouping of datacenters which share similar fault domains is important when deploying VPC-level constructs within the user space. Some components within a cloud deployment are tied to the underlying region such as VPCs (exception—GCP VPCs are global). However, many components such as instances, NAT GW, and others are availability zone-level constructs. This means that they are normally deployed within a subnet, which is tied to an availability zone. If that availability zone fails, so does the subnet(s) within the availability zone, and all of the components and instances deployed within that subnets tied to that availability zone.

Route Tables

A route table is a mechanism whereby traffic can be directed to a specific location. Very similar to route tables within a traditional router, however with the exception route propagation from gateways such as the virtual private gateway (more on that below), the route table that is assigned to each subnet is a *static* route table, with routing destinations configured by an end user. An example route table is shown in Fig. 7.8.

Internet Gateway

Bridging the connectivity between the public internet and the VPC, an internet gateway is a region-level construct that can be attached to a VPC and will perform a 1:1 NAT function between the private addresses of the VPC that are associated with instances, and public IPs known as Elastic IPs (AWS) or static public Ips (Azure and GCP) note that only AWS has the Internet Gateway construct, for GCP you can use the Cloud NAT functionality, and Azure allows instances to have public IPs directly assigned to instance interfaces which will then give an instance internet connectivity.

Elastic IP or Static Public IP

Public addresses owned by the cloud provider can be used as an outside NAT address for private VPC addresses that are assigned to instances. This static 1:1 NAT function happens at the Internet Gateway or Cloud NAT for AWS and GCP, and directly on the instance in Azure.

Network Interfaces

Assigned directly to instances, network interfaces in the cloud world have been decoupled from the physical services to which they are attached. Network interfaces in the cloud can usually have one primary IP address and many secondary IP addresses. These are also an availability zone-level construct, with the ability for in some cases to be detached from one instance and then reattached to another instance in the same subnet for which the network interface resides.

Virtual Private Gateway

Connecting to the cloud from on-premises can be done either through VPN or a physical fiber connectivity; we cover more on hybrid connectivity below; however, the virtual private gateway or VGW is how both of these types of connectivity are brought directly into the VPC. The VGW is a similar construct to the IGW, in that it is a managed service; however unlike the IGW, the VGW does not perform network address translation. Any CIDR range configured within the VPC can be advertised from the VGW to your on-premises networks.

Transit Gateway

A very familiar construct to the VGW, in that it is a gateway that can be attached to a VPC, and it allows for cloud connectivity. However, the similarities end there. The Transit Gateway is a mechanism where you can attach to many VPCs (up to 5000) and bridge connectivity between each of these through each VPC's TGW attachment. Further to allowing VPCs to communicate with each other, Transit Gateway also allows you to connect to on-premises via VPN termination, or direct connect, directly attached to the transit gateway. We cover more on interconnectivity of VPCs and on-premises further later.

NAT Gateway

Similar to an IGW, however NAT-GW is an availability zone construct that can be deployed in a single subnet. When traffic is then destined for the public internet via the NAT-GW, it will translate the source address to its own address, using port address translation (PAT) and forward the traffic on to its original destination. The NAT-GW subsequently allows many instances to reside behind a single public elastic IP.

7.2.2.2   Services Inside and Outside the VPC

When deploying IT systems on the cloud, you have a wide selection of services that you can choose to use. Some of these services are built or deployed natively inside the VPC, whereas others may be outside of the boundaries of the VPC. An example of some of the available services that public cloud providers offer includes compute, block storage, object storage, databases, security services, big data, machine learning, serverless, and many others.

Services inside a VPC are privately accessible by deployments that might be using compute in that same VPC directly, or these services may be reachable via connectivity to that VPC, such as when using AWS Direct Connect/Azure Express Route/GCP Cloud Interconnect, or VPN to connect to a VPC. Services that are outside the VPC mostly are traditional services that were built to be publicly accessible initially, and as cloud architectures evolved, it became necessary to be able to connect to these public services via private means from within the VPC. Object storage and noSQL databases are one such classic example of public services that have traditionally had public IP addresses and have been reachable from the public internet, but are now used by applications within a VPC, and have a need to be connected to privately, from within the VPC (Fig. 7.9).

Service network connectivity for cloud can be categorized into four main types, public services, native VPC constructs, network interface-based attachment to the VPC, and private connectivity for public services. We dive into each of these in the sections below.

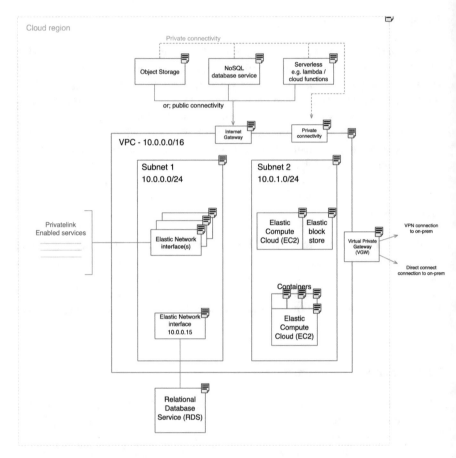

Fig. 7.9 Connectivity to services in the cloud

Public Services

Usually, legacy cloud services were being built with public connectivity in mind. AWS's object storage service—S3—for the longest time was only accessible via public IP addressing via the public AWS internet backbone. S3 serves a good portion of the public internet's content, and therefore it made a lot of sense to be publicly accessible. Other services such as NoSQL database services and serverless also exist within the public cloud realm and can be connected to publicly. When connecting to a public service, the service is reachable via a fully qualified domain name which would resolve to multiple A-records that are direct public IP addresses of the service itself.

Native VPC Constructs

Services that are built upon virtual machines inside a VPC are directly attached to these such as block storage or natively operate inside a VPC directly. These include compute and block storage attached to VMs but also include services such as DNS private hosted zones and DNS resolution in the VPC. Cloud provider container and Kubernetes services such as EKS and ECS are also built upon the cloud providers compute deployment (VMs) and operate inside the VPC but also have networking interface attachments within a VPC.

Network Interface-Based Attachment to the VPC

Services such as the relational database services, big data, and load balancing use network interfaces that operate across VPC boundaries. With this mechanism, the cloud provider can build a service inside a service-owned VPC and then connect to the user-plane VPC through a network interface that exists inside the user-plane VPC. The management of such a service is generally taken care of by the cloud provider; however, the network interface attachment in the user-plane VPC allows for data connectivity from everything within the VPC or connected to the VPC to be able to connect to this service. This mechanism is also used for the control plane for services that are deployed natively inside the user-plane VPC such as containers or kubernetes yet still require connectivity back to the cloud provider managed control plane.

Private Connectivity for Public Services

From legacy that is cloud, is that most services were built on the public cloud platform first, with public connectivity, and public IP addressing. As VPC became more popular, and users of the cloud realized they could have the same private domains as their on-premises datacenters, public connectivity to services in the cloud from applications that were deployed in the cloud soon fell out of popularity. The natural movement of connectivity shifted toward public-when-necessary and private-as-default.

Services such as AWS's privatelink took many of the publicly deployed services and made it possible to connect to these privately within the VPC. With privatelink acting as an intermediary, many of the once public cloud services could now directly borrow IP addresses from the user-space VPC and allow anything within the VPC to connect to each said service. At time of publishing, AWS privatelink supports private VPC connectivity to over 95 AWS services that could previously only be reached via the public AWS backbone and public IP addresses. All three cloud providers have an equivalent to privatelink (GCP is call private connect, and Azure is called privatelink).

7.2.2.3   Future of User-Space Services and Connectivity *Within* the Virtual Private Cloud

Looking toward Network 2030, we can expect a large number of services to continue to be released by cloud providers; however, the lower-level networking constructs of the VPC probably won't change very much, as these are the foundation for cloud provider managed services to connect to applications either publicly or privately and do not have much need to change at the VPC level, just like on-premises routing and switching concepts haven't changed foundationally a great deal in the last 10 years or so.

Where we will see innovation is the deployment of overlay networks or even upper layer networking constructs like service mesh which provides greater functionality than the lower VPC networking layer, for example, mutual TLS authentication, routing policies, service discovery, telemetry, and additional load balancing functionality, all built into the application or container deployment. Overlay networks have been used and will continue to be used to overcome the shortfalls of VPC-level capabilities, and service mesh is being widely used in microservices architectures and will only continue to grow in popularity.

The main downside to using upper layer networking constructs like overlay networks or service mesh in the cloud is that often the VPC gets forgotten about and architectures like these tend to lead to deploying all VMs or instances into a single VPC to minimize lower-level VPC complexity and rely on the upper layer to perform most of the network tasks. This can result in a single hyperscale VPC or very large VPC when deploying large environments. A hyperscale VPC is one that contains thousands of instances or virtual machines vs. a highly distributed architecture that could contain hundreds of instances across thousands of VPCs—which is another common pattern. Azure and GCP can only support a VPC size of 65,000 and 15,500 VMs, respectively, in a single VPC, and AWS does not call out the upper size of a VPC; however limitations such as number of network interfaces per region (5000), or routes in a route Table (1000), or subnets per VPC (200), will affect the size that a VPC can comfortably grow to. Note that some of these limits can be raised, but the main issue is that having all of your instances in a single VPC opens you up to reaching VPC limits much sooner than if you had a multi-VPC architecture.

Going forward to Network 2030, limits of the underlying VPC will surely be raised beyond the levels that they are today, but a lot of folks will be relying on upper layer capabilities to perform complex network functions and not rely on the lower level VPC, causing the single VPC architecture to become more popular.

The other end of the scale from the hyperscale VPC is VPC sprawl or the highly distributed VPC architecture, where every business unit of an organization, or every application deployed in the cloud gets its own VPC. Organizations can then have thousands of VPCs, each relying on VPC-level connectivity for inter-application communication. Network 2030 could see the reverse of hyperscale VPC deployments (single large VPCs) which is hyper-distributed multi-VPC architectures, where 1000s or 10,000s of VPCs are deployed essentially like containers, connected to whatever it is they need to be connected to, perform their function, and then shut down. The only downside of this multi-VPC architecture is that cloud providers charge for data transfer between VPCs, so having many VPCs and a lot of application communications that is inter-VPC by nature can become cost prohibitive quite quickly from a data transfer perspective. We cover more on inter-VPC connectivity in the next section.

Overall, we'll probably see a healthy mix of folks using a few VPCs as they start using the cloud, then moving to several as they start to grow their deployment, then 10s, then 100s, and probably a consolidation back to 10s depending on the amount of communication interdependencies between their applications in different VPCs, and the user of the cloud's appetite to deploy application-level networking like service mesh. Providers like GCP will see larger VPCs just given the global nature of a VPC in GCP, as opposed to AWS and Azure customers being only able to deploy VPCs in a single region and requiring another when they expand to another region. There is no single right way to build out your VPC architecture, but one thing is clear, VPC is here to stay until network 2030 and beyond.

7.2.3   Network Connectivity

Starting off with one VPC when thinking about using the cloud and allowing your application folks to deploy systems in this single VPC is an easy transition from the physical world, given that a VPC almost looks like a virtual datacenter. But given the attractiveness of using the VPC as a boundary for your applications, and the VPC sprawl that follows, the connectivity between your VPCs becomes an important part of your cloud architecture.

Users of cloud more often than not have on-premises deployments in pre-existing datacenters, and this is where cloud connectivity goes beyond the border of the cloud provider and connects into the on-premises world.

The next few sections cover VPC connectivity within a cloud provider's deployment and then on-premises to cloud connectivity for hybrid cloud architectures; before we dive into these, I'd like to cover a brief section on an important construct in networking—the communications matrix.

When building networks, it is always a good idea to break down items into their smallest possible component, and that happens to be the network flow required between a set of applications (or the user and the application). To build a path through a network, you need to know the source of the traffic, the destination, the port number or application type, and probably whether the traffic needs to be encrypted or not. Any decent network engineer who manages a network should be able to stitch together connectivity requested by application folks, as long as they've been given these details.

That said, and keeping your communications matrix in mind (or how you would like to connect your IT systems together), here are some of the options for connecting both VPCs and Vnets together in the cloud, and connecting your cloud to on-premises, depending on the type of connectivity, your application communications matrix requires the following:

7.2.3.1 VPC Peering

VPC peering is a mechanism for connecting two VPCs together in a point-to-point fashion. Cloud providers use overlay networks or encapsulation to provide user-level separation at the VPC, which gives them the ability to connect two VPCs together by bridging the encapsulated domains together. This VPC point-to-point connectivity is extremely useful if there are applications across different VPCs that need to talk to each other. In a point-to-point fashion, this is very useful; however as VPC deployments grow and many applications from many VPCs need to communicate, this becomes unwieldy to manage, especially since this point-to-point connectivity requires any VPCs to have unique IP addressing, and appropriate routing to connect the VPCs together. Figure 7.10 shows a VPC point-to-point connection, allowing applications in VPC A with IP addresses in the range 10.0.0.0/16 to communicate with VPC B that will have applications with IPs in the range 192.168.0.0/16. As a packet leaves an instance in VPC A that is destined for VPC B, it will follow its subnet route table to the logical entity that is the peering connection, and onward to the destination VPC, in this case VPC B.

Where peering becomes too complex to manage is where users deploy ten or more VPCs that require interconnectivity. While a user might start out with only one or two VPCs that require connectivity, this often snowballs into most if not all VPCs needing connectivity. The architecture that results from all VPCs needing connectivity is the full mesh architecture, and in this case, the number of peering connections is equal to n being the number of VPCs, and number of peering connections = $n(n - 1)/2$. This means that 10 VPCs peered together in a full mesh will equal 45 peering connections, and each route table for each VPC will need a

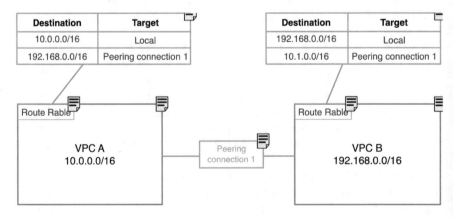

Fig. 7.10 VPC peering

route to each VPC, meaning 45 manually created routes need to be deployed in each
VPCs route table. Due to the complexity of the full mesh deployment, users of the
cloud will sometimes default to a hub and spoke peering architecture, with centrally
hosted shared services in a central VPC, and peering connection to each application
that resides in a spoke. Figure 7.11 shows both the full-mesh and hub-and-spoke
peering architectures commonly used.

It is worth noting that hub-and-spoke and full-mesh network architectures have
been around for a long time before cloud, and VPC peering is merely borrowing
these mechanisms to describe architectures of VPC connectivity.

7.2.3.2   Privatelink

Mentioned earlier as a mechanism for enabling connectivity to services from a VPC
privately, privatelink can also enable connectivity to VPCs and applications running
in those VPCs.

> **All three of the main cloud providers have a version of Privatelink or
> Private service connectivity; each has variations that we won't cover
> here, but the overall premise is the same across all three.**

Privatelink's first use case is a mechanism for bringing public cloud services
such as object storage, no SQL databases, or serverless functions into a VPC privately. This private connectivity to public services means that the VPC does not
need public internet access for any of the VPC deployed applications to use them.
To take this one step further, cloud providers also offer privatelink capabilities for
software-as-a-service (SaaS) applications that are deployed in another VPC, or what

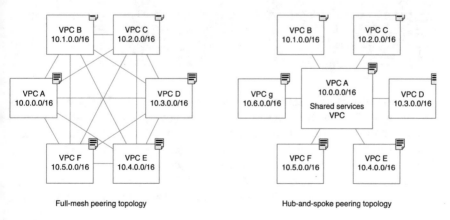

Full-mesh peering topology   Hub-and-spoke peering topology

Fig. 7.11   Full-mesh and hub-and-spoke peering topologies

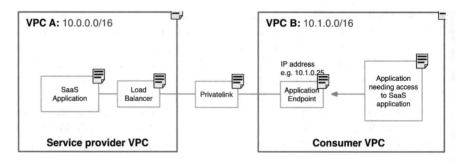

Fig. 7.12 Privatelink connectivity for services in another VPC

is called a service provider VPC. This is especially useful if users are building an application that needs to be reached by many VPCs. Figure 7.12 shows how an application can be deployed in a service provider VPC and then accessed through a consumer VPC. Privatelink with this use case greatly reduces the reliance on large-scale VPC peering connectivity, given that the application can be dropped into many VPCs at the same time, even VPCs that might have overlapping IP addressing with each other.

7.2.3.3 Transit GW

To further help with large-scale VPC connectivity, Transit Gateway, currently only available through AWS, is a mechanism for connecting many VPCs together in a single region (the TGW itself is a region-level construct); TGW can also connect to your on-premises via VPN and Direct Connect which we cover next. Transit Gateway is proving to be the backbone for a lot of VPC area networks in the AWS Cloud.

> There is less of a need for Transit Gateway within GCP given that their VPC can span multiple regions and is essentially global, and Azure has the ability to connect VPCs together using their VPN gateway service.

Transit Gateway architectures are enabling global scale for VPC-level networking and have allowed customers to bridge existing on-premises networks such as their SDWAN deployments into their cloud backbone network. As Transit Gateway continues to add additional capabilities such as increased SDWAN vendor support, it will continue to be the heart of a lot of VPC deployments.

Figure 7.13 shows Transit Gateway as the hub of connectivity for multiple VPCs (up to 5000) and also connectivity via Direct Connect and VPN to on-premises sites.

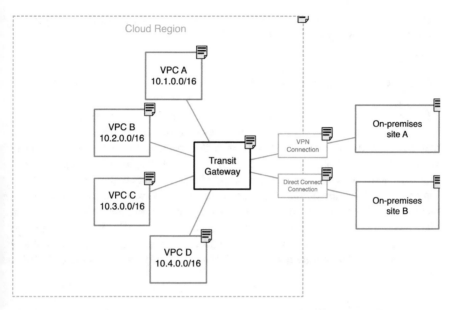

Fig. 7.13 Transit Gateway connectivity for VPCs and on-premises

GCP does have the advantage of a single VPC globally; however, if we think about single VPC limits such as the 15,500 VM limit per VPC, it can often make sense to split a VPC architecture into multiple VPCs, and Transit Gateway helps when a user's topology exceeds ten or more VPCs.

7.2.3.4 Managed Virtual Private Network Connectivity

For on-premises connectivity, and sometimes connectivity to other virtual machines, all three cloud providers have the ability to attach VPN connectivity to a VPC; this is often referred to as managed VPN and is normally IPsec site-to-site connectivity from a VPC to another IPsec capability appliance. Transit gateway architectures can help site-to-site VPN connectivity to many VPNs, without having to build connectivity to each individual VPC; most cloud providers also have the ability to attach client-to-site VPN connectivity to a VPC also.

7.2.3.5 Build Your Own Virtual Private Network

Prior to large-scale manage VPN solutions like the AWS Transit Gateway, or even Azure's Virtual WAN which we dive into below, users of cloud would often build instances or virtual machines that would run site-to-site VPN software such as Openswan or Strongswan; this gave the instance the ability to connect traffic from on-premises through a VPN connection over the public internet and into the

VPC. The VPN instance would then act as a virtual router of sorts, routing traffic from on-premises to the VPC and back again.

7.2.3.6   Direct Fiber to the Cloud

All three of the main cloud providers have the ability for companies to provision fiber from their on-premises datacenters to a cloud provider's edge location, which can then in-turn connect over the cloud providers backbone into the cloud providers region. This direct path to the cloud enables users of the cloud to bridge their on-premises and the cloud in a more consistent and sometimes higher bandwidth fashion than just using the public internet to connect to the cloud. Each provider gives users the ability to connect to either the cloud providers public services directly in what looks like dedicated public peering to the cloud or connect privately to specific VPCs that run the user's private cloud workloads; this connectivity is normally through a public cloud providers Point of Presence (PoP) and is shown in Fig. 7.14.

7.2.3.7   SDWAN Connectivity

The SDWAN architecture is covered in Chap. 4, and given that the cloud has fast become a user's equivalent of a datacenter, it makes a lot of sense for users to extend their on-premises connectivity using SDWAN up into the cloud. The main downside of extending SDWAN to the cloud is that the SDWAN edge device needs to run on a virtual machine within a VPC, and this is something that has taken a long time for vendors (either firewall, router, or SDWAN vendors) to optimize their appliance's performance for, which is mainly due to on-premises appliances having the hardware itself within the SDWAN providers control where Application Specific Integrated Circuits (ASICs) can be used, which can give these appliances much greater packet per second (PPS) performance vs. using standard non-optimized

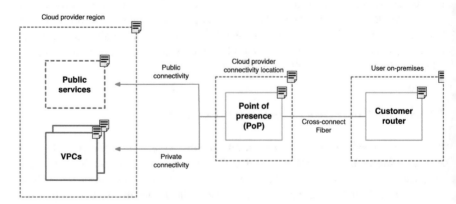

Fig. 7.14 Direct fiber connectivity to the cloud

software on x86-based virtual machines in the cloud, in the cloud, SDWAN providers software vs. previously the hardware needs to be specialized to utilize multiple CPU cores and also integrate into cloud constructs such as autoscaling for horizontal scale-out of SDWAN edge appliances vs. the previous mentality of "build a bigger hardware box" that appliance vendors have used in the past.

The first method for extending an SDWAN deployment into the cloud is to simply deploy an active/passive set of instances in a VPC and direct any traffic to and from the cloud via these instances. In this topology, shown in Fig. 7.15, there would normally be an orchestrator (not shown) that would also detect a failure of the active instance and proceed to call the cloud providers API to shift routing from the active SDWAN appliance to the secondary.

There are three downsides with Fig. 7.15's active/standby topology. The first is that having to call an API to shift a route within the cloud can take anywhere from 3 to 30 s, with a third-party orchestrator needing to do the detection and API call. The second is that the virtual edge appliance is limited by a single instance's worth of bandwidth, meaning that all traffic will traverse one instance, which causes this instance to become a bottleneck. If an appliance is not utilizing multiple cores on a virtual machine, traffic normally cannot exceed around 1.5 Gbps per instance, which, depending on the customers deployment, could be far lower than is needed. The third issue is that each VPC that wants to become a part of the SDWAN deployment will need SDWAN appliances deployed within. This means that when a user starts to see VPC sprawl, they will need to deploy up to two SDWAN appliances in each and every VPC, which can become cost prohibitive.

The fix to the aforementioned shortfalls of having an SDWAN deployment per VPC is to leverage other cloud-native constructs such as Transit Gateway that can give you equal cost multipath over multiple SDWAN appliances and integrate with autoscaling to horizontally scale the fleet of SDWAN appliances as they need traffic and scale the fleet back down when they do not, offering cost savings when the edge infrastructure isn't needed. Lastly, using a centralized construct such as Transit Gateway allows you to build a centralized Transit VPC that can service all of your

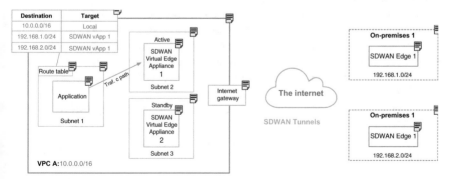

Fig. 7.15 SDWAN deployments extended to a single VPC

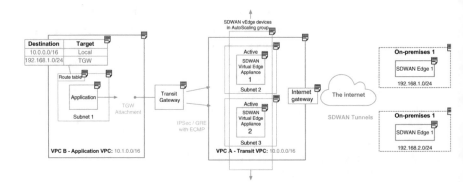

Fig. 7.16 Deploying a scalable fleet of SDWAN appliances with Autoscaling and Transit Gateway

other VPCs, regardless of the breadth of your VPC sprawl. This architecture is shown in Fig. 7.16.

7.2.3.8 Virtual WAN

While Amazon seems to be the only provider who has built a centralized data-plane router that allows many thousands of VPCs to connect to each other (Transit Gateway) and has added integration with SDWAN vendors to connect to on-premises SDWAN deployments with this virtual router construct, these solutions tend to be more of the roll-your-own-type solutions. This means that users need to manage the transit VPC, Transit Gateways, and the instances within the transit VPCs.

Azure has taken the idea of on-premises networking integration one step further with their Virtual WAN solution, shown in Fig. 7.17, which allows for the automated deployment of SDWAN appliances within a Virtual WAN construct that provides the hub for site-to-site VPN, Express Route, remote users, SDWAN spokes, and Vnets to all connect together in one virtual hub deployed in each region. This service also takes advantage of the Azure backbone network, allowing connectivity such as Express Route or VPN to on-ramp at one Azure edge location, and off-ramp at another, traversing the Azure backbone in-between. While still in its early stages, the promise of using a cloud provider's backbone in place of using a national or international MPLS IP VPN deployment or other WAN technologies such as VPLS is immensely attractive for users of cloud that have many disparate sites requiring connectivity.

Azure may have been the first to allow this type of service for users of cloud to easily take advantage of their cloud backbone, yet GCP has followed suite with its Network Connectivity Center, which with some differences, essentially opens up their backbone for users to take advantage of and integrate with SDWAN appliances as well.

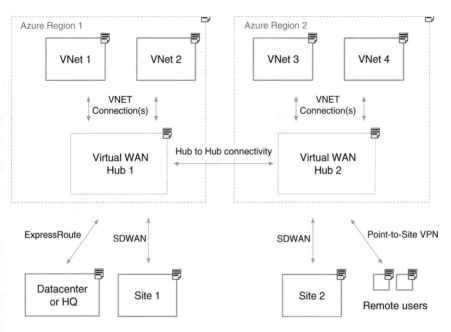

Fig. 7.17   Azure's Virtual WAN service [3]

7.2.3.9   Future State of Cloud Network Connectivity

We've seen the growth of network connectivity options within the cloud over the last several years move from public-based connectivity to a vast expansion of private options within, and connecting to, the virtual private cloud (VPC).

As companies have chosen to use cloud as their primary datacenter, the technologies that were previously needed to connect to their datacenter from their on-premises cloud such as fiber connectivity, VPNs, and SDWAN have now been needed to be extended to the cloud.

Looking toward Network 2030 and cloud network connectivity, we can expect that even more constructs such as SDWAN and other virtual network functions such as firewalls, IDS, IPS, optimization, and visualization tools are not only available within the cloud but have native integration into the underlying cloud infrastructure, enabling users to select which of these components they'd like to be deployed within their cloud connectivity deployment and have these deployed and managed in a one-click (or one API call) fashion.

We're only now starting to see the beginning of the marriage between traditional hardware vendors and cloud deployments. It is certainly an interesting space where you have an appliance vendor such as a firewall company, who has been securing networks for 15 or 20 years, with a large list of features that their customers know and love, and then cloud which abstracts the hardware away from the software itself; with this marriage, you'll have to see cloud providers give the appropriate hooks for these vendors to integrate into their platform like we are starting to see with AWS's TGW SDWAN integration, Azures Virtual WAN, and GCP's Network Connectivity Center.

We'll also need to see appliance vendors solve the performance gaps that they have now; they cannot rely on ASICs and high PPS appliances with specific hardware capabilities. The shift will be one in which multi-threaded applications take advantage of the width of hardware in terms of available CPU threads that a software application can use, especially in the case where tunneling protocols such as IPsec have been found to bind themselves to a single RX queue on a network interface, which in turn binds itself to a single CPU thread, limiting the available bandwidth that can be sent via an IPsec tunnel through an instance. Multi-threaded solutions will be needed to give customers the full amount of throughput available on a virtual machine, and integration into services such as AutoScaling such that appliances are able to horizontally scale out and scale in infrastructure as needed, which is what customers have come to expect from their cloud deployments and the elasticity that is so fundamental to everything that is cloud.

7.3 APIs

For the last part of this chapter, it makes sense to mention the very thing that has enabled cloud from the beginning—Application Programming Interfaces, or APIs. We covered earlier that cloud providers needed to abstract the services that they offer to customers such as VPC from the physical hardware that these services are built upon, that mechanism is the availability of APIs.

Almost every step that a customer needs to perform when building out infrastructure on the cloud is available through the use of an API call. This means that there is an underlying interface that can be called to perform a function. These APIs are available to perform functions on services within a user's account. This section glosses over what is arguably one of the most important functions of using cloud computing; however, given that APIs are widely accepted and understood within the IT community as a whole, we'll only spend a very brief moment to mention API connectivity. It is also worth mentioning that the API calls mentioned herein are referenced to infrastructure deployed on the cloud, and not the APIs or API Gateways that you might build as part of your application deployment.

An example in Fig. 7.18 shows the AWS VPC API call required to create a virtual private cloud in AWS, given the CIDR range 10.0.0.0/16, and also asking for an IPv6 range too. We are showing the actual API call here, however if a user were to

```
https://ec2.amazonaws.com/?Action=CreateVpc

&CidrBlock=10.0.0.0/16

&AmazonProvidedIpv6CidrBlock=true
```

Fig. 7.18 CreateVPC public a = API call

use either the cloud providers console which allows the creation of a VPC through a graphical interface. A user could also use the command-line interface which would allow creation of the VPC through a command on a terminal, or through CloudFormation or another infrastructure as code definition such as a JSON file describing the user's cloud infrastructure they would like to deploy. Regardless of these interaction options, under the hood, the same CreateVPC API call would probably be initiated.

As shown in the sample API call, the destination of the request in this case is ec2. amazonaws.com, which is a public endpoint on the public internet and reachable only if public internet access is available for the entity performing the API call; this is also shown in Fig. 7.19.

With the release of privatelink, the public API endpoint can also be placed into a VPC directly, such that any applications within that VPC, or even any applications on-premises or that are privately connected to the VPC, can also now connect to the API endpoint which is now considered private for anything within that private VPC realm, as per Fig. 7.20. This could be the deployment of constructs such as an SDWAN controller within a VPC that might want to call the ec2 API to shift routes, or failover instances.

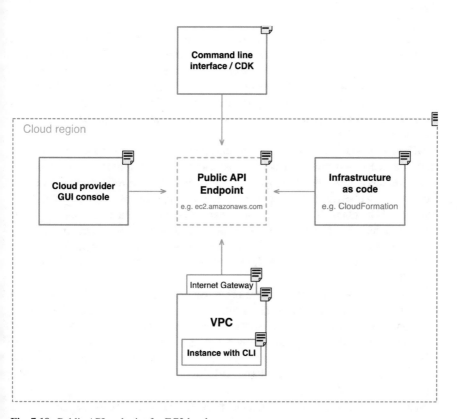

Fig. 7.19 Public API endpoint for EC2 level constructs

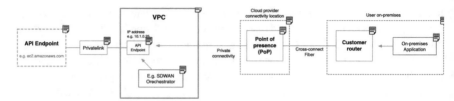

Fig. 7.20 EC2 API via privatelink for private connectivity

```
{
"Type" : "AWS::EC2::VPC",
"Properties" : {
"CidrBlock" : String,
"EnableDnsHostnames" : Boolean,
"EnableDnsSupport" : Boolean,
"InstanceTenancy" : String,
"Tags" : [ Tag, ... ]
    }
}
```

Fig. 7.21 JSON VPC snippet

For the future of virtual infrastructure APIs as we approach Network 2030, the key is that each and every component or service released by public cloud providers should and will have their own API that can be invoked, even for managed services, which would ultimately call the infrastructure APIs when being deployed within a user's VPC such as with a cloud provider managed Kubernetes service, which is built upon virtual machines or virtual infrastructure within the users VPC environment.

When users start to build their IT systems on cloud providers, they will generally start with the cloud providers console, as it provides a graphical feedback on the infrastructure that is being deployed, making it very easy to follow and get started. However, as we've discussed, these environments can soon grow at a very fast pace, and it would be untenable to deploy or manage these via human means and a graphical user interface. Therefore, it becomes a necessity to use mechanisms like CloudFormation or Terraform, tools that describe infrastructure as code, in the form of a JSON or YAML. An example JSON snippet for an Amazon VPC can be seen in Fig. 7.21.

Seasoned cloud users would therefore have a full library of architectures in the form of JSON or YAML deployments that they can quickly use to launch architectures on cloud providers infrastructure. These architectural playbooks are essentially a part of a build once and deploy many idea that users can take advantage of when scaling their IT systems on the cloud, and as long as cloud providers keep deploying services, and keep providing APIs for these services, users will be able to take advantage and grow out their repertoire of architectures for each and every scenario needed. Additional to this, cloud providers almost since the beginning have been offering full stack deployment tools to help with everything needed for an application deployment, including the VPC infrastructure; these tools are too numerous to mention them all, but some examples are Amazon Elastic Beanstalk, GCP App engine, and Azure App Service.

7.4   Conclusion

This chapter has hopefully given you the background on cloud networking needed to help foresee the future of cloud networking for both inside the cloud and connecting to the cloud. We're currently at an interesting intersection of on-premises solutions and users wanting to take advantage of all that cloud has to offer such as global reach, scalability in minutes, and elasticity. Traditionally, these constructs haven't been available when building networks and IT systems in the past, so there is some catching up to do, but hopefully by the time Network 2030 rolls around, cloud will be a seamless part of the story, enabling users to do what it is users are best at, and not have to worry about deploying, securing, and managing their cloud network.

References

1. https://aws.amazon.com/compliance/shared-responsibility-model/
2. https://docs.aws.amazon.com/outposts/latest/userguide/outposts-networking-components.html
3. https://docs.microsoft.com/en-us/azure/virtual-wan/virtual-wan-about

Chapter 8
Integrated Space-Terrestrial Networking and Management

Daniel King and Ning Wang

8.1 Introduction

Exponential increases in Internet speed have facilitated an entirely new set of applications and industry verticals underpinned by evolving fixed network infrastructure. The costs of deploying new fixed fibre networks are a limiting factor. As 5G and Internet infrastructure build-out continues, we must now look up both figuratively and physically, for our next networking opportunity. In the future, space communication will play a significant role in providing ubiquitous Internet communications in terms of both access and backhaul services [1].

Legacy satellites, probes, and space-based objects like the International Space Station (ISS) rely mostly on radio technology for communication. Using radio, it would take approximately 2.5 s to send data to the Moon and back to Earth, and between 5 and 20 min depending on planet alignment. In 2014 the ISS tested OPALS (Optical Payload for Lasercomm Science) system developed by NASA, and this achieved a data rate of 50 Mb/s. By 2015 gigabit laser-based communication was performed by the European Space Agency (ESA) and called the European Data Relay System (EDRS) [2]. The ESA system is still operational and extensively used.

In 2020 we observed a slew of next-generation meshed satellite constellations [3]—OneWeb, SpaceX (Starlink), Viasat-4, and TeleSat—with Amazon (project Kuiper) and Facebook also developing space-based communication projects. These new space networks will be capable of providing global gigabyte Internet via

D. King (✉)
Department of Computing and Communications, Lancaster University, Lancaster, UK
e-mail: d.king@lancaster.ac.uk

N. Wang
Institute of Communication Systems (ICS), Engineering and Physical Science, University of Surrey, Surrey, UK
e-mail: n.wang@surrey.ac.uk

© The Author(s), under exclusive license to Springer Nature
Switzerland AG 2021
M. Toy (ed.), *Future Networks, Services and Management*,
https://doi.org/10.1007/978-3-030-81961-3_8

Earth-to-space lasers instead of radio, and, instead of bouncing signals between Earth and space and back to Earth, the signal can be transmitted in space using space-based laser communication. These new satellite constellations are positioned in a Low Earth Orbit (LEO) approximately ≤2000 km altitude. They number from thousands to tens of thousands, in a grid-like pattern, and will provide continuous Internet coverage. The constellation will orbit the Earth on the order of 100 min, travelling at roughly 27,000 km/h.

These new satellite constellations will form a mesh network infrastructure in space that will connect to existing network infrastructure on the ground and provide lower latency. The potential for lower latency for long-distance connectivity stems from building "nearly shortest" paths (after incurring the overhead for the uplinks and downlinks) instead of circuitous terrestrial fibre routes.

These new networks will provide connections of 100 Mbps to residential users, and multiple Gbps to enterprise users, across vast rural areas and provide competitive low-latency bandwidth in metro areas, thus significantly offloading Internet traffic from traditional terrestrial infrastructures. Current and near-future space-based networks include Telesat with 120 satellites and 40 grounds stations, OneWeb with 720 satellites and 70 ground stations, and SpaceX with planned 42,000 satellites and 120 ground stations [4]. Also, several additional satellite Internet projects are proposed for operational deployment by 2025.

Future space networks will also need to cooperate with the existing terrestrial network infrastructure, exploiting heterogeneous devices, systems, and networks, thus, providing much more effective services than traditional Earth-based infrastructure and greater reach and coverage than proprietary and isolated space-based networks.

This chapter discusses the current state of the art for space-terrestrial network integration and highlights specific use cases and technical challenges. A fundamental challenge will be the future seamless integration of space networks with the current terrestrial Internet infrastructure, to maximise the benefits for Earth-based and space-based infrastructure. To limit the discussion's scope, we mainly focus on the Low Earth Orbit (LEO) satellite system, which can provide low end-to-end latency compared to its GEO (Geostationary Earth Orbit) counterpart. The shared vision in this scenario is that multiple (up to tens of thousands) LEO satellites can be interconnected to form a network infrastructure in space that will be further integrated with the ground's network infrastructures. On the other hand, the critical challenge, in this case, is the frequent handover between the two networks caused by the constellation behaviours at the LEO satellite side, which is considered to be the most notable feature, which incurs a wide range of technical challenges in the context of space-terrestrial network integration. The rest of this chapter aims to describe different strategies for such network integration and the specific technical issues that need to be addressed.

Many new satellite constellations will use best-of-breed commercial-off-the-shelf (COTS) technologies and Free Space Optical (FSO) subsystems [5], enabling the on-demand deployment of satellite infrastructure. FSOs are designed to provide high-bandwidth, optical wireless network access to end users by using satellites with high bitrate interfaces, which cover large areas of the Earth. These constellations will provide a global space backbone network with optical links since satellites can support terrestrial residents regardless of topographical limitations if a line-of-sight (LOS) of an earth-to-space and space path LOS exists. Therefore, this new infrastructure offers high-quality data services even to isolated areas. Inter-satellite links (ISLs) are designated for routing data traffic hop-by-hop through satellites towards the ultimate destination satellite with up-and-down links between aircraft and dedicated and fixed-ground stations on the surface of the Earth [6]. Usually, such links will have exceedingly high data rates. Thus, ISLs are used for intercontinental communications. The receivers can be stationary, such as those placed on top of buildings, mountains, towers, and so forth. The receivers can also be in motion, such as those installed in aircraft, ships, and ground vehicles.

It is envisaged that future integrated space and terrestrial networks (ISTNs) will be comprised of the following key components:

Satellite   A Low Earth Orbit (LEO) satellite has a lower physical orbit compared to legacy satellite systems, potentially bringing a short-latency benefit at the expense of constellation complexity. Medium earth orbit (MEO) and geostationary earth orbit (GEO) satellites can provide more physical stability, but they come with a relatively longer transmission delay than LEO systems. The current satellite systems mostly provide relay function; however, in the future, satellite systems may build up a meshlike network to provide routing and forwarding function. LEO satellites should be organised as a routing system and work as routers covering data-plane and control-plane functions.

Ground Station and Terminal   Ground stations and terminals are physical terrestrial devices that act as gateway or interfaces between terrestrial and space networks through radio communications. The networking mechanisms and protocols used in space networks are different from those in the traditional IP framework in the terrestrial infrastructures. Hence, ground stations and terminals have been responsible for protocol translations and creation/maintenance of tunnels for data packets to traverse different network environments. It is also worth mentioning that, while ground stations use dedicated gateways between the space network and the terrestrial infrastructures today, it is envisaged that in the future network/user devices will be able to communicate direct with satellites, allowing Internet traffic to be exchanged between user devices without necessarily always going through ground stations.

- Controller (SDN architecture-based): The satellite network system may also employ a hierarchical architecture. Some of the satellites play the role not only of a router but also a controller.
- Mobile edge computing (MEC) server: MEC has been a terminology used mainly in the context of 5G where local computing and storage capabilities can be embedded at the mobile network edge to provide low latency data/computing services to locally attached end users. It is envisaged that in emerging space and

terrestrial networks, LEO satellites can also interconnect MEC servers in the satellite constellation once equipped with computing and data storage capabilities.

8.2 Use Cases and Design Options for Integrated Space and Terrestrial Networks (ISTNs)

Given the capital and operational costs of launching and managing space-based infrastructure, it will be critical to identify the key use cases that are commercially viable and operationally possible. The likely use cases and scenarios for ISNs include the following:

8.2.1 Using ISTNs for Backbone Internet

The first use case is to use networked LEO satellites to provide transit service as backbone infrastructure in space [8], while the second use case is to use individual satellites as access nodes to provide enhanced service coverage especially in rural areas. We split the first use case into two different scenarios of using a LEO satellite network as a backbone. The decoupled scenario is based on the availability of peering ISLs in space, in which case the routing infrastructure can be completely decoupled from its terrestrial counterpart. In comparison, in the coupled scenario there is no peering link between neighbouring LEO satellites, and hence each LEO satellite can be independently deemed as an "overlay" node on top of the terrestrial network infrastructure. The main reason for this situation is the current difficulty in establishing ISL links between satellites due to limitations in the design of antennae. So, without loss of generality, we elaborate on specific features based on both scenarios.

8.2.1.1   The Decoupled Scenario

This is a more classical view of the internetworking between a LEO satellite network and the terrestrial infrastructure. Thanks to the availability of ISLs, it is possible to deploy separate routing mechanisms among satellites which do not need to rely on the terrestrial routing infrastructure. The default scenario here is that once user data packets have been injected into the space network, they will only need to return to the ground when reaching the last-hop satellite which is normally the closest to the destination. The delivery of the packets is based on dedicated routing mechanisms in the space network which can be completely different from that on the terrestrial infrastructure as shown in Fig. 8.1.

8.2.1.2   The Coupled Scenario

One typical design rationale behind this scenario is uncertainty about the readiness of ISLs based on laser commutations. Without the availability of such links, one typical scenario will be the one that is shown below, where a LEO satellite is integrated with the terrestrial infrastructure *on per-hop basis*. As such, it is not appropriate to run a dedicated routing protocol between the satellites, but instead each satellite is supposed to be an "integral" component of the overall framework on the ground running a common routing protocol. Another view can be that the introduction of these satellites offers the opportunity to create "shortcut" paths compared to routes across domains. Another key difference compared to the decoupled scenario, shown in Fig. 8.2, and the role of downlinks/uplinks between satellites and the ground infrastructure. Links in the decoupled scenario are only used for access purposes, while in the coupled scenario such links will take both roles of access and transit, in which case the bandwidth capacity needs to be adequate for such purposes.

8.2.2   Using ISTNs for Access Networks

The benefit of using LEO satellites to provide access service is mainly due to its ubiquitous access coverage, even in remote areas such as oceans, mountains, or deserts where it is difficult or even impossible to deploy any fixed Internet access infrastructures. A typical use case can be described as follows: passengers on a cruise ship in the Atlantic Ocean would like to access video content offered from a content provider in mainland Europe. Today, this use case is addressed by installing a satellite dish on the ship and using onboard Wi-Fi to provide Internet connectivity to the passengers to reach the content source or content node. In the future, individual users onboard will be able to use their own mobile devices to access Internet through the LEO satellites that have local coverage of the area. Even though the

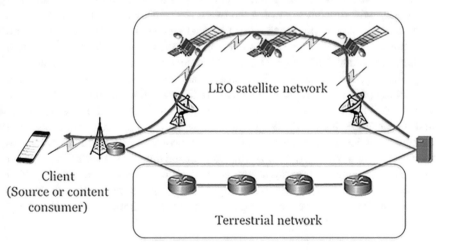

Fig. 8.1 Decoupled scenario

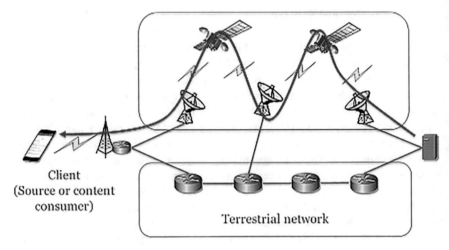

Client
(Source or content
consumer)

Terrestrial network

Fig. 8.2 Coupled scenario

last-mile access is already provided by LEO satellites, in order to stream video content from the data centre on the land, still it is necessary to build a content delivery path from the content source to the users, involving either a complete chain of LEO satellites or a combined path consisting of both terrestrial routers and LEO satellites.

8.2.3   Fundamental Design Options

In this section, we address basic networking challenges for integrating space and terrestrial network infrastructures including addressing and routing paradigms. It is worth noting that routing optimisation across LEO satellite networks has been extensively studied in the literature, but how to seamlessly harmonise or even unify the routing infrastructures between the two types of networks has been much less investigated. Below we highlight two different strategies. As mentioned above, from the classical viewpoint, a small number of dedicated ground stations or satellite terminals are strategically deployed at specific network locations which act as gateways that separate the LEO satellite network and the terrestrial infrastructure (e.g. performing protocol translation functions). In this case, the common practice today is that individual LEO satellite networks run their own individual protocols locally and "encapsulate" them with the ground stations acting as gateways. However, with emerging wireless access communications, it is envisaged that ground network elements and user devices will be able to communicate directly with satellites without necessarily relying on traditional ground stations at (a limited number of) fixed locations. Such a feature will have implications for the design of solutions to natively integrate space networks and their terrestrial counterparts with much more "blurred" network boundaries.

8.2.3.1   Design Option I: Integrated Architecture Based on IP Routing

In this option, the first strategy is to directly apply the IP routing paradigm to the space network and let it "natively" interface with the terrestrial infrastructure; see Fig. 8.3. This option can be seen as a relatively conservative scheme with minimum disruption to the existing network. However, due to the constellation behaviours, the space-terrestrial link can be very dynamic, which will lead to potential disruptions such as frequent and simultaneous broken link events and unwanted routing protocol convergence between directly interfaced space-terrestrial networks without "gateway" functions provided by fixed-location ground stations. Thus, the relative mobility between the space and terrestrial network infrastructure is certainly one of the key features to be investigated, and some preliminary studies within this topic (i.e. applying IP routing principles in LEO satellite networks) have recently been carried out with the consideration of constellation behaviours. For example, in [9], a preliminary study of applying BGP in the satellite network is provided and the results indicate that up to 45% of available satellite connectivity is wasted due to unstable eBGP sessions. To address such issue, in [9] a scheme named NTD-BGP is proposed aiming to preserve the eBGP sessions between the space and terrestrial routers across mobility events. However, this approach has difficulty fitting the inter-AS scenario where the terrestrial router moves into a new satellite AS while NTD-BGP requires the BGP speakers to always establish the eBGP session using a fixed loop-back address. Thus, if the dedicated loop-back address is not advertised to the new satellite AS, the eBGP session would be unable to be established.

Apart from the traditional static IP address configuration, dynamic address configuration may be an alternative option. For example, by utilising relative fixed

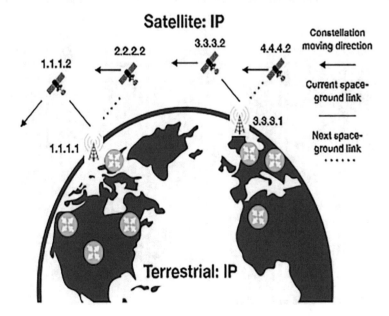

Fig. 8.3 Envisioned addressing and routing system

geography information such as longitude and/or latitude, the IP addresses can be
bound to predefined regions instead of the physical router interfaces. Thereby, from
the terrestrial router's point of view, IP addresses of the space peers become "static".
However, such an addressing scheme could lead to convergence issues for IGP/
iBGP within the space network domain, a problem which can potentially be
addressed by proactive route calculations according to predictable satellite constellation behaviours.

8.2.3.2 Design Option II: An Integrated Architecture Based on Evolved/New IP

In this option, the strategy is to design a comprehensive, integrated addressing and
routing system for both the space and terrestrial networks based on evolved/new
IP. The new addressing and routing system in this case should be able to natively
overcome the issues caused by the topology dynamics including the dynamicity
within the satellite constellations and the space-terrestrial links; see Fig. 8.4. The
ultimate ambition of this option is to integrate any network especially future network architectures, rather than being restricted to legacy IP-based networks.
Therefore, for this option, the requirement for compatibility, scalability, robustness,
and mobility support should be satisfied from basic design. Although developing
such a new architecture is particularly challenging and will also face significant
pressure from the deployment side as there may be strong impact on the current
network system, the reward can potentially be significant.

The most prominent benefit is that end-to-end communication is natively supported (rather than requiring tunneling or protocol translation between the

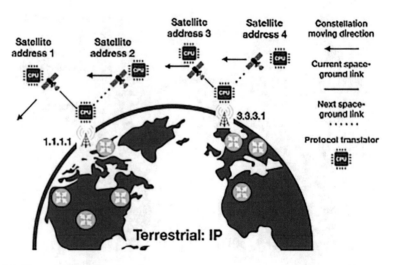

Fig. 8.4 New satellite addressing scheme

networks) since the space and terrestrial networks are running the same addressing and routing framework. Any method is open for discussion to achieve this goal. This may include, but is not limited to, introducing additional fields in IP packet headers and/or routing tables, or even location-free schemes. In principle, routing in space should simultaneously take into account both the satellite address and the IP address, instead of completely relying on satellite addresses by encapsulating the IP address. Therefore, the user devices running on legacy IP can freely switch their connections between the space and terrestrial network depending on the network performance. As such, advanced Internet services, such as caching and video acceleration, can also be supported in the satellite network. Finally, it is worth noting that, since there is no packet header encapsulation in this case, when a packet is being delivered through the terrestrial network, the header field for satellite addresses can be reserved for other purpose or eliminated using mechanisms such as variable-length IP addressing schemes.

8.3 Challenges of Integrated Space and Terrestrial Networks (ISTNs)

Many new satellite constellations will use best-of-breed commercial-off-the-shelf (COTS) technologies and Free Space Optical (FSO) subsystems [7]: this will help enable more rapid on-demand deployment of satellite infrastructure. FSOs are designed to provide high-bandwidth, free-space optical network access to end users by using satellites with high bitrate interfaces, which cover large areas of the Earth. These constellations will provide a global space backbone network with optical links since satellites can support any geographical residents regardless of topographical limitations and whether a LOS space path exists. Therefore, this new infrastructure offers high-quality data services even to isolated areas [8]. ISLs are designated for routing data traffic hop-by-hop through satellites towards the ultimate destination satellite with up-and-down links between the aircraft or a ground station on the surface of the Earth.

Several general infrastructure challenges have been identified for successful ISN deployment and operation. These include the following:

- As LEO satellites orbit the Earth at relatively high-speed, the space-based path latency and bandwidth will fluctuate as routes shift across the satellite topology.
- Future LEO satellites will support multiple link types, air interfaces, and frequencies, including high-bandwidth free-space optical links and low-speed radio interfaces.
- Atmospheric conditions and weather severely degrade communication between satellites over space-ground links, significantly reducing throughput or requiring new routing paths to be selected.
- The ISTN links will become bandwidth-constrained, and it will be necessary to compute alternative paths around those congested links. Dynamic path selection

based on current and predicted demands will need to be factored in; thus traditional Dijkstra techniques for path routing will not be sufficient.

Existing Internet architecture and protocol mechanisms will likely apply to converged space-based and Earth-based network infrastructure; however, there will be limitations. This section outlines some of the challenges, requirements, and potential strategies to pursue for future ISNs.

8.3.1 Routing and Forwarding

Routing and signaling across emerging next-generation satellite networks is far from static; satellite-to-satellite connectivity changes frequently, space-based link latencies, and links from space-to-ground will change regularly. Satellites will also have to contend with predictive routing capabilities, as links will only be established when optical alignment is possible. Given that meshes of 100s and 1000s of satellites are also expected [9], techniques that use per-hop Dijkstra calculation will be extremely inefficient.

Next-generation space networks are not static. The satellite that is overhead a particularly ground station changes frequently, the laser links between space-based satellites change often, and link latencies for satellite to ground links will vary based on atmospheric conditions.

Several control plane challenges have been identified for space-based networks [10], and these include:

- New link acquisition, predicted link availability, and link metric dynamicity: As the acquisition and tracking of satellites and links change, there is a need to adjust basic link and TE metrics (delay, jitter, bandwidth) and update the existing routing traffic engineering database.
- Space-based path computation: Selection of the best path across ISLs and direct uplinks and downlinks, consideration of cloud cover, air turbulence, and external object occlusion.
- Temporal routing: Consideration of the time-varying topology of the space network will necessitate frequent routing updates.
- Predictive routing: Time-scheduled routing paths based on expected satellite orbits and air-interface alignment.
- Rerouting of paths: Which may be required in the event of projected space-based debris orbits that prevent line-of-sight between adjacent nodes, interface and node failures, and adverse weather which may affect space-to-ground communication points.
- Resilience: Overall, the network must be resilient to failures and capable of routing with low latencies, even when traffic levels are significant enough to oversubscribe the preferred paths.

Several of the challenges outlined above, and other aspects such as variable addressing, security, and privacy, have been highlighted in the recent ITU-T NETWORK 2030 Study Group 3 (Future NET2030 Architecture Framework) [11] and future Internet architecture, protocols, and applications [12].

8.3.2 Network Control and Addressing

Integrating the space-based infrastructure with an existing network might be achieved using traditional Internet routing techniques and identifying the extraterrestrial portion of the network as a specific domain (such as an IGP area or an AS). The space-domain might run a traditional routing control plane, likely logically within an Earth-based representation which programs the path via an SDN-programming technique. However, this approach would not be capable of computing paths based on the unique space connectivity dynamics. Furthermore, if the space-domain was connected to traditional Earth-based Internet domains (including ASes via BGP), it might create unwanted route flapping, causing routing instability.

Due to the unique characteristics of the space-based nodes (which may have multiple interfaces and lines of sight to next-hop satellite nodes or ground stations, may fluctuate), other network control methods may be needed.

8.3.3 System Resilience

Legacy satellites might typically operate independently from their orbiting counterparts. However, next-generation space-based infrastructure will be utilising multiple links between satellite nodes and ground-stations, which leaves potential network paths susceptible to the consequences of node and link failures or anomalies. Loss of node payload, communication link, or other subsystem components might render the entire satellite node inoperable.

In a satellite network, there are several types of failures a routing system might be concerned with; these include:

- Failures of components in the forwarding plane—e.g. ISL communication failure.
- Control plane malfunction, if the central controller is destroyed or disconnected, or the distributed control plane suffers a catastrophic failure or attack.
- Misconfiguration of satellite node or ISL forwarding, or degradation of satellite orbit and loss of communication sight to neighbouring nodes.

In general, satellite node failures or components of the forwarding plane are problematic, but as the latest generation of space infrastructure is highly meshed, routing around node failures is feasible. Once a failure occurs, the centralised controller, or distributed control plane, would have to respond and update the forwarding state in devices to route traffic around the failed nodes or links. As failure may be seen as an extreme case of an unexpected change in traffic level, a traffic re-optimisation mechanism would likely be required.

8.4 Advanced Features for Integrated Space and Terrestrial Networks (ISTNs)

8.4.1 Multilayer Networking

The Low Earth Orbit (LEO) satellite uses a lower physical orbit, which provides latency benefits, but this orbit will incur more dynamic connectivity and oscillating link characteristics [9]. The Medium Earth Orbit (MEO) and Geostationary Earth Orbit (GEO) satellites provide more physical stability and reduced dynamicity of the links as the satellites remain static. The current GEO satellite system mostly provides relay function; however, in the next generation, satellite systems could interact providing multilayer routing and forwarding functions between satellite layers, akin to multilayer networking in terrestrial networks [13]; see Fig. 8.5.

8.4.2 Traffic Engineering

Traffic engineering (TE) has been well investigated for more than two decades in the context of the traditional terrestrial Internet. However, TE has not been systematically understood in the integrated space and terrestrial network environment, especially given the district characteristics of the two types of networks and also the mega-constellation behaviours of LEO satellites. It is generally understood that the inter-satellite link capacity is not compared to the optical fibre links in the terrestrial Internet. As such, the traffic injected into the space network has to be selective. Policies can be enforced either based on the traffic type and their QoS requirements or based on other contexts such as the distance between source and destination pairs. For instance, in [9] it has been argued that routing through a chain of LEO satellites will outperform the usage of terrestrial Internet in terms of end-to-end delay if the distance of the source and destination is beyond 3000 km. It is also worth noting the capability of TE in the space network also largely depends on the specific routing mechanisms that are deployed, which has been the case in terrestrial network environments, e.g. IP/MPLS/SDN. As mentioned above, the capability of TE in integrated space and terrestrial network infrastructures will also depend on the routing mechanisms deployed in the two network environments, either with separated protocols (the case today) or with a unified protocol suite.

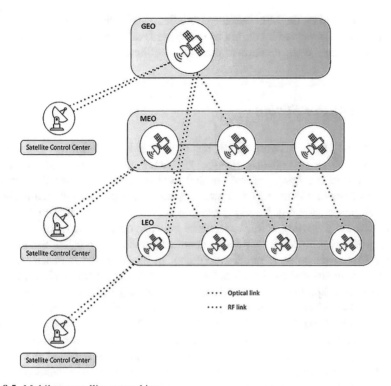

Fig. 8.5 Multilayer satellite networking

8.4.3   Quality of Service

In theory, using the additional path and bandwidth capabilities from the space network should be able to improve Quality of Service (QoS) and resilience offered from the terrestrial network infrastructures. However, without systematic network engineering solutions, QoS/resilience requirements will not be automatically met in practice. First of all, at the service management level, how to establish provider-level service level agreements (SLAs) that include QoS and resilience requirements can be negotiated between the terrestrial network operators and space network operators needs to be investigated. Secondly, in order to enforce the actual QoS-awareness (e.g. end-to-end QoS-constrained paths), routing optimisation, resource allocation, and traffic admission control mechanism need to be in place, especially by taking into account the constellation mobility of the LEO satellite infrastructure which may cause stability issues. Finally, concerning the assurance to the service performance, mechanisms need to be in place for handling potential traffic disruption caused by the LEO satellite constellation behaviours, including potential transient loss of connectivity during handover, and varied propagation delay performance depending on the location of the LEO satellites.

8.4.4 Resource Slicing

In the context of 5G, network slicing has been deemed as a promising feature for operators to provision network resources and functions to tailor for heterogeneous requirements of emerging applications and services. While the business model for network slicing on the traditional network operator side has been relatively clear, a more complex scenario of involving satellite operators has not yet been previously elaborated. As a starting point, a terrestrial network operator can rent virtual network resources provided by a satellite operator to build a dedicated backhaul link for connecting its point of presences (PoPs). In this case the terrestrial network operator can create end-to-end slices for supporting different application types, and the backhaul component of a selected subset of slices (e.g. eMBB (Enhanced Mobile Broadband) for video content delivery) can leverage on the satellite capability.

On the other hand, a satellite operator could also slice its own satellite link resources and lease to multiple terrestrial network operators for backhauling or extended access services, by applying intelligent beamforming techniques to cater for different geographical areas. As shown in Fig. 8.1 (for simplicity only one satellite is shown, but it can be a chain of LEO satellites), sliced satellite link capabilities can be leased to terrestrial network operators (e.g. mobile operators) in order for them to build their own service-tailored slices provided that the sliced satellite capability is able to fulfil the targeted service requirements. For instance, as shown in Fig. 8.6, once terrestrial network operator A has deployed a MEC-based content prefetching/caching network function within its network slice (Slice A.1) for transmitting 4K/8K video content, then it can use leased satellite capability for backhauling 4K/8K video in that slice. From the business point of view, we can envisage a cash flow from end customers (subscribers of terrestrial network slices) to the terrestrial network operators and further to the satellite operator.

8.4.5 Content Caching

Considering future satellite constellations are comprised of high-capacity links, providing ubiquitous global coverage, it will enable operators to provide content closer to the end user. Therefore, the opportunity for content caching becomes compelling. The satellite infrastructure and topology also provide the capability for multi-casting traffic [14], thereby facilitating the distribution and updating of cached content to various locations.

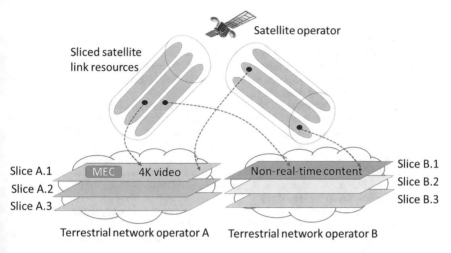

Fig. 8.6 Integration of terrestrial and satellite networks

Supporting in situ content caching will facilitate a range of future Internet applications and service many end users and their demand for low-latency content to be provided without multiple transmissions, and minimising delay. However, it is worth noting that the dynamic and time-varying nature of the satellite topology and limited space-based resources will have to be considered when deploying cache capable nodes and the type of path and node placement algorithms used.

8.5 Summary

In this chapter we shed lights on the envisaged scenario of future integration of space and terrestrial network infrastructures and the associated technical challenges. We specifically focused on the LEO satellite scenario which is expected to provide low latency communications compared to GEO/MEO counterparts. The key technical issue in integrating LEO satellite networks with the terrestrial infrastructure is the constellation behaviours that trigger dynamic but predictable infrastructure mobility between the two networks. This effectively is the fundamental cause of many derived technical challenges in realising the space-terrestrial network seamlessly. In this chapter we highlight different design strategies, including both basic addressing, packet routing, and forwarding strategies and a variety of advanced features that can be envisaged in the integrated space-terrestrial network environments in the future.

References

1. J. Liu, Tsinghua University. 2030 from earth to space – one world network (2020)
2. First SpaceDataHighway Laser Relay in Orbit. ESA (2016), http://www.esa. int/Our_Activities/Telecommunications_Integrated_Applications/EDRS/ First_SpaceDataHighway_laser_relay_in_orbit
3. I.D. Portillo, B.G. Cameron, E.F. Crawley, A technical comparison of three low earth orbit satellite constellation systems to provide global broadband. Acta Astronaut. **159**, 123–135 (2019)
4. G. Curzi, D. Modenini, P. Tortora, Large constellations of small satellites: A survey of near future challenges and missions. Aerospace **7**, 133 (2020). https://doi.org/10.3390/ aerospace7090133
5. E. Hudson, Broadband high-throughput satellites, in *Handbook of Satellite Applications*, ed. by J. Pelton, S. Madry, S. Camacho-Lara, (Springer, Cham, 2017). https://doi. org/10.1007/978-3-319-23386-4_95
6. H. Kaushal, G. Kaddoum, Optical communication in space: challenges and mitigation techniques. IEEE Commun. Surv. Tutor. **19**(1), 57–96 (2017)
7. I.K. Son, S. Mao, A survey of free space optical networks. Digit. Commun. Netw. **3**, 67–77 (2017)
8. Space Exploration Technologies, SpaceX non geostationary satellite system Attachment A: technical information to supplement Schedule S (2016), https://licensing.fcc.gov/myibfs/ download.do?attachment_key=1158350
9. M. Handley, Delay is not an option: low latency routing in space, in *Proceedings of the 17th ACM Workshop on Hot Topics in Networks (HotNets '18)*, (Association for Computing Machinery, New York, 2018), pp. 85–91
10. D. King, A. Farrel, Z. Chen, An evolution of optical network control: From Earth to Space. *2020 22nd international conference on transparent optical networks (ICTON)*, Bari, Italy, 2020, pp. 1–4
11. Network 2030 Network2030 Architecture framework document
12. Z. Chen et al. New IP framework and protocol for future applications. NOMS'20 (2020)
13. Z. Liu, J. Zhu, J. Zhang, Q. Liu, Routing algorithm design of satellite network architecture based on SDN and ICN. Int. J. Satell. Commun. Netw. **38**, 1304 (2019). https://doi.org/10.1002/ sat.1304
14. S. Liu, X. Hu, Y. Wang, G. Cui, W. Wang, Distributed caching based on matching game in LEO satellite constellation networks. IEEE Commun. Lett. **22**(2), 300–303 (2018)

Chapter 9
Network Slicing and Management

Sameh Yamany and Luis M. Contreras

Abbreviations

AAA	Authentication authorization and accounting
ADAS	Advanced driving-assisted systems
AI	Artificial intelligence
BSS	Business support systems
DN	Data network
E/AO	Edge/access operator
eCPRI	Enhanced common public radio interface
ENNI	External network-network interface
FRR	Fast reroute
IoT	Internet of Things
ISR	Interrupt service routine
KPI	Key performance indicator
LSO	Lifecycle services orchestration
LSVR	Link state vector routing
MEC	Multi-access edge computing
ML	Machine learning
MSS	Maximum segment size
NFVO	Network function virtualization orchestration
NS	Network slicing
OSS	Operations support systems
QFC	Quantum flow control
QoS	Quality of service

S. Yamany (✉)
VIAVI Solutions, Boulder, CO, USA
e-mail: sameh.yamany@viavisolutions.com

L. M. Contreras
Telefónica I+D/CTIO, Madrid, Spain
e-mail: luismiguel.contrerasmurillo@telefonica.com

© The Author(s), under exclusive license to Springer Nature
Switzerland AG 2021
M. Toy (ed.), *Future Networks, Services and Management*,
https://doi.org/10.1007/978-3-030-81961-3_9

QUIC   Quick UDP internet connections
RCS   Rich communication services
RoE   Rf over Ethernet
RTT   Round trip time
SLA   Service level agreement
SLO   Service level objectives
SP   Service provider
TCP   Transmission control protocol
TSN   Time-sensitive networking
UE   User equipment
UTRAN   UMTS terrestrial radio access network
UNI   User network interface
VIM   Virtual infrastructure manager
VNF   Virtual network function

9.1   Introduction

Network slicing is a paradigm through which different virtual resource elements of common shared infrastructure (in both connectivity and compute substrates) become allocated to a specific customer who perceives the resulting slice as a fully dedicated, self-contained network for it. The resources are virtualized through a process of abstraction of lower-level elements, providing great flexibility and independence when allocating specific elements to the customer. This process permits the exercise of advanced actions such as scalability, reliability, protection, relocation, etc., along the network slice lifetime, without impacting the customer service. All these possible actions represent an incredible asset for a novel way of service provisioning with respect to the conventional mode of network operation.

Network slicing, despite not being a new concept [1–3], acts as a foundational concept and systems to current 5G/future networks and service delivery, with the goal of providing dedicated private networks tailored to the needs of different verticals based on the specific requirements of a diversity of new services such as high-definition (HD) video, virtual reality (VR), V2X applications, and high-precision services [4]. Network slicing is supported by the technological paradigms of software defined networking (SDN) and network function virtualization (NFV) in an integrated manner. These three concepts, SDN, NFV, and slicing, encompass the overall trend of network softwarization, governing the transformation of operational networks. All of them will form the basement for the evolution of the network towards 2030.

9.1.1 *Value Provided by Network Slicing*

Network slicing allows the provision of tailored end-to-end logical networks on top of a common and shared physical network infrastructure, being offered to external customers in an on-demand manner. With this, a customer can make use of a complete logical end-to-end network for its specific service with full guarantees. The physical network infrastructure can now be consumed following a Slice-as-a-Service (SAAS), opening up new business opportunities for telecom operators, where the network is transformed into a production system.

The versatility offered by a dynamic consumption of network resources (both computing and networking) facilitates the emergency of different business models with the participation of distinct stakeholders, either in a Business-to-Business (B2B), Business-to-Consumer (B2C), or business-to-business-to-consumer (B2B2C) fashion. Apart from that, the operators can also leverage network slicing for their purposes, making it easier to distinguish internal service concerns. With that, several types of network slices can be assumed in terms of management, control, and usage [5]. Thus, it is possible to differentiate in:

- *Internal slices*, which are the network slices where the operator keeps the overall control and management of the slice. These are commonly used for the internal services of the operator.
- *External slices*, which are the ones offered to external customers. Those customers will perceive the allocated slices as a dedicated network. This category can be classified into two further types:
  - Network slices for external customers that are yet managed by the operator, which performs the control and management of the slice. In this case, the external customer simply runs its service on top, without further control or management capabilities for the allocated slice.
  - Network slices for an external customer that are also managed by them, performing the control of the allocated resources and service functions. The control capabilities could be limited to some point by limiting the set of operations and/or configuration actions allowed.

Because of this variety of slice types, distinct operational implications are observed from the specific customer need. The primary point is to control the allocated abstract resources to the customer (e.g., the possibility of steering the traffic by directly programming policies on the network elements involved in the traffic forwarding). The absence of such control implies that the slice simply accommodates the customer service, with the customer not being able to reconfigure it, consuming the slice as a kind of static network. This is the case of the external slices managed by the operator, where different customers have similar service needs that can be fitted in the same slice (e.g., customers requiring a generic enhanced Mobile Broadband—eMBB—service), assuming that slice is correctly dimensioned to bear the load of the distinct customers supported.

On the other end, when control capabilities are granted to the customer, it can flexibly manage the allocated resources (and service functions, if any), for example, by reconfiguring paths in the slice adapting them to the changing conditions of the customer service traffic. At the time of enabling control capabilities for the customers, this should be carefully enabled because the different actions from the set of customers supported in the same network can be conflicting. Even though the specific configuring actions of a given customer are performed on the abstracted resources of its corresponding slice, since those virtualized resources all pertain to a common physical resource shared among slices, contradictory configurations can collide with actions from one customer negatively impacting on the slice of another customer. The way of avoiding such an impact is to provide isolation among slices of a different customer that require slice control. That isolation can be achieved at different levels and usually implies a strict and dedicated allocation of resources per customer (see, for instance [6], for isolation options at transport network). This is the case of the external slices managed by the customer, where each slice should be essentially dedicated per customer.

Figure 9.1 graphically illustrates the different types of slices described, showing the control and management capabilities in each case.

9.1.2   New Business Proposition

Network slice transforms a set of the infrastructures (network, cloud, data center) components/network functions, infrastructure resources (i.e., connectivity, compute, and storage manageable resources), and service functions to meet the new demand of emerging business opportunities created by industry verticals (e.g.,

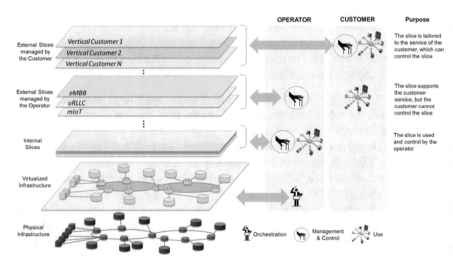

Fig. 9.1 Types of network slices according to management and control levels of responsibility

advanced driving-assisted systems or ADAS, connected healthcare, smart energy grids, smart cities, connected factory automation, connected retail industry, smart ports, etc.).

Network slice behavior for each business opportunity is realized via network slice instances (i.e., activated slices, dynamically and nondisruptively reprovisioned). Network slices considerably transform the networking perspective by abstracting, isolating, orchestrating, softwarizing, and separating logical network components from the underlying physical network resources and monetizing the same infrastructure based on the vertical services each slice support.

9.1.3   The Current State of Network Slicing

Slicing is a move towards on-demand segmentation of resources and deployment of virtual elements to enhance services and applications on a shared infrastructure. Therefore, slicing should be considered from multiple viewpoints, technical and business ones. Many groups, including ITU-T, ETSI, IETF, 3GPP, and ONF in addition to open source and research projects, are currently considering slicing as a tenet of their assets.

Slicing itself is not new [7]; it has been considered in the past, and it is progressively being included in the 5G standards. Early forms of network slicing included the ability to define, deploy, and operate user-defined network instances in isolation through the node operating systems and resource control frameworks part of programmable networks for IP service deployment [1].

Standards Development Organizations (SDOs) and some other industry associations have been looking at the network slice concept from different angles and perspectives. For example, ITU-T Slicing (2011) [8] defined slicing as a Logically Isolated Network Partitions (LINP) composed of units of programmable resources such as network, computation, and storage. More recently, ITU-T IMT2010/SG13 (2018/2019) [9, 10] describe the concept of network slicing and use cases of when a single user equipment (UE) simultaneously attaches to multiple network slices in the IMT-2020 network. In IETF, network slicing is defined in [11] as managed partitions of physical and/or virtual network and computation resources, network physical/virtual and service functions that can act as an independent instance of a connectivity network and/or as a network cloud.

In 3GPP, the current network slicing architecture is defined in the following technical specifications:

- Charging management; network slice performance and analytics charging in the 5G System (5GS) [12]
- Charging management; network slice management charging in the 5G System (5GS) [13]

ETSI E2E network slicing [14] defined a next-gen network slicing (NGNS) framework as a generalized architecture that would allow different network service providers to coordinate and concurrently operate different services as active NS. ETSI Zero Touch Network and Service Management Industry Specification Group (ZSM ISG) is specifically devoted to the standardization of automation technology for network slice management [15].

9.1.4 Network Slicing and 5G Mobile Networks

As mentioned in previous sections, network slicing is not a new concept within the mobile telecommunications world. For example, mobile virtual network operators (MVNOs) exploit slicing in legacy networks. Typically, this is accomplished by reserving a set of subscribers IMSI9 for the MVNO and slicing at the subscriber management/billing layer. Network sharing [16] can also be considered a precursor for slicing. For example, the MOCN shares a single RAN between different operators' CN, and MORAN shares a single RAN with separate frequency allocation per operator and GWCN.

With the new control and user plane separation in 5G, particularly with the 5G CN SBA, a much finer granularity of slicing is allowed. The functions in the network become logical functions that may be instantiated in physical locations as service requirements and capabilities demand. This is further enhanced by Network Function Virtualization (NFV) that permits the logical functions to be instantiated on a virtualization abstraction layer hardware supported on COTS hardware.

In this new context, a network slice is defined as "a logical network that provides specific network capabilities and network characteristics," [16] and a network slice instance is defined as "a set of Network Function instances and the required resources (e.g., compute, storage, and networking resources) which form a deployed network slice" [16]. Consequently, the slice instance will also determine the preferred control/user plane splits, function locations, and required telemetry to provide assurance of the SLA. Example types include slices for emergency services networks (ESNs), mMTC, and enterprises. Moreover, an MVNO may be provisioned as having access to a subset of slices of the required types.

Network slicing effectively requires disaggregation of the 5G Core Network (CN) and RAN in the service/tenant domain. Ideally, the slices are independent and isolated from the point of view of SLA assurance, as this simplifies resource management and service of SLA. However, this arrangement requires a sacrifice of efficiency. Additionally, there are limits to the isolation that is attainable, for example, when it comes to meeting stringent latency and bandwidth requirements on the air interface. Resource allocation to the slices is generally dynamic and potentially contingent on priority, e.g., for ESN, and the concept of a broking service to manage this contention has been proposed [16].

9.2 NS Primer

The GSM Alliance (GSMA) specifies a Generic Network Slice Template (GST) [17] as a universal and generic blueprint to be used for the deployment of a network slice instance (NSI). This blueprint focuses initially on 5G networks but it can be extended generically for expressing any kind of network slices. The GST is a compendium of attributes intended to characterize a particular customer service (Table 9.1).

According to 3GPP, a specific NEtwork slice type (NEST) is generated by filling the GST attributes with values. In other words, the NEST is a filled-in version of the GST that allows the operator and the customers to agree on the Service Level Objectives (SLOs) and other characteristics of the slice. The customer uses the NEST to request the provisioning of an NSI able to satisfy a particular set of service requirements. The NEST is processed by the 3GPP Management System [18], which maps the values to the requirements of the slice being deployed. Figure 9.2 shows the overarching 3GPP Management System architecture for slice management and control.

9.3 NS Architecture Elements

The vision of many industry verticals being enriched by new services with different mixes of low latency, ultra-reliability, massive connectivity, and enhanced Mobile Broadband is most valuable if delivered simultaneously in the same network, precisely what network slicing promises to deliver. However, the right architecture for an E2E network slice has to consider the different network components that constitute the E2E service chain. For example, an enhanced Mobile Broadband (eMBB) slice has to consider the radio segment of the network, the access network components, the transport network, and the mobile core network. Successfully orchestrating the network slice characteristics between all these network components will require a coherent architecture definition of the slice and its parameters at each segment of the slice service chain.

Figure 9.3 illustrates the example of deploying an access network to serve an eMBB and an ultra-reliable low latency application at the same time. Whereas the former necessitates the use of a double split (options 2 and 7) architecture between the 5G core (NGC) and antenna, the latter needs to place the core functions closer to the edge to meet the tight latency requirements.

Emerging 5G services demand different SLAs for eMBB, URLLC, and mMTC applications. While eMBB challenges the bandwidth inefficiency of existing fronthaul technologies, URLLC applications require ultra-reliable low latency networks, and mMTC demands a network that can manage a vast number of endpoints in a power-efficient manner. These new challenges have led to the consideration of new ways of splitting critical baseband and radio functions.

Table 9.1 Generic slice template attributes

#	Attribute	Description	#	Attribute	Description
1	Availability	Not described in current version	19	Positioning support	Support of geo-localization methods
2	Area of service	Area of access to a network slice	20	Radio spectrum	Radio spectrum
3	Delay tolerance	Slice does not require low latency	21	Root cause investigation	Capability of providing degradation root cause
4	Deterministic communication	Support of determinism for periodic traffic	22	Session and service continuity support	Continuity of a PDU session
5	Downlink throughput per network slice	Achievable DL data rate at slice level	23	Simultaneous use of the network slice	Merging capabilities with other slices
6	Downlink throughput per UE	Achievable DL data rate at user level	24	Slice quality of service parameters	QoS parameters for the network slice
7	Energy efficiency	Bit/Joule for the slice	25	Support for non-IP traffic	Indication of other type of traffic supported
8	Group communication	Support of multicast, broadcast, etc.	26	Supported device velocity	Maximum speed supported the slice
9	Isolation	Segregation level from other slices	27	Synchronicity	Synchronicity of communication devices
10	Maximum supported packet size	Maximum packet size in the network slice	28	Terminal density	Devices per km^2
11	Mission critical support	Priority respect to other slices	29	Uplink throughput per network slice	Achievable UL data rate at slice level
12	MMTel support	Support of multimedia services (e.g., IMS)	30	Uplink throughput per UE	Achievable UL data rate at user level
13	NB-IoT support	Slice supporting NB-IoT	31	User management openness	Possibility of managing users in the slice
14	Customer network functions	List of NFs provided by the customer	32	User data access	Access to Internet or VPNs
15	Number of protocol data unit (PDU) sessions	Maximum number of concurrent sessions	33	V2X communication mode	Support of V2X

(continued)

Table 9.1 (continued)

#	Attribute	Description	#	Attribute	Description
16	Number of terminals	Maximum number of simultaneous terminals	34	Latency from user plane function (UPF)	Delay from UPF to application server
17	Performance monitoring	Indication of KPIs and KQIs to monitor	35	Network slice specific authentication and authorization (NSSAA) required	
18	Performance prediction	Capability for predicting network status	36	Multimedia Priority Service support	

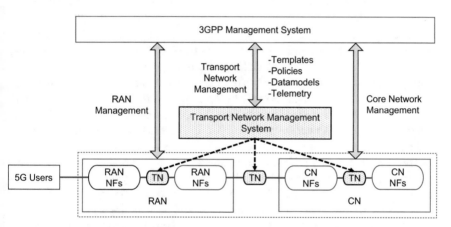

Fig. 9.2 3GPP slice management and control

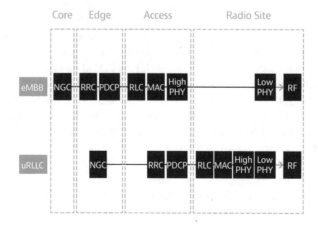

Fig. 9.3 Slicing for different verticals

Proper network design requires careful analysis of various SLAs associated with the above functional split options and use cases. They are characterized by latency, frame loss ratio, and time error metrics. Considering the diversity of the use cases and SLAs, 5G transport networks can be economically viable only if they are designed on a single converged physical network. In addition, network slicing enables the deployment of multiple services with distinct SLAs on a single physical network.

Starting with the radio segment, there are many design decisions in the new 5G New Radio (NR) made with this desire to create network slices to address numerous use cases simultaneously. In this case, flexible numerologies and bandwidth parts are vital in enabling network slicing. The higher numerologies have larger subcarrier spacing, and hence shorter symbols and slots can be scheduled more rapidly. As the slots are self-contained in terms of how data are scheduled on them, this is key to supporting low-latency applications. For the numerology with 120 kHz subcarrier spacing (the highest numerology that supports data), the slot lasts a mere 125 μs. This gives extraordinarily frequent opportunities for scheduling low-latency data in conditions where the channel conditions support it and can deliver the low latency required. But it would be wasteful if an entire carrier had to be given up to numerology with the shortest slots. Splitting a block of the spectrum into two carriers could overcome this but would reduce the spectral efficiency with the need for guard bands. Thus, another key part of delivering network slicing is the use of bandwidth parts. The ability to use different numerologies within the same carrier means that services with vastly different characteristics can be supported while preserving spectral efficiency. While the use of high subcarrier spacing and short slots is consistent with low latency, it is also subject to inter-symbol interference. This will reduce the reliability of the transmission and could have an impact on ultra-reliable low latency communication (URLLC). But even when high-order numerologies are used where the delay spread is significant and indicates against the use of such short symbols, there are mitigations in the 5G standard. First, the most robust modulation and coding schemes may be sufficient to overcome inter-symbol interference. But packet duplication can also assist in this regard. This mechanism creates a second RLC entity along with a second logical channel on the radio bearer when more reliability is required. The data is then transmitted twice—once in each RLC entity. This significantly raises the chance that each packet will be delivered successfully. Mini slots described earlier are also a key feature for low latency. Resource blocks in the resource grid can be reserved for mini slots and not be used for transmissions scheduled as part of regular slot scheduling. Mini slots can start at any time in the period and don't need to be aligned to other slot boundaries. This is ideal for low-latency applications as the transmissions can begin as soon as they are required by the low-latency application with no constraints on delay. Mini slots can be used in the DL and UL and can be as short as one symbol in length.

The 5G standard makes it possible for data to be grouped into different logical channels. Logical channel prioritization (LCP) allows these logical channels to be assigned different prioritizations. This supports network slices that can respect the relative importance and latency requirements of the slices. Restrictions can be placed on the logical channels so that they must be restricted to being scheduled on specific combinations of configured cells, numerologies, or PUSCH transmission durations. Thus, logical channels have a well-defined hierarchy of priority consistent with their importance, and the latency can be controlled by arranging for the latency to be low for slices that need it. In combination with the other flexible uses of the 5G infrastructure, these various features facilitate network slicing to deliver the rich mix of heterogeneous services that 5G will provide.

The next segment to architect for network slicing support is the access network. Traditionally converged access networks took advantage of WDM technology. However, while sufficient for the initial deployment, massive deployment of fiber, 4G, and 5G radios necessitates an economic and ubiquitous technology such as Ethernet. For example, in 4G and 5G mobile networks, to allow for the convergence of legacy CPRI-based and new Ethernet-based network technologies, the RoE standard can be deployed in fronthaul networks. An Ethernet-based technology can be most fruitful if its statistical multiplexing gains are effectively used. Taking advantage of this multiplexing gain can be realized only with a careful analysis of the transport network latency requirements. Time-sensitive networking is the ultimate goal of a cost-effective and massively scalable converged-access network.

At the transport layer, the NEST is processed to extract requirements that directly apply to the creation and instantiation of the transport slice. This consists of identifying parameters that directly or indirectly impact the selection of transport resources to form the E2E slice connectivity. Some attributes can be directly translated into transport network requirements (e.g., "slice throughput"), while others have indirect implications (e.g., "latency from (last) UPF to the application server") with respect to the location of the application (e.g., edge or central cloud) and the resources to be committed in the transport network to reach it. Thus, from the perspective of the transport network control entities, it is essential to account for mechanisms for assisting such translation, resulting in the instantiation of the transport network slice.

The transport network is in charge of enabling connectivity between the end users and the service functions composing a given E2E service as requested by a customer. According to the characteristics of the supported service, such connectivity can have distinct properties or characteristics and pursue different Service Level Objectives (SLOs).

The dynamicity in creating those services could be more significant if compared with the existing ones nowadays, allowing great flexibility in the provisioning of services with the need of handling from ephemeral to long-lasting services, of very different characteristics in terms of requirements, on top of the common transport substrate. To support the transport slicing capabilities, it seems convenient to define a new component, the Transport Network Slice Controller (T-NSC), in charge of control the provision of the transport slices and the management of their life cycle. The T-NSC has an awareness of slicing at the transport level. The Transport Slice Controller [19] will support both a Northbound and a Southbound Interface (NBI and SBI, respectively). Through the NBI, different customers request transport slices adapted to the specific needs of each particular service. Thus, the T-NSC acts as the single entry point to the transport network for these requests, resolving potential conflicts and/or incompatible requests. On the other side, once the requests are processed, the T-NSC instructs a number of per-technology network controllers to proceed with the proper actions associated with the previous requests. To summarize, the NBI supports the transport slice description, while the SBI enforces the transport slice realization.

It is essential to highlight that the NBI is considered to be technology-agnostic, that is, only the transport slice characteristics are expressed through that interface. The T-NSC performs the mapping of the technology-agnostic view to realizing the slice using a specific transport technology. In that mapping process, the T-NSC interacts with each particular technology controller involved in the slice provision. This implies the NBI to be common to all kinds of customers, even though not all of them would consume all the NBI capabilities. On the other hand, there can be expected the support of a variety multiple of SBI, one per each of specific transport technologies present in the network.

The final segment to orchestrate in an E2E network slice is the edge and core networks. As we mentioned, network slicing needs to deliver the QoS service level agreements (SLAs), meeting the requirements of the industry verticals without needing to custom design and deploy dedicated networks for each of the vertical use cases. The Next Generation Core (NGC) was architecture to support network slicing by including the Common Control Network Functions (CCNF) to enable the automation of deploying and managing a network slice with specific QoE and SLA characteristics.

Common Control Network Functions such as a single common access and mobility management function (AMF) are allocated to terminate a UE's NAS connection for all slices. This single AMF proxies session management messages to and from Session Management Functions (SMF) in the different network slices. For dedicated network functions—such as in the case of user plane—each data connection of the UE is served by a SMF+UPF belonging to the same assigned slice. The UE can have multiple PDU sessions in a slice to different data networks or multiple PDU sessions to the same data network via different slices, via the combination of slice identifier and APN. A UE can establish and maintain connections to a maximum of eight slices in parallel.

Figure 9.4 shows a service deployment example for 5G system network slicing focusing on the internals of NGC slicing, i.e., the UE, radio, RAN, and transport slicing aspects are not shown in this figure.

To enable a NGC to properly create, manage, and communicate data about a certain slice through its components, 3GPP defined single-network slice selection assistance information (S-NSSAI) to be used to identify a network slice and it consists of the following:

- A slice/service type (SST) refers to the expected network slice behavior in terms of features and services. Examples of SST include eMBB, URLLC, and MIoT.

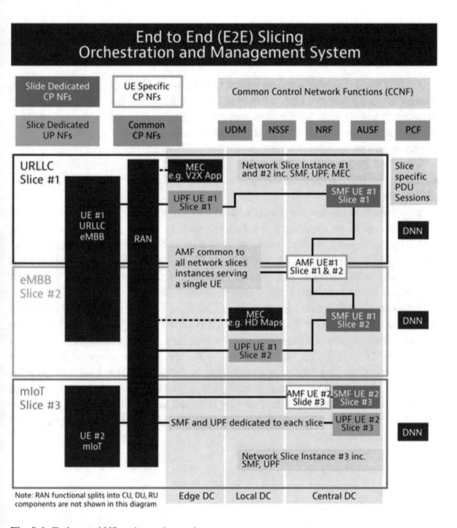

Fig. 9.4 End-to-end NS orchestration and management

Table 9.2 Standardized SST values

SST value	Slice/service type and its characteristics
1	Slice suitable for the handling of 5G enhanced mobile broadband (eMBB)
2	Slice suitable for the handling of ultra-reliable low latency communications (URLLC)
3	Slice suitable for the handling of massive IoT (MIoT)

- A slice differentiator (SD), which is optional information that complements the slice/service type(s) to differentiate among multiple network slices of the same slice/service type.

An S-NSSAI can have standard values or non-standard values. An S-NSSAI with a non-standard value identifies a single network slice within the public land mobile network (PLMN) with which it is associated. Standardized SST values enables global interoperability for slicing scenarios spanning multiple PLMNs, as roaming use cases can be supported for the most commonly used slice/service types.

Table 9.2 (standardized SST values) (referenced from [20]) shows the list of standardized SST values as defined in 3GPP release 15 and their respective characteristics.

The S-NSSAI can be associated with one or more network slice instances. The NSI ID serves as an identifier for a network slice instance. The NSSF may return NSI ID(s) to be associated with the network slice instance(s) corresponding to certain S-NSSAIs. A PDU session, which is the 5G system (5GS) association between the UE and a data network that provides a PDU connectivity service, is associated with one S-NSSAI and one DNN (Data Network Name).

Network slice instance selection for a UE is normally triggered as part of the registration procedure by the first AMF that receives the registration request from the UE. The UE provides the requested network slice selection assistance information (NSSAI) for network slice selection in 5G NR-RRC message, if it has been provided by NAS. The NSSAI is a collection of S-NSSAIs. The AMF retrieves the slices that are allowed by the user subscription and interacts with the NSSF to select the appropriate network slice instance (e.g., based on allowed S-NSSAIs, PLMN ID, etc.). Figure 9.5 (network slice selection example) shows an example of 5G network slice selection.

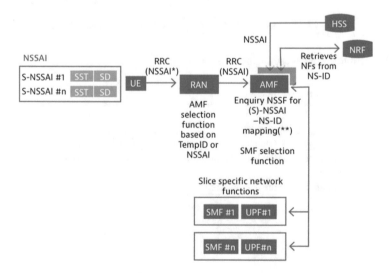

Fig. 9.5 Network slice selection example

9.4   Network Slicing Characteristics

Network slices are based on the allocation of resources to accomplish the service expectation from the customer. This fact imposes some specific characteristics present in NS, which differentiate them from other service offerings. The following subsections describe some of the most relevant characteristics to take into account.

9.4.1   Scalability of Slices

The consideration of scalability, when applied to network slices, is twofold. On the one hand, it can refer to the scalability of each particular slice. On the other hand, it also applies to the overall system scalability, considering the total number of slices that can be supported.

In the former, scalability is associated with the use of resources by a particular slice along its lifetime. The resources allocated for the slice are not only the ones nominally needed for honoring the service requested but also necessary to allocate resources that could provide protection and availability in case of network failure events affecting the slice. Thanks to the network softwarization process, the scaling can be dynamic, which implies that a given slice can scale up or down along the time, including compute and networking resources, adapting and fitting to the real demand at every moment. This dynamicity requires the support of an accounting system that could permit business models such as the pay-as-you-grow schemas.

Moreover, the latter scalability approach applies to the overall scalability exhibited by the operator in terms of how many slices can be deployed (potentially depending also on the type of slices). In principle, this depends on the number of resources available and the consumption pattern (or allocation) of those resources per slice. However, the larger the distinction of services (for instance, by means of a very granular set of SLOs supported), the higher the diversity of slices to be managed. Despite the fact that a more tailored consumption of resources can help accommodate a larger number of slices, this can lead to scenarios of high complexity because of the very different slice profiles to maintain. A coarser characterization of slices, with the idea of reducing the number of managed profiles, can have the advantage of simplifying the life cycle management of the complete set of slices in a network. Such coarser grouping of slices potentially implies the support in a common type of slice of many customer slices with slightly different SLAs, all of them being satisfied in principle by the slice profile with which the coarse grouping is defined. The customer slices could even be hosted on the same network slice instance, except for the customers requiring some control capability (to avoid conflicts in the configuration of resources). For those customers, it is always necessary to provide a dedicated slice.

9.4.2   Arbitration Among Slices

Once the slices are deployed, a number of events can imply the need to implement arbitration mechanisms among the slices deployed in the network. Scaling events, as commented before, will dynamically change the number of allocated resources. Additionally, network failures can force to migrate services around the network (because of shortage of computing or networking resources in a given area), then motivating a reassignment of resources per slice.

These kinds of situations show the need to make available mechanisms to provide arbitration among slices when competing for resources, in order to efficiently use the available network and compute assets and enforce negotiated SLAs. Arbitration is assumed to be an internal capability of the operator, being transparent to the customer that only influences the arbitration results through the SLAs.

Importantly, the arbitration can happen not only among slices of different customers but also in the slice realization of a single customer request. This is because there is not always a correspondence of 1 to 1 between customer slice request and the realization of the slice in the network. In some cases, a single customer request is directly mapped into an existing slice, which accommodates similar slices of some other different customers. In other cases, a single customer request could require the realization of more than one slice in the network (interconnected either in a recursive or serial manner).

Thus, arbitration can be applied at the time of slice provision (or activation, if the slice is not activated at the time of provisioning), but also during the slice lifetime, when service or network situations force to do so. The arbitration process can produce conflicts to other slices, affecting the service of other customers' slices in the network. When orchestrating, managing, and controlling the slices, it is then necessary to consider the SLA thresholds and intervals of SLO compliance since it is the basis of the calculation for the different trade-offs to apply on the need of reconfiguration of an existing slice. For instance, the requirement of strict slice isolation could be temporarily overcome by reallocating some resources to another vacant slice. However, if any kind of degradation or interference from one slice to another is closed to happen, the resources should be guaranteed for those slices with the requirements of strict isolation in its SLA.

Finally, in the case of massive failure, it could be required some prioritization among slices. The prioritization criteria could vary, mostly influenced by commercial aspects, even though other considerations, such as the applicable regulation, could determine a particular behavior. Alternatives are to prioritize in terms of the type of service, premium customers, percentage of slices affected, critical SLAs negotiated, or associated penalties, etc. For instance, in [23] an isolation index is proposed as feasibility criteria for basing prioritization of transport slices in the need of slice reconfiguration.

9.4.3 Slice Temporality

The duration of the slices can significantly vary among services, as reflected in the examples provided in [22]. Network softwarization, complementing both SDN and NFV techniques with automated orchestration capabilities, facilitates a very fast invocation and deployment of services (in the form of slices) when compared to traditional networks. Consequently, the burden for service creation is alleviated, opening the opportunity to satisfy slice demands of short duration (e.g., for a few hours, days, or weeks) in an easy way. The aforementioned automation also permits periodic actions, for instance, bandwidth calendaring slices, making them available in specific periods of time (e.g., weekends).

Another aspect to take into account is the frequency with which the different slices become requested. For instance, if a specific slice is requested very often, it could not be worthy to reallocate resources to other slices when the frequent slices are no longer operational since the time for a newer request of such kind of a slice will be short. Then it can be more practical to maintain the resources reserved until a new request is received.

Both duration and frequency are variables that should be considered at the time of orchestrating the slices in order to determine the potential impacts of the slice deployment. This is not only at the instant where the slice request is received. Both can assist in predicting the availability of resources in the future, so instantiation decisions can be based on overall resource availability in a given time frame (essentially the duration expected for the new slice request). This introduces further constraints in determining the availability of resources for satisfying a slice request in its expected lifetime.

In any case, long- and short-lasted slices will coexist in the same shared infrastructure. The planning, accounting, and billing of used resources should be adapted to this new situation.

9.5 Network Slicing Management

End-to-end slices are a service management issue. The enablers of management and orchestration are the usage of open and standard interfaces for interacting with different purpose nodes and systems, as well as the definition of normalized models for service and devices. An overview of network management and orchestration can be found in [23]. The result of the orchestration process is the allocation of the resources, as well as the management of their life cycle, including service assurance and fulfillment along the lifetime of the slice.

The customers can require distinct levels of control for the slices they have requested to the provider. Extreme cases can be, on one hand, customers that do not require any capability of control and management of the allocated assets (just pure communication service) and, on the other hand, customers requiring full control of their slices. A gradual level of control needs could be found in between.

Then, the operator should provide configuration and administration capabilities to the customers according to the levels of control that they request. These capabilities could come by simply exposing some interfaces for that required control actions (e.g., APIs), up to granting direct access to the resources (e.g., IP address to access the element console). The more abstracted way, the less invasive for the operator.

3GPP [24] defines a number of management functions managing network slices in support of communication services. These functions are known as:

- *Communication service management function* (CSMF), which is responsible for translating the communication service-related requirements into network slice-related requirements
- *Network slice management function* (NSMF), which is responsible for the management and the orchestration of an instance of a network slice
- *Network slice subnet management function* (NSSMF), which performs the same task as the NSMF, but at a sub-instance level

Figure 9.6 shows the relationship among the 3GPP slice management functions. These functions have been also mapped to the ETSI NFV orchestration framework in [1], as represented in Fig. 9.7. The slice management functions are considered to be part of the broader OSS/BSS components. Both NSMF and NSSMF can be considered as functionally similar in this case.

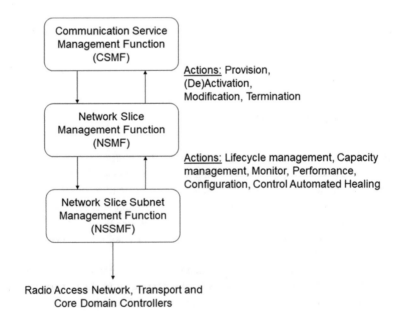

Fig. 9.6 3GPP slice management functions

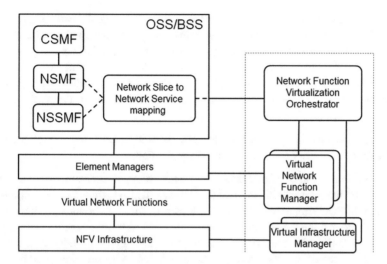

Fig. 9.7 Mapping of the 3GPP network slicing concept to the ETSI MANO framework

9.5.1 Multi-Provider Slice Orchestration

Multi-provider orchestration facilitates the instantiation of virtualized network functions in computing infrastructures from distinct operators, which constitute different administrative domains. The service is then formed by composing resources (in terms of compute and the functions running on them, as well as the required connectivity) hosted in different providers.

Multi-domain slices will be based on this multi-provider orchestration capabilities, where slice orchestration is built as an extension of them. For enabling such a multi-operator slice provisioning context, a number of business and technical aspects have to be considered, as follows:

- Multi-provider business coordination: In a multi-domain scenario, it is necessary to define how the variety of involved stakeholders interplay during the provisioning and instantiation of a multi-domain slice. Such coordination can assist on the trading of resources (compute, networking) and services (in the form of virtualized functions offered as a service) which are combined and composed to form the end-to-end slices.
- Intra- and inter-provider SLAs: An end-to-end orchestrator (acting either in hierarchical or peer-to-peer mode) should take care of decomposing the end-to-end SLA in different SLAs per provider, which are responsible of enforcing it internally per domain.
- Billing principles: In multi-domain environments different business models can occur, through bilateral agreements, brokering mechanisms, etc. Pricing schemas, related to the particular economic relationship among providers, accompany that business models.

- Specification of services and their advertisement to external customers: In multi-domain ecosystems (e.g., federations), the providers participating on them may make use of capabilities outside its own domain. This is also applicable to resources, functions, and slice offerings and solutions available on neighboring domains. Since a common way of advertising such capabilities is the usage of catalogues, a multi-domain synchronization or integration of the particular service catalogues.

Complementary to the business aspects, there are also a number of technical implications necessary to be taken into consideration:

- Multi-domain slice orchestrator: The multi-domain functionality should be restricted to some specific entities per domain. Those entities will be responsible of the interworking across domains, interacting with functionally similar entities on the other domains. These entities, referred to as multi-domain orchestrators, are responsible of the abstraction of the provider assets (i.e., underlay infrastructure and networking resources, functions offered in a virtualized manner, etc.) for later on advertising them to other providers.
- Decomposition of end-to-end slices on per-domain slice parts: The customer solely interacts with a provider that receives the slice request. This provider, which can be assumed to be the origin provider for that customer, is responsible of decomposing the slice taking into account its own capabilities plus the capabilities available in other providers. This implies to implement a specific logic for slice decomposition as applicable to the different domains.
- Multi-domain environment configuration: Manual configuration of available providers is a basic option for establishing the multi-domain ecosystem. Alternatively, automatic procedures can be considered for the discovery of neighboring administrative domains. The network softwarization trend will accelerate the adoption of this kind of solutions.
- Support of common abstraction models across domains: It is basic to support a common way of handling and interpreting the description of the resources available (i.e., for either network, compute, and storage) as well as the capabilities present on each of the providers involved.
- Standard interfaces, protocols, and APIs for the multi-domain interworking: Standardized mechanisms facilitate the integration among providers, reducing the time and cost of achieving such interworking.

9.5.2 Slice Operation

Slice operation implies the definition and usage of monitoring and telemetry capabilities to report the status of the network slice. The supervision procedures will not differ in essence from those followed for a conventional network since the network slice constitutes a logic network. However, there is a need for the abstraction of the physical performance parameters in order to derive from them the particular indicators for a given slice. This refers to all kinds of resources being part of the slice from a pure infrastructure perspective, that is, networking and compute resources. Moreover, when service functions are part of the slice, it is necessary to consider parameters associated with functions providing a complete overview of the slice.

All those indicators should be in principle common to the ones observed in conventional networks. At the time of processing, correlating, and determining operational situations, the same type of actions could be expected. This, however, can impact the scalability of the supportive systems (i.e., OSS/BSS) that assist on network operation, since now with slicing can be multiple incarnations of logical networks if compared with the traditional way. Essentially, there will be a need to implement a single operational procedure per (customer) slice, which can be solved in two ways: the first one, by first collecting all the information for a later customization per customer, and the second one, by separating in origin the parameters per slice and later on proceed with the processing and correlation. Following one or another way can impact the OSS/BSS capabilities, as well.

It is relevant to remark that due to leveraging virtualization and abstractions, physical resources supporting the slices can change along the slice duration time. Thus, it is necessary to keep a record and track of indicators even in the case that the actual resources changes, through dynamic association and aggregation of them. This fact affects as well the definition of the parameter itself since it should be generic to resources of the same kind.

Those parameters, usually associated with raw values of the physical resources, have to be processed in order to extract the specific information per slice. This becomes especially critical when the network slice is offered to customers, since the generated information should be only the one required per customer to guarantee privacy, prevent information leakage, etc.

Such information is sensitive in several aspects. On the one hand, it can be particular to the specific service of the customer, and then revealing functional characteristics of the service that could be considered as a secret for the customer. On the other hand, that information will form the basis for assessing the compliance of the committed SLAs between the customer and the operator.

9.5.3 Slice SLA Management

If a mobile operator knows the specific information about what type of vertical service terminals, like a vehicle, or IoT device, are in use, where they are located, and what that device is doing, the operator can start to implement different network slices and apply different service level agreements (SLAs) to different scenarios. Each vertical being served by a network slice will have specific needs such as how much coverage is needed, the amount of network bandwidth needed, how sensitive the device is to communication delays, and how available the IoT device and application need to be. What is the energy need for the device in that vertical? Does the device have a full-time power source, such as an autonomous car, or does it have a finite battery life such as a smart meter? What is the vertical's sensitivity to delays in communication? Every vertical will be different. For example, the needs of the mobile health vertical using IoT devices in pacemakers will be vastly different from a utility company needing information from smart meters. Smart water meters are not sensitive to delay, availability, or bandwidth requirements but do require deep coverage as they are often located in underground locations and are high-energy efficient to lengthen battery life. Conversely, a pacemaker application does not require deep coverage but is sensitive to delays and network bandwidth because decisions need to be made quickly. Therefore, the SLAs will be vastly different for these two examples.

For pacemakers, each device will need to be managed individually to a stringent degree by the application provider with no delays in network communications. For smart meters, an application vendor might sample a geographic region to see if there are any meters in the area and determine if any of the meters are malfunctioning. In this case, delay and individual monitoring are not important. For every vertical, automated identification of a device will be critical for network slicing to succeed on a mass scale. The Internet of Things will add billions of devices to the Internet and stands to trigger the next industrial revolution. However, IoT's demands are challenging for current cellular networks: These applications require high data rates and the lowest possible latency as "things" need to communicate with each other continuously. 2G networks were designed for voice, 3G for voice and data, and 4G for broadband Internet experiences. 5G, with its network slicing capabilities, is going to be the biggest facilitator of IoT. 5G will be the first network designed to be scalable, versatile, and efficient in terms of energy consumption. In other words, each device and network created that is based on a network slice will use only what it needs and when it needs it, instead of consuming anything and everything that's available.

5G promises to deliver latency and improve data rate and coverage. It will follow what is called a "non-orthogonal multiple access" model that supports multiple users to share limited bandwidth channels. This allows myriad devices to share data in a timely manner without having to wait for other devices to use a bandwidth channel and release it. Consequently, we will be able to add indefinite numbers of devices to the Internet without having to worry about scalability issues. With 5G, there will be computing capabilities fused with communications everywhere, so billions of devices don't have to worry about computing power because 5G networks will bring computer processing to devices that need it. 5G networks will be faster but also a lot smarter.

5G networks are designed not only to better interconnect people but also to interconnect and control machines, objects, and devices. They will deliver new levels of performance and efficiency that will support a broad set of industries. 5G is not just about multi-Gbps peak rates, but it also brings ultra-reliable low latency, high reliability, and massive IoT scale. Allowing industries to tap into these new capabilities will help to facilitate the next industrial revolution. Mobile operators can ensure the

Category	Application Example	Availability	UL Bandwidth & Data Size	DL Data Size	Frequency	Power	Delay Sensitivity	Coverage
Automotive	Connected Car	HIGH 99.9%	HIGH 10Mbps 200bytes	HIGH 200bytes	HIGH Continuous	LOW	HIGH 1ms	NORMAL
Industrial Control	Switch on/off, device triggered to send	HIGH 99.999%	HIGH 50Mbps 0-20bytes 50% of cases require UL response	HIGH 20bytes	MED 1 day (40%) 2hrs (40%) 1 hour (15%) 30mins (5%)	LOW	HIGH 1ms	DEEP
Utilities/ Meters	Smart Water Meter	LOW 99%	LOW 50kbps 20bytes with cut off of 200 bytes	LOW 50% of UL data size	MED 1 day (40%) 2hrs (40%) 1 hour (15%) 30mins (5%)	HIGH 10yrs & 4800mAH	LOW 5sec	DEEP
Security	Smoke alarm detectors, power failure otification, tamper notifications	HIGH 99.9%	LOW 50kbps 20bytes	LOW 0 ACK payload size is assumed to be 0bytes	LOW Every few months, Every Year	LOW	MED 1sec	DEEP

Fig. 9.8 CIoT SLA criteria [NS-33]

success of CIoT and 5G through the management of the devices and how those devices interact with the network (Fig. 9.8).

9.5.4   Optimization and Assurance

As models become more sophisticated, and as network slicing becomes more of the normal choice to provide competing requirements within the same infrastructure, more aspects of the network parameterization choices and the corresponding performance will be required as inputs and outputs for these models. Some of these will include choices such as the numerology, choice of cyclic prefix, RACH channel configuration, and configuration of BWPs. These will consider the variations in UE capabilities of the device population including which numerologies are supported and what channel bandwidth can be accessed on which carriers, so that the parts of the spectral resources that can be accessed by which devices is modeled. Thus, the overall composite capacity achievable for the mix of devices and subscribers over multiple network slices will be resolved by the model. The most sophisticated models will not only model the dynamics of UEs as they move around a network on a specific spectral resource; they will complement this with model components that include the interactions between network layers. Network layer modeling will include which carriers and BWPs are used to deliver service to each class of subscribers in the various locations. This will also include the interworking between LTE and 5GNR carriers in dual connectivity mode, respecting the system parameters that control the management of spectral resource layers. Such powerful models will capture more faithfully the complex system of interactions that is the NR and LTE radio interface and will underpin optimization of the many parameters and choices that can be tuned in the next-generation networks. The resulting networks will utilize the physical resources more efficiently and deliver the best balance of coverage and capacity for each class of subscriber on each network slice. These AI and ML and models that possess ever more powerful predictive capability will be in the ascendancy, as initiatives such as disaggregation in the RAN, service-based architecture in the core, and the O-RAN Alliance open up the network into more discrete components. As these components become more programmable, so will the data that they expose to fuel the next generation of advanced AI models. These open programmable networks and the transport networks that fuel them, when combined with sophisticated models, optimization, and prescriptive analytics, will be a potent combination. Drawing on many areas of advanced technology, with autonomic monitoring, self-regulation, and intelligent adaptability, we will have the most sophisticated hybrid digital and physical systems ever created by humans. From these, new artifacts will emerge a communication system that delivers a richness of experience with breadth and depth well beyond what we dare to imagine today.

9.6 Network Slicing Targeting Year 2030

At the writing of this book, 5G mobile network deployments are taking place worldwide. While still there are different challenges to get the maximum benefits from what a 5G network promises, there are already discussions about what the forthcoming network generation will demand. The typical life cycle of network generations predicts that by 2030 we should be starting the prospect of upgrading 5G and deploying 6G networks around the world. With such advanced network, the potential of new services and capabilities will have an impact on how important network slicing functions. From new holographic type communications that will require high speed in the range of 100 Gbps and low latency in the range of 5–7 ms to high precision communications with on-time and throughput guarantee, each will challenge how efficient a network slice programming and assurance becomes.

The expected requirements for the Beyond 5G (B5G) and potential 6G wireless technology include:

- Extreme high data rate and capacity with peak data rate >100 Gbps
- Extreme coverage including 3D precision in potential upper atmosphere and space
- Extreme low energy and cost, with devices battery life time for decades or no need for charging
- Extreme low latency below 1 ms
- Extreme high reliability with guaranteed QoE, security, privacy, and resilience
- Extreme connectivity with densification of 10M devices per square kilometer

All of these requirements will undoubtedly challenge the network slicing concept in terms of automation, accurate mobility, reliability, energy efficiency, high flexibility, security, and trust.

Another significant development that will impact network slicing adoptions and expectations is the high acceleration in the acceptance, evolution, and generalized deployment of disaggregated solutions, such as O-RAN. An O-RAN, or Open Radio Access Network (O-RAN), is a concept based on interoperability and standardization of RAN elements, including a unified interconnection standard for white-box hardware and open source software elements from different vendors. O-RAN architecture integrates a modular base station software stack on off-the-shelf hardware which allows baseband and radio unit components from discrete suppliers to operate seamlessly together.

O-RAN underscores streamlined 5G RAN performance objectives through the common attributes of efficiency, intelligence, and versatility, and network slicing will be one technology vehicle used to achieve many of these attributes. Open RAN deployed at the network edge will benefit network slicing verticals such as autonomous vehicles and the IoT and will support network slicing use cases effectively and enable secure and efficient over-the-air firmware upgrades.

The enticing Open RAN concept of flexible interoperability also brings challenges for test and integration. To fulfill the O-RAN promise of reduced OPEX and total cost of ownership (TCO), operators must take responsibility for multi-vendor, disaggregated elements and make sure they perform together to maintain QoE standards. With Open RAN reducing the barrier to entry for dozens of new players, interoperability is a paramount concern for both the O-RAN ALLIANCE and OpenRAN group. Open Test and Integration Centers (OTIC) worldwide have been established as a collaborative hub for commercial Open RAN development and interoperability testing to address this challenge. The operator-led OTIC initiative benefits from the support of global telecom organizations with a shared commitment to verification, integration testing, and validation of disaggregated RAN components. Network slicing within an O-RAN ecosystem is being researched and tested in many of these OTIC labs at the writing of this book.

9.7 Conclusions

This chapter describes and details how network slicing considerably transforms the networking perspective by abstracting, isolating, orchestrating, softwarizing, and separating logical network components from the underlying physical network resources. As such, they are intertwined to enhance Internet architecture principles.

In the foreseeable future (e.g., by 2030), different forms and factors of network slicing are expected to become the norm, realized through diverse operational modes and taking multi-tenancy and precision slicing to an extreme, and as such, slicing impact is broad in terms of networking (i.e., feature-rich, capability-rich and value-rich) and deep from both a vertical (multilayer) perspective as well as a horizontal (end-to-end and multi-domain) view. A future-thinking perspective on cloud network slicing takes customer/tenant-provider recursive relations to an extreme combined with flexible tenant-driven choices on the network protocol stack and actual software instances under its responsibility.

References

1. A. Galis, S. Denazis, C. Brou, C. Klein, *Programmable Networks for IP Service Deployment* (Artech House Books, Norwood, 2004), p. 450
2. A. Galis et al., Management and service-aware networking architectures (MANA) for future internet, in *Invited Paper IEEE 2009 Fourth International Conference on Communications and Networking in China (ChinaCom09)*, (IEEE, Geneva, 2009)
3. Rochwerger, J. Caceres, R. Montero, D. Breitgand, A. Galis, E. Levy, I. Llorente, K. Nagin, Y. Wolfsthal, The RESERVOIR model and architecture for open federated cloud computing. Internet Scale Data Centers **53**, 4 (2009)
4. ITU-T Network 2030, New services and capabilities for network 2030: description, technical gap and performance target analysis, https://www.itu.int/en/ITU-T/focusgroups/Network2030/Documents/Deliverable_NETWORK 2030.pdf
5. L.M. Contreras, D.R. López, A network service provider perspective on network slicing. IEEE Softwarization (2018), https://sdn.ieee.org/newsletter/january-2018/a-network-service-provider-perspective-on-network-slicing
6. L.M. Contreras, J. Ordonez-Lucena, On slice isolation options in the transport network and associated feasibility indicators, in *IEEE International Conference on Network Softwarization (NetSoft)* (2021)
7. ITU-T Technical Report, FG-NET2030-Arch Network 2030 – Architecture Framework
8. ITU-T Y.3011, http://www.itu.int/rec/T-REC-Y.3001-201105-I
9. ITU-T IMT 2020 Technical Report: Application of network softwarization to IMT-2020 (O-041), https://www.itu.int/en/ITU-T/focusgroups/imt-2020/Pages/default.aspx
10. ITU-T Recommendation on network slicing ITU-T Y.3112
11. A. Galis, J. Dong, K. Makhijani, S. Bryant, M. Boucadair, P. Martinez-Julia, IETF draft "Network slicing", https://tools.ietf.org/html/draft-gdmb-netslices-intro-and-ps-01
12. 3GPP TS 28.201, https://portal.3gpp.org/desktopmodules/Specifications/SpecificationDetails.aspx?specificationId=3692
13. 3GPP TS 28.202, https://portal.3gpp.org/desktopmodules/Specifications/SpecificationDetails.aspx?specificationId=3684

14. M. Kiran, K. Smith, J. Grant, A. Galis, UCL, ETSI E2E network slicing reference framework and information model, Xavier Defoy-Interdigital published in October 2018 by ETSI as a standard specification document, https://www.etsi.org/deliver/etsi_gr/NGP/001_099/011/01.01

15. ETSI Zero Touch Network (ZTN), https://www.etsi.org/technologies/zero-touch-network-service-management

16. VIAVI Solutions, *Understanding 5G*, 2nd edn, 2021

17. GSMA NG.116 Generic Network Slice Template v4.0, https://www.gsma.com/newsroom/all-documents/generic-network-slice-template-v4-0/

18. 3GPP TS 28.531, Management and orchestration; Provisioning, V16.3.0 (2019)

19. A. Farrel, et al., Framework for IETF Network Slices, draft-ietf-teas-ietf-network-slices-01 (work in progress) (2021)

20. 3GPP TS 23.501, System architecture for the 5G System (5GS); Stage 2 (Release 15), V15.12.0 (2020)

21. L.M. Contreras, J. Ordonez-Lucena, On slice isolation options in the transport network and associated feasibility indicators, in *IEEE International Conference on Network Softwarization (NetSoft)* (2021)

22. 5G-TRANSFORMER deliverable 1.4, 5G-TRANSFORMER final system design and Techno-Economic analysis (2019), http://5g-transformer.eu/wp-content/uploads/2019/11/D1.4_5G-TRANSFORMER_final_system_design_and_Techno-Economic_analysis.pdf

23. L.M. Contreras, V. López, R. Vilalta, R. Casellas, R. Muñoz, W. Jiang, H. Schotten, J. Alcaraz-Calero, Q. Wang, B. Sonkoly, L. Toka, Network management and orchestration, in *5G System Design: Architectural and Functional Considerations and Long Term Research*, ed. by P. Marsch, Ö. Bulakci, O. Queseth, M. Boldi, (Wiley, London, 2018)

24. 3GPP TR 28.801, https://portal.3gpp.org/desktopmodules/Specifications/SpecificationDetails.aspx?specificationId=3091

Chapter 10
Routing and Addressing

Yingzhen Qu, Adrian Perrig, and Daniel King

10.1 Introduction

The current Internet, which has evolved for more than 50 years, is facing a set of unique challenges, both technically and commercially. The exponential growth of the Internet and emerging demands from connected devices, increased mobility, security and resilience are met through incremental updates. Routing protocols have been critical networking technologies, and continuous development of routing protocols is essential to provide network services, which are the building blocks for new applications and services.

Figure 10.1 classifies widely used routing protocols into different categories.

Distance vector protocols are based on the Bellman-Ford algorithm, and are also referred to as routing by rumor, as they rely on neighbor-based information. Routers iteratively calculate the best routes to others as routing information propagates through the network. Common distance vector protocols include Enhanced Interior Gateway Routing Protocol (EIGRP) (https://tools.ietf.org/html/rfc7868) and Routing Information Protocol (RIP) (https://tools.ietf.org/html/rfc2453).

In link state protocols, each router floods its connectivity information to all other routers and locally calculates the shortest paths to them using Dijkstra's algorithm. Any change of link status (e.g., an interface shutdown) will be advertised to all

Y. Qu (✉)
Futurewei Technologies Inc., Santa Clara, CA, USA
e-mail: yingzhen.qu@futurewei.com

A. Perrig
Department of Computer Science, Network Security Group, ETH Zurich, Zürich, Switzerland
e-mail: adrian.perrig@inf.ethz.ch

D. King
Department of Computing and Communications, Lancaster University, Lancaster, UK
e-mail: d.king@lancaster.ac.uk

© The Author(s), under exclusive license to Springer Nature
Switzerland AG 2021
M. Toy (ed.), *Future Networks, Services and Management*,
https://doi.org/10.1007/978-3-030-81961-3_10

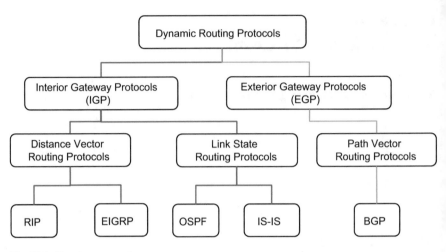

Fig. 10.1 Routing protocol category

routers in the network, allowing them to recalculate the shortest paths and maintain an up-to-date view of the entire network topology. Examples of link-state protocols include Open Shortest Path First (OSPF) (https://tools.ietf.org/html/rfc2328) and Intermediate System to Intermediate System (IS-IS) [1].

A path vector protocol, such as the Border Gateway Protocol (BGP) [2], iteratively builds up an Autonomous System (AS) path for each destination network. An autonomous system is composed of a set of routers under a single entity's administrative control. One of the advantages of path vector protocols is each destination network has a path dynamically added to it. Therefore, a loop is detected if the AS finds its own AS number in the path received. Routing protocols can be broken down just one more level as interior gateway protocol vs. exterior gateway protocol. Interior gateway protocols are used within an autonomous system, typically not routed between autonomous systems from the ground up, and interior gateway protocols are designed to fast route convergence, e.g., how fast to route around a link failure in a network. Exterior gateway protocols are used to route traffic between autonomous systems from the ground up, and they were designed to hold large amounts of routes, e.g., the routing table of the Internet. Another important thing is the ability to perform routing policies. For example, if there are two Internet providers, a routing policy can be defined for given prefix to prefer an ISP over another. This is done by allowing a preferred ISP ingress into the autonomous system. To access external resources to the autonomous system, the autonomous system needs to build a neighborship with another autonomous system. This is how routing on the Internet works. Of course, there are way more complicated things that make the Internet work. But at the networking level, fundamentally, that's how the Internet is shared and how it works.

Recent research and investigation for the future of the Internet identified several technology objectives, including contextual addressing, application-aware networking, increased stability and security, faster convergence, and decreased operational costs.

At the core of the Internet are routing protocols, including OSPF, IS-IS, and BGP, facilitating how Internet routers communicate with each other to distribute information that enables them to select routes for Internet connectivity. Existing routing protocols likely need to be enhanced. New routing protocols may also be required to meet the new requirements for the emerging requirements and long-term Internet evolution goals.

10.2 Addressing

Internet Protocol (IP) addressing facilitates how one device attached to the Internet is distinguished from every other device. They are used to direct requests to an appropriate destination (destination address) and indicate where replies should be sent (source address). Due to the rapid growth of the Internet and exponential increase of connected devices, several short-term fixes have been developed for coping with Internet addressing demands. The continued growth and deployment of the Internet of Things (IoTs) and new network types such as space-networking will place new requirements on existing addressing schemes.

10.2.1 IP Address

An Internet Protocol address (IP address) is a number assigned to each device connected to a network using the Internet Protocol (IP). An IP address is used to both identify a host and the location of the host.

There are two versions of the Internet Protocol commonly used today. The original version is Internet Protocol version 4 (IPv4), which was deployed on the ARPANET [3] in 1983 and still carries the most traffic on the Internet today. An IPv4 address is a 32-bit number and is typically written in dot-decimal notation, such as 192.168.1.1. As the number of devices on the Internet increases, the IPv4 addresses have been exhausted at the IANA level since 2011. A new version of IP (IPv6) using 128-bit number was standardized in 1998 [4], and the deployment of IPv6 started in the mid-2000s.

10.2.2 Name-Based Network

Compared with using IP address as both the name and addressing, another way is to access data by name regardless of the location. Information-centric networking (ICN) evolves the Internet architecture by introducing uniquely named data, so data is independent from location, application, and means of transportation, enabling in-network caching and replication [5].

The Locator/ID Separation Protocol (LISP) as defined in RFC 6830 (https://tools.ietf.org/html/rfc6830) was published by IETF as an experimental RFC in 2013. LISP separated IP addresses into two numbering spaces: Endpoint Identifiers (EIDs) and Routing Locators (RLOCs). IP packets addressed with EIDs are encapsulated with RLOCs for routing and forwarding in the network. The EID-to-RLOC mapping is stored in a mapping database.

Figure 10.2 shows a typical deployment of LISP in the global Internet [6].

10.2.3   Current Internet Addressing Techniques

The introduction of IPv6 supports of the next generation of wireless, high-bandwidth, multimedia Internet applications, as well as growth in the global number of users and devices. Continued IPv6 deployment provides expanded scale, reduced operational costs by utilizing simpler network models enabling new service and application innovations.

General benefits for IPv6 also include reduction in the deployment of network address translation technologies, increasing address capacity for wireless peer-to-peer (P2P) applications. However, several Internet addressing challenges exist, and new requirements are being introduced based on predicted services and future Internet architecture.

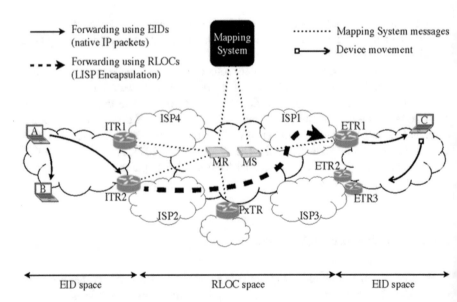

Fig. 10.2 A typical deployment of LISP

10.2.4   Requirements for Addressing in the Future Internet

Several emerging addressing requirements have been identified for future Internet; these include:

- Beaconing for source, destination, and route discovery
- Semantic and contextual addressing
- Flexible addressing
- Inherent security: via identifiers which allow applications and users to validate source and destination

Furthermore, addressing should enable upper layer protocols to identify end points unambiguously, and be agnostic to the underlay technologies and hardware. Thus, it allows the exploitation of new transmission technologies, and decouples users, applications and services from lower-layer hardware. This would also facilitate the interconnection of emerging future Internet devices to existing Internet infrastructure.

10.2.5   Addressing Semantics

Within a limited domain, it is possible to set an address with some special semantic, so service providers or network operators can apply local policies or have certain service bound depending on the sematic. For example, a semantic may denote different device types, or connectivity requirements (https://datatracker.ietf.org/doc/draft-king-irtf-challenges-in-routing/).

The following list shows how address semantics may interact with routing:

- New semantics to IP addresses may have implications for how network routing is performed.
- Semantic techniques might not be supported by existing routing protocols and so would require changes.
- Semantic techniques might enable advanced routing features or offer benefits in scaling and management of routing systems.

10.3   Routing

10.3.1   Network Path Selection

Typically, two approaches may be used for network path selection:

- Firstly, a priori assessment by having the feasible paths and constraints computed in advance
- Secondly, real-time computation in response to changing network conditions

The first scenario may be conducted offline and allows for concurrent or global optimization and several factors to be applied. As network complexity increases, the required computing power may increase exponentially, especially when evaluating large search spaces.

The second approach must consider the speed of calculation. The response processing may delay the service setup, especially if the path selection request is in response to a network failure. Path selection constraints may be applied to reduce complexity. However, the path computation's accuracy and optimality may be negatively affected.

In both scenarios, the amount of information that needs to be imported and processed can become very large (e.g., in large networks, with many possible paths and route metrics), which might impede the scalability of either method.

In the last decade, significant research has been conducted into future Internet architectures. During this research, several techniques emerged, highlighting the benefits of path awareness and path selection for end hosts during this research, and multiple path-aware network architectures have been proposed, including SCION [7] and RINA [8].

When choosing the best paths or topology structures, the following criteria may need to be considered:

- Method a path, or path set, is to be calculated, e.g., a path can be selected automatically by the routing protocol calculated the best path or imposed by a central entity, for example, for traffic-engineering reasons.
- What criteria are used for selecting the best path, e.g., classic route preference, or administrative policies such as economic costs, resilience, and security, and if requested, applying geopolitical considerations.

10.3.2 Traffic Engineering

A fundamental capability of the Internet is to route end-user traffic from the source to the ultimate destination. Transit routers along the path will implement control and optimization techniques to steer traffic along the path from the source to destination. Routers may utilize traffic engineering (TE) techniques to apply scientific principles to the measurement, characterization, modeling, and path selection control and ensure the end-user traffic is forwarded and end-user application requirements are met.

Another objective of Internet TE is to facilitate reliable network operations, which can be achieved by providing mechanisms that enhance network integrity and embrace policies that emphasize network survivability in the event of failures, thus, reducing susceptibility to network outages arising from errors, faults, and physical failures and force majeure events, occurring within the network infrastructure.

Traffic engineering techniques can be applied using distributed or centralized control plane techniques. Both scenarios would utilize the key TE components, including path steering, policy, and resource management:

- Path steering: This is the ability to forward packets using more information than just knowledge of the next hop.
- Policy: This allows for the selection of next hops and paths based on information beyond basic reachability.
- Resource management: This controls how different resources can be shared among different services.

10.3.3   Predictive Routing

Predictive routing means the change in the state of a router/host can be predicted; hence the routing algorithm can make route changes before or as an event occurs. There are new categories of applications that may benefit from predictive routing: such as cars driving on a highway or robots moving in a factory. These are applications where packet loss or delay is potentially very harmful, but the device's movement can be either predefined or predicted.

An alternative approach that alleviates the effects of slow routing protocol convergence is embodied by protocols with packet-carried forwarding state, such as SCION [7] or Segment Routing [9]. In such protocols, forwarding information that is carried in the packet header does not rely on router's (inter-domain) forwarding tables and thus avoiding inconsistent forwarding table state due to asynchronous update mechanisms. Moreover, the nature of the path exploration process in SCION (referred to as beaconing) which creates path segments does not require any convergence for connectivity—instead, additional paths are created over time that become available. Basic end-to-end connectivity, however, is established based on the initial path segments that are disseminated.

In general, the network infrastructure is fixed subject to the impacts of failure, maintenance, and upgrades. However, there is a new class of network emerging based on the use of mobile network infrastructure components such as large constellations of low earth orbiting satellites. These have the property that while the network infrastructure is dynamic, and the best paths are constantly changing, the best path is predictable in advance. This allows a new approach to routing based on current knowledge of the future disposition of the infrastructure rather than on preconfigured "static" paths, or dynamically discovered paths.

10.3.4 ManyNets and Routing for Space-Based Networks

While there is no relation between wireless mesh network routing challenges and protocols developed in IETF MANET WG, routing in space with LEO satellite constellations presents domain specific routing challenges.

A system, where complete global connectivity is provided through LEO satellites, which includes inter-satellite connectivity using Free Space Optical (FSO) transmission, introduces unique set of challenges w.r.t routing in space and possible traffic engineering [10]. This is because (as noted earlier) of the continuous changes to the network paths as the nodes in the orbit are on the move. There is no routing protocol today which does shortest path computation when all the nodes in the network are continuously moving. However, one characteristic of this network is the movements of satellites are completely predictable and this can be factored for new route computation methods. This also introduces unique set of Fast ReRoute (FRR) challenges which are not applicable for terrestrial networks. However, it is worth noting at this time the applicability and possible deployment of such a system is constrained by free space optics (FSO) limitations. These limitations concern with the inter-satellite link capacity, which is currently in the order few Gbps [11, 12], while the sub-sea fiber optical cable provides bandwidth in the order of 10's of Tbps.

The resulting low Earth orbit (LEO) constellations will not only bridge the digital divide by providing service to remote areas, but they also promise much lower latency than terrestrial fiber for long-distance routes. Unlocking this potential is nontrivial: such constellations provide inherently variable connectivity which today's Internet is ill-suited to accommodate. In fact, the use of the BGP protocol to integrate the satellite network in today's Internet unfortunately encounters several major challenges:

- The highly dynamic nature of ground station to satellite links creates.
- Scalability limitations for BGP, especially due to weather disruptions.
- During early phases of deployment, connectivity will fluctuate so often that slow routing convergence with BGP could make the partially deployed constellation unusable.
- The higher cost and lower bandwidth of satellite network links complicates their use for all data traffic, thus complicating the management of differentiated traffic.

There have been proposals to address these challenges. Giuliari et al. propose an optimal solution based on the SCION path-aware-networking architecture, and given this clean-slate baseline, they then develop a more pragmatic solution based on a CDN-like architecture [12].

10.3.5   Mobility

Mobility needs to provide ubiquitous connectivity to mobile users, independent of type and location of devices, access technologies, etc. A mobile node must be able to continue to communicate with others when access location or technology changes when moving and still providing efficient content delivery and trustworthiness.

There have been researches and proposals on mobility for years. One current approach to mobility issues is that they are resolved by the applications themselves using technologies such as MPTCP, QUIC (https://datatracker.ietf.org/doc/html/rfc9000), etc. at the transport layer. Another approach is the Mobile Ad hoc Networks (MANETs) (https://datatracker.ietf.org/wg/manet/about/), which is to provide a network layer solution to support node motions, including IP routing protocol functionality suitable for wireless routing applications.

For the Internet of Everything (IoE), the collaboration of IoE-based devices with current Internet protocols is challenging, specifically in terms of mobility and scalability.

Mobility scenarios in cellular networks pre-REL15 [13] involves only access layer, i.e., UE's mobility from one NodeB to another NodeB with the same or different Mobility Management Entity (MME). However, 3GPP REL15 [13] presents various mobility scenarios which involves IP address changes with or without service continuity as described in various Session and Service Continuity (SSC) modes. In the scenario, where IP address change causes disruption to session continuity, to maintain service continuity, various solutions are specified in [13], involving changes to transport layer protocols at UE. While other category of such solution involves network-assisted service continuity with multiple PDCP sessions and stitching these sessions in backhaul network to prevent the services interruption at the UE without any or with minimal packet loss.

However, there are not widely accepted/deployed solutions in network layer yet for new service requirements described in Gap Analysis of Network 2030" [https://www.itu.int/en/ITU-T/focusgroups/net2030/Documents/Gap_analysis_and_use_cases.pdf]. With the development of new applications in NETWORK2030 with uRLLC requirements, it is desired to support mobility in network layer, which avoids the session interruption and minimizes the packet loss and latency.

10.3.6   Domain-Specific Routing Protocols and Algorithms

New routing protocols are being developed in the IETF for data centers, e.g., RIFT and LSVR. These are protocols specifically optimized for use in certain types of domain and topologies. Such protocols trade general applicability for high performance in the target domain. Soon, there could be more domain-specific cases that require new routing protocols or algorithms, such as routing for satellite communications.

10.3.6.1 Industrial Internet and the Internet of Things

Industrial internet refers to the interconnected networks of sensors, robots, etc. Internet of Things (IoT) network consists of control systems, embedded systems, etc., and in consumer market, IoT is essentially the technology to build smart home and smart cities and enable applications including healthcare, disaster recovery, etc.

New technologies and standards are being developed at a rapid pace to form different IoT ecosystems and networks. From routing perspective, the typical common requirements among these networks are the following:

- Low power consumption. Typical IoT devices are powered by batteries with limited processing power and memory, and this means they need to be conservative on power consumption when sending data packets or control packets. Routing protocols designed for such IoTs should be quiet without sending too many control packets, and then resulted data packets should not have big encapsulation header.

- High availability. Applications such as disaster recovery require the network to provide nondisruptive service in case of network failure, power outage, natural disaster, etc.

- Mobility. IoT devices should be able to connect and communicate with the network or other devices without location and access technology limitations, whenever and wherever.

- Large number of connections. The number of various IoT devices to be connected to the network will be in thousands or millions, so routing protocols are required to connect these huge number of heterogeneous systems.

There are two key issues that future network designers need to contend with in IoT networks. Firstly, the high path quality is needed, which requires the routing system to establish the path and allocate the resources, including the case where it may need to configure the network to strategically replicate and eliminate packets to maximize their chances of successfully traversing the network [14]. Additionally, many IoT devices are designed to meet extreme physical size, cost, and lifetime power budgets. The protocols that these devices use require extreme regard to resource conservation and may not be able to use the "standard" network protocols which are optimized for characteristics such as generality and performance.

10.4 Routing Security and Resilience

Ensuring the security of routing mechanisms continues to be a challenge. Routing attacks include route hijacking, which diverts traffic to an adversary-controlled domain, and denial-of-service attacks, which can prevent communication from happening altogether. Over the past four decades, numerous researchers studied secure routing in a variety of network types and settings. We briefly highlight the core challenges and several proposed approaches.

An overview of routing security is available as a taxonomy for secure routing protocols by Hollick et al. [14], which emerged from a recent Dagstuhl seminar on secure routing [15]. The taxonomy establishes the following general services that need to be protected: identity service, routing service, topology service, and transport service. An adversary can have a variety of capabilities, resources, and goals—the security section lists different categories of capabilities and resources as defined in Sect. 10.5 of this document. In the context of routing, the main goals are to violate the following security properties: availability of routing and forwarding, authenticity of routing information, confidentiality/privacy of routing and topology information, and anonymity of entities (e.g., mobile users could be located via the routing protocol). In terms of security properties of the forwarded packet data, the routing system should prevent the redirection of traffic flows through entities that intend to eavesdrop or alter packet traffic—if communication is already passing through a malicious entity, it is the responsibility of the data plane to ensure traffic secrecy and integrity.

Routing protocols, especially IGPs, have been running in a relatively benign environment. With the development of new applications, it is critical for the network to provide nondisrupted service especially to high-value traffic. As more hosts/IoTs are added to the network, security is becoming more and more critical.

Secure intra-domain routing protocols have been largely neglected compared to inter-domain settings, as one assumes a benign environment under single administrative control in these settings. In existing intra-domain protocols, however, adversaries can launch several attacks: availability, denial-of-service, or traffic redirection. The typical approach for securing link-state intra-domain routing protocols is to attach a cryptographic signature to link-state updates, as is done for instance in secure OSPF. Within a single administrative domain, the entity identification problem is simplified, as the network administrator can establish and distribute cryptographic keys and certificates among networking devices and systems.

Inter-domain secure routing continues to be a challenge up to today. While S-BGP and its successor BGPSEC have been developed over the past 20 years, they have seen limited deployment due to several reasons: worse scalability than BGP (due to the inability for prefix aggregation and the need for periodic dissemination of routing updates), operational challenges (obtaining and handling certificates, updating router software and possibly even hardware), limited security benefits (new attacks are made possible), slower convergence than BGP, and disruption of policy mechanisms (ASpath alteration/prepending). A beacon of hope is the resource public-key infrastructure (RPKI), which provides the prefix and AS certificates in BGPSEC, as it enables route origin validation, which is easier to deploy than full BGPSEC and in itself addresses several attacks [16]. Unfortunately, the RPKI introduces a circular dependency with routing, as route message verification requires RPKI certificate validation and RPKI certificate validation requires a route to a server to fetch the RPKI certificate database. Moreover, the RPKI also opens up vulnerabilities to misbehaving RPKI authorities, where a misconfiguration or malicious action can result in rendering an address range unreachable [7].

It appears that an Internet redesign is needed to resolve the thorny issues to secure BGP. The SCION secure Internet architecture has thus redesigned the routing and PKI infrastructure from ground up to achieve high levels of security [7]. By avoiding inter-domain forwarding tables on routers and utilizing a path exploration system that does not rely on convergence, many attacks and vulnerabilities are prevented by design. The control-plane PKI in SCION is constructed such that the distribution of cryptographic credentials follows the transmission of routing messages, thus avoiding circular dependencies between routing and certificate distribution. The definition of trust roots within each isolation domain ensures operational sovereignty and prevents external entities to affect operation due to misconfigurations or misbehavior. As a consequence of its design, SCION can prevent all known routing attacks.

Current routing protocols are built and operated on the assumption of a high degree of trust. IGPs are typically running within a controlled and secured domain and BGP connected with trusted neighbors. For future networks, there are three possible solution directions (not exclusive of each other):

• Making existing routing protocols more secure by adding new authentication mechanisms/algorithms, etc.
• Securing and authenticating the information distributed by routing systems (such as by RPKI mechanisms applied to BGP)
• Using a new secure routing protocol, e.g., SCION

In case of link or node failure, routing protocols should be able to continue to provide an acceptable level of service. This could be achieved through local repair techniques, such as Loop-Free Alternate (LFA) Fast Reroute (FRR) [17, 18]. Meanwhile, routing protocols should reconverge fast and bring the network back to a stable state.

Mutually Agreed Norms for Routing Security (MANRS) (https://www.manrs.org) is a global initiative, supported by the Internet Society, and is made up of network operators to improve global routing security.

MANRS provides critical fixes to reduce the most common routing issues, outlining four simple but concrete actions for network operators:

• Anti-spoofing: Prevent traffic with spoofed source IP addresses. Network operators should enable source address validation and prevent packets with incorrect source IP address from entering and leaving the network.
• Filtering: Prevent propagation of incorrect routing information. Implementing prefix filters within a network can help protect against threats such as prefix hijacking and route leaks.
• Coordination: Facilitate global operational communication and coordination between network operators, maintain globally accessible, and up-to-date contact information.

- Global validation: Facilitate validation of routing information on a global scale. Network operators need to ensure that their network's routing information is publicly available including the announcements that the network originates and the routing policy.

Figure 10.3 shows a summary of proposed actions by MANRS (https://www. manrs.org)

10.5   Emerging Routing Protocols

Internet paths often require evaluating and assessing route metrics, including latency, jitter reliability, bandwidth, and congestion. Depending on the number and overall path length, computing paths is often processor and time-consuming. The design of effective path evaluation strategies is a balancing act between accuracy, computation time, and be more specific.

Several emerging routing techniques are being developed to address the scaling concerns and path selection complexity for the future Internet and support emerging domain-specific technologies. These new routing techniques are discussed in the following subsections.

A CLOS network [19] is a kind of multistage circuit-switching network. It was invented by Charles Clos to solve the problem of explosive growth of telephone network. Figure 10.4 shows a Clos network, where each leaf is connected to every spine node, and vice versa.

Modern data centers, especially large-scale data centers, host tens of thousands of end points, and this puts on new challenges on network architecture and routing protocols. For operational simplicity, many data centers have chosen BGP [2] with a CLOS topology as the most appropriate routing protocol and architecture as described in RFC 7938 (https://tools.ietf.org/html/rfc7938).

Fig. 10.3 Proposed actions for service providers by MANRS

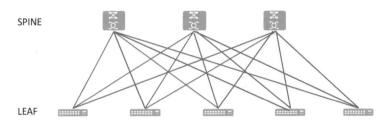

SPINE

LEAF

Fig. 10.4 A CLOS network architecture

10.5.1 RIFT

RIFT (Routing In Fat Trees) is a novel routing protocol defined by IETF (https://datatracker.ietf.org/wg/rift/about/). It mainly targets Clos [19] and fat-tree network topology-based data centers and is optimized with minimization of configuration and operational complexity.

RIFT is mixture of both link-state and distance-vector technologies and can be described as "link-state towards the spine" and "distance vector towards the leaves." Here are the major characteristics of RIFT:

- Northbound link state routing with flooding reduction, lower levels are flooding their link-state information in the "northern" direction, so that each level obtains the full topology of levels south of it.
- Southbound distance vector routing, each upper node generated a default route to the "southern" direction.
- Link state is advertised one-hop southbound and then reflected one-hop northbound. This is when a node detects that default route encompasses prefixes for which one of the other nodes in its level has no possible next-hops in the level below, and it has to disaggregate it to prevent black-holing or suboptimal routing through such nodes.
- Optional zero touch provisioning (ZTP), only top tier nodes need to be configured.
- Packet formats are defined in Thrift [20] models.

Figure 10.5 is a simplified illustration of the RIFT protocol (https://datatracker.ietf.org/meeting/103/materials/slides-103-rtgarea-rift-update).

10.5.2 LSVR

The link state vector routing (LSVR) working group (https://datatracker.ietf.org/meeting/103/materials/slides-103-rtgarea-lsvr-update) at IETF is proposing a new solution which leverages BGP link-state distribution and the Shortest Path First (SPF) algorithm and targets Massively Scaled Data Centers (MSDCs).

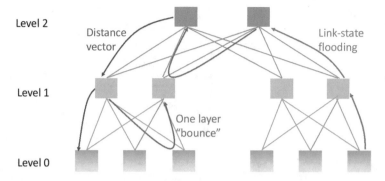

Fig. 10.5 Illustration of RIFT: routing in fat trees

BGP has been chosen as the single routing protocol to simplify routing in MSDCs and their interconnections. While BGP offers operational simplicity and scalability, it lacks some advantages that IGP can provide, such as being a hop-by-hop routing protocol, and it is missing the fabric topology information as IGP. As the size of data center grows, the CLOS tiers increase as well as configuration complexity.

Link state vector routing, also known as BGP-SPF, using BGP as base protocol, and add the best of IGP characteristics, and the following are the main advantages:

- Complete fabric topology at each node, and path computations using SPF, TE, CSPF, LFA, etc.
- Faster convergence compared with classic BGP
- Simplicity—incremental from base BGP, operational and troubleshooting
- Incremental updates, no flooding and selective filtering
- Reliable transport using TCP

The key idea of BGP-SPF is that BGP runs Dijkstra algorithm to calculate best path instead of the BGP Bestpath decision process. BGP-SPF supports various peering models, such as peering in single-hop or route-reflector, as long as all BGP speakers in the BGP-SPF domain can receive link-state NLRI and hence perform consistent distributed route computing. To be backward compatible, BGP-SPF introduces the BGP-LS-SPF SAFI for BGP-LS SPF operation and extends BGP-LS (https://tools.ietf.org/html/rfc7752) with new attribute TLVs.

Figure 10.6 shows the building blocks of LSVR protocol:

10.5.3   SCION

The SCION (Scalability, Control, and Isolation On Next-generation networks) inter-domain network architecture has been designed to address security and scalability issues and provides an alternative to today's BGP. SCION combines a globally distributed public key infrastructure, a way to efficiently derive symmetric keys

Fig. 10.6 Components
of LSVR

Best Path	•Dijkstra SPF algorithm •Routing table calculation
BGP	•Reliable node/link state distribution
BGP-LS Extension	•Router node/link state encoding

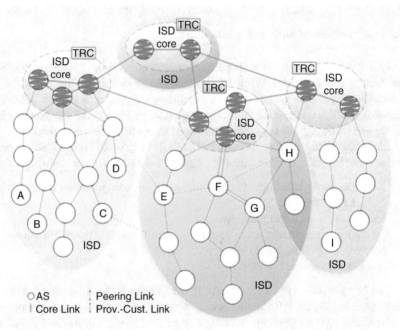

Fig. 10.7 SCION architecture overview

between any network entities, and the forwarding approach of packet-carried forwarding state. Instead of relying on inter-domain lookup tables, the AS-level forwarding path is encoded in the header of the packet. Each router verifies a message
authentication code with a symmetric cryptographic key before forwarding. Figure
10.7 depicts the ISolation Domains (ISDs), which group a set of ASes into independent routing domains. The ISD is governed by a set of core ASes that provide connectivity to other ISDs, a role that is typically held by the largest ISPs that also
provide global connectivity in today's Internet. The Trust Root Configuration (TRC)
of each ISD enables setting of the local roots of trust, in essence the set of public
keys that are used to verify public-key certificates. The partition into ISDs also
enhances the scalability of the SCION routing system, as the beaconing mechanism

creates intra-ISD and inter-ISD path segments among a reduced number of entities compared to BGP.

The SCION Internet architecture provides a fundamentally clean-slate approach to multipath communication: at the control plane, the routing system discovers a variety of AS-level path segments (which can also differ in the interface or links connecting neighboring ASes), which are globally disseminated through a path server infrastructure; at the data plane, cryptographically protected packet-carried state encodes the AS sequence and the AS-to-AS interfaces in the packet header.

End-hosts fetch path segments from the path server infrastructure and construct the exact forwarding route themselves by combining those path segments. The architecture ensures that a variety of combinations among the path segments are feasible, while cryptographic protections prevent unauthorized combinations or path segment alteration. The architecture further enables path validation, providing per-packet verifiable guarantees on the path traversed.

SCION's intrinsic multipath communication provides a natural defense against distributed denial of service (DDoS) attacks. An attacker must congest all paths instead of only one, which increases the needed attack capacity and complicates the attack since access to all paths must be prevented. Further, an AS can choose not to publicly announce some of its path segments at the path servers, but still share them with select communication partners "out of band." The ability to use such "hidden" path segments as part of multipath communication guarantees the existence of a fall-back path that is not publicly known and therefore cannot be clogged through a DDoS attack.

10.6 Conclusions

A set of requirements and several major innovations have been outlined in this chapter for future Internet routing and addressing. Fundamentally, the recurring theme for future routing and addressing is the adherence to a key set of principles which must be preserved and applied to the future architecture of the Internet, and these include:

- Heterogeneity support

  Given existing knowledge of Internet evolution, we must assume the requirement for heterogeneity to be much higher than it is today. Multiple types of devices and applications, network nodes, and protocols will coexist. Hence, the capability to support heterogeneity should remain and potentially be an enforced requirement.

- Massive scale, throughput, and scalability

  With the deployments of massive IoT devices and large-scale data centers, the number of devices needs to be supported by routing protocols that keep increasing. Protocol design and extension have to put scalability into consideration.

New applications and services also add new challenges, such as application specific service requirements in terms of latency and throughput. Routing protocols should have the capability to provide diverse routing services to meet the needs.

- Autonomic networking

  Generally, device and connection management require a series of manual processes and expert knowledge, although more recently the advent of device and service models, along with well-defined APIs, is facilitating the rise of network automation.

  Future services and increasing need for traffic-engineering will require the network to be more flexible and adaptive. As new algorithms and protocols are then proposed, it will also require additional configuration steps and models. Future Internet network devices will need to be adaptive but also more autonomic, with the capability to auto-configure and self-heal.

- Security and resilience

  Internet has changed people's everyday life dramatically, and this also means an ever-increasing dependency of the Internet. To provide reliable and secured services is key requirement for networks. Routing security, as an essential piece of a secured Internet, continues to be a challenge. Network resilience is to maintain an acceptable level of services against failure or damage, and for routing protocols, this means to reconverge fast when failure happens on top of various local repair techniques.

  Fundamentally, future Internet must also stay true to the principles that have made the existing Internet so successful, which include openness and decentralization.

References

1. Intermediate System to Intermediate System intra-domain routing information exchange protocol for use in conjunction with the protocol for providing the connectionless-mode network service (ISO 8473). International Standard 10589: 2002, 2nd ed, 2002
2. Y. Rekhter, T. Li, S. Hares, A border gateway protocol 4 (BGP-4). RFC 4271 (2006), https://www.rfc-editor.org/info/rfc4271
3. *A History of the ARPANET: The First Decade (Report)* (Bolt, Beranek & Newman Inc., Arlington, VA, 1981)
4. Internet Protocol, Version 6 (IPv6) specification, RFC8200, TBC
5. IRTF Information-Centric Networking Research Group (ICNRG)
6. D. Saucez, L. Iannone, O. Bonaventure, D. Farinacci, Designing a deployable internet: the locator/identifier separation protocol. IEEE Internet Computing Magazine 16(6), 14–21 (2012)
7. A. Perrig, P. Szalachowski, R.M. Reischuk, L. Chuat, *SCION: A Secure Internet Architecture* (Springer, New York, 2017)
8. J. Day, *Patterns in Network Architecture: A Return to Fundamentals* (Prentice Hall, Hoboken, 2008)

9. C. Filsfils, S. Previdi, L. Ginsberg, B. Decraene, S. Litkowski, R. Shakir, Segment routing architecture, RFC 8402 (2018), https://www.rfc-editor.org/info/rfc8402
10. D. King, A. Farrel, Z. Chen, An evolution of optical network control: from earth to space, in *22nd International Conference on Transparent Optical Networks (ICTON)*, (ICTON, Bari, 2020), pp. 1–4. https://doi.org/10.1109/ICTON51198.2020.9203098
11. P. Miller, Ka-Band – the future of satellite communication, http://www.tele-satellite.com/TELE-satellite-0709/eng/feature.pdf
12. Giacomo Giuliari, Tobias Klenze, Markus Legner, David Basin, Adrian Perrig and Ankit Singla. Internet backbones in space. In ACM SIGCOMM Computer Communications Review ACM New York, 50(1), 2020.
13. TS 23.501 System architecture for the 5G System (5GS)
14. A. Herzberg, M. Hollick, A. Perrig, Assessment of the effective performance of DPSK vs. OOK in satellite-based optical communications, ICSO 2018. Dagstuhl Rep. **5**(3), 28–40 (2015). https://doi.org/10.4230/DagRep.5.3.28
15. D. Cooper, E. Heilman, K. Brogle, L. Reyzin, S. Goldberg, On the risk of misbehaving RPKI Authorities, in *Proceedings of ACM HotNets-XII*, (ACM, New York, 2013)
16. R. Lychev, S. Goldberg, M. Schapira, Is the juice worth the squeeze? BGP security in partial deployment, in *Proceedings of ACM SIGCOMM* (2013)
17. M. Shand, S. Bryant, A framework for loop-free convergence, RFC 5715 (2010), https://www.rfc-editor.org/info/rfc5715
18. Topology independent fast reroute using segment routing, https://tools.ietf.org/html/
19. C. Clos, A study of non-blocking switching networks. Bell Syst. Tech. J. **32**(2), 406–424 (1953). https://doi.org/10.1002/j.1538-7305.1953.tb01433.x
20. Apache Software Foundation, Thrift interface description language, https://thrift.apache.org/docs/idl

Chapter 11
Quality of Service (QoS)

Toerless Eckert and Stewart Bryant

11.1 Introduction

Current QoS in TCP/IP networks has not seen significant improvement or an increase in its adoption for the last 10–20 years. The authors think that this is to a large extent because of the limitations of the current forwarding plane QoS functionality, which makes it difficult, if not impossible to offer scalable, low-cost QoS service differentiation. For this reason, this section primarily discusses the (high-speed) forwarding plane aspects of QoS and only touches on control, management, and policy plane where deemed necessary.

In this section, QoS is used to refer to the packet level service experience offered to traffic between an ingress and one (unicast) or more (multicast) egress point(s). QoS is also used to refer to the experience received by sequences of packets called flows and includes the aspects of throughput, congestion behavior, loss, reordering, latency, and jitter across the packets of such a flow.

11.2 Current QoS

11.2.1 Fundamentals (IPv4, IPv6, MPLS)

The fundamental QoS experienced by IP is the so-called datagram service. In hop-by-hop forwarding of IP packets, every hop can discard the packet when there are no resources to forward the packet—or for other (error) reasons such as Time to Live (TTL) expiry.

T. Eckert (✉) · S. Bryant
Futurewei Technologies, Santa Clara, CA, USA
e-mail: toerless.eckert@futurewei.com; sb@stewartbryant.com

© The Author(s), under exclusive license to Springer Nature
Switzerland AG 2021
M. Toy (ed.), *Future Networks, Services and Management*,
https://doi.org/10.1007/978-3-030-81961-3_11

309

The IPv4 packet header [1] contains an 8-bit "Type of Service Octet" to indicate to the network the QoS that the packet is requesting. The semantic of this field was originally the one specified in RFC 1349 [2], but was changed in 1998 to indicate a 6-bit "Differentiated Services Code Point" (DSCP) (see RFC 2474 [3]) that is indicating the so-called Per-Hop-Behavior (PHB) of the desired service of the packet and 2-bit of "Early Congestion Notification" (see RFC 3168 [4]). IPv6 [5] uses the same DSCP and ECN semantic for the TOS octet as IPv4.

Each MPLS [6] label stack entry carries a 3-bit traffic class (TC) field (formerly known as the EXP field) that indicates the TC of the packet [7]). This is similar in function to the DSCP in IP, but due to its size of only three bits, it is more limited. As it is included in every label stack entry, it can be different in each entry in the label stack. However, the TC field is typically the same across all label stack entries when using it to indicate a PHB in the sense of the DSCP.

When packet drops are undesirable (as it is in most cases), the datagram service has to be augmented by either a resource reservation mechanism or a congestion control mechanism or both. Even when packet drops can be avoided through such mechanisms, temporary or permanent congestion may lead to increased queuing latency of all packets sharing the same fate along a path, which may also be undesirable.

11.2.2   Best Effort

The most basic end-to-end QoS in IP is called "best-effort" (BE). It is the default QoS in most TCP/IP networks, and it is the only QoS in the Internet: end-to-end transport protocols running on top of IP, such as TCP, SCTP, and RTP, or any congestion sensitive protocol such as QUIC running on top of UDP relies on packet drops in the network to perform "Congestion Control" (CC), reducing their sender rate until no packet drops occur—and increasing it (in that face of available data to send) as long as no packet drops start to occur.

The goals of CC when using BE are to maximize the goodput of the network, not to starve any traffic flows and ideally to even provide some fair share of bandwidth across traffic flows. Goodput is defined as the number of bits per unit of time forwarded to the correct destination, minus any bits lost or retransmitted [8].

There is no commonly agreed upon definition of "fair share" of bandwidth for Internet/BE QoS. In the absence of further policies, the goal of CC designs is typically to provide "equal" share of bandwidth to end-to-end traffic flows competing on a congested interface. As a result, applications and subscribers using more flows in parallel will receive more network resources, which can be very "unfair."

Service providers often implement additional fairness policies for BE traffic, most often without clear documentation, and without any ability for subscribers or applications to discover those policies or to proactively support better fairness.

In one case, congestion points in subscriber access service providers, such as mobile network edge routers connecting the service providers subscriber mobile endpoints to the Internet, bandwidth under congestion could be shared in a weighted fair fashion across subscribers, where the weight of a subscriber's traffic could be decided by the subscription tier. In another case, traffic from applications known to behave "unfairly", for example, by monopolizing bandwidth, may be throttled.

None of these more advanced policies can be built by Internet service providers solely from standardized TCP/IP protocols or rely solely on standardized TCP/IP node requirements. Matching traffic to subscriber requires the matching of IP address(es), but those are the only subscriber-controlled fields allowing the identification of the subscriber. Even when this is used, this mapping information is only available to the direct access service provider. Matching traffic to application type is only possible through "Deep Packet Inspection" (DPI), but this does not work with today's mostly end-to-end encrypted application traffic.

As a result, managing resources for Internet traffic relies primarily on ad hoc methods, often quite convoluted, non-scalable, and non-proactive. In the wake of the Covid-19 pandemic in 2020, for example, several over-the-top (OTT) streaming providers reduced bitrates for their services at the requests of national or transnational governments because the Internet itself is not able to implement policies that prefer more critical traffic over entertainment content: "the European Union has asked Netflix and YouTube to reduce their service from high definition to standard quality" [9].

11.2.3 Congestion Management (CC, AQM, ECN, PCN, L4S)

End-to-end congestion management for TCP/IP networks is an ongoing subject of research, standardization, and deployment challenges and has still many open issues.

One of the most persistent problems is "bufferbloat" (https://www.simula.no/file/pi2conextpdf/download) or severely increased path latency (as much as seconds) due to full queues: it occurs when forwarders only discard packets when the outgoing interfaces queue is full and CC of the transport stacks does not compensate for the problem.

Congestion avoidance (CA) attempts to avoid bufferbloat by targeting low occupancy queues even under maximum load. With today's IP, CA can rely on enhanced CC algorithms in the transport protocol, active queue management (AQM) in forwarders, additional signalling between forwarders and transport stacks (ECN/PCN), or a combination of these mechanisms.

Without any further help from forwarders, CA in the transport protocol can only rely on recognition of increasing latency and reacting with throttling traffic. Likewise, a reduction in path latency can be used as a trigger to increase the rate sent into the network. CC techniques that use delay this way are known as "Delay Variation" (DV) methods. However, when flows that rely on DV-CC share fate with flows relying solely on packet drops CC, the DV flows will not receive a fair share of bandwidth because they will throttle first. Instead, they would have to revert to the less desirable and more "aggressive" packet drop based CC to compete.

Forwarders can reduce bufferbloat through the so-called active queue management. One of the first AQM mechanisms was to probabilistically drop packets even under low queue occupancy with algorithms such as "Random Early Discard" (RED); see RFC 2309 [10]. Note that RED is not recommended to be used by default [11] because it lacks the ability to auto-configure itself to appropriate parameters. Newer AQM mechanism, including "Proportional Integral Controller Enhanced" (PIE) [12], resolves these issues.

Forwarders implementing AQM can avoid discarding packets in support of CA by leveraging "Early Congestion Notification" (ECN) [4]. When a transport stack supports ECN, it sets the "ECN Compatible Transport" (ECT) ECN Bit. When a packet is to be discarded by AQM, and has the ECT bit set, the forwarder instead sets the "Congestion Experienced" (CE) ECN bit and forwards the packet. The transport stack can then use the CE bit to trigger appropriate CA responses, for example, the same response as a packet loss, except that no packet retransmission is required and no time is wasted discovering a packet loss.

Beyond AQM, congestion can proactively be avoided by "Pre-Congestion Notification" (PCN) [13]. Instead of relying on measuring and managing a queue or latency through it, PCN utilizes rate measurements of the traffic sent and raises notifications when the average rate exceeds, for example, some threshold of 90% of the maximum rate permissible on an interface. PCN can therefore reduce queuing latency even more than ECN because it does not have to rely on any queue build-up. PCN can be deployed by utilizing the two ECN bits of the IP header for signalling, which leverage transport stack operations that expects those bits to be triggered by ECN.

Beyond ECN signalling, the amount of queuing under congestion depends on the speed with which the transport protocol signals congestion from receiver to sender, which in turn depends on the setup of ECN signalling and the RTT and rate of flows and paths. The term scalable congestion control characterizes advanced CC algorithms that not only aim to achieve low latency in queuing but also that can scale across different RTT and path/flow rates. For instance, "Data-Center TCP" (DCTCP) [14] averages two congestion signals per round trip whatever the flow rate is.

To effectively allow the introduction of low-latency, scalable congestion control algorithms into the network in the presence of higher-queuing latency causing legacy CC algorithms, the "Low Latency, Low Loss, Scalable Throughput" (L4S) Internet Service [15] proposes an architecture in which both type of traffic share the available bandwidth dynamically, ideally giving each flow the same fair share of bandwidth, whether it is using legacy or scalable CC, and independent of how many legacy or scalable CC flows there are. This is proposed to be achieved via two queues, one for classic and one for scalable traffic and coupling ECN congestion feedback, for example, via extensions to PIE, called Dual PI^2 [16].

In summary, hop-by-hop forwarder support for congestion management via AQM, ECN/PCN, and L4S style composite AQM is crucial to help make congestion-controlled traffic work better, but it still has fundamental issues to solve and limitations caused by IP and the limited QoS options in the IP header.

The most fundamental limitation of IP is the availability of only two bits for ECN, which requires the delivery of information about the congestion level along the path stochastically across a sequence of packets therefore limiting the speed of reaction. Research such as LIVE [17] has shown how this can be solved by providing, for example, eight bits of congestion feedback per packet. Likewise, classical ECN does not provide indication of available bandwidth, thus requiring CC in transport protocols to periodically probe the network to detect an increase in the available bandwidth. For many real-time and interactive applications, fast up-speeding is crucial, such as the need for almost immediate sending with highest possible speed for use in switched conferences. This current CC limitation would most readily be resolved by additional data in the network packet header.

The core issue of CC is the absence of a framework for "fairness." For instance, assume that "each flow gets the same bandwidth" was an appropriate definition of "fair." Initially, "TCP Friendly Rate Control" (TFRC) [18] was written to create such a standard for CC behavior with roughly that definition of fairness as the goal, but that method does not work well across the variety of different CC algorithms nor media and application requirements. For instance, the short- vs. longer-term burstiness of different applications and their transport stack can vary widely and this makes CC between those flows difficult to become "fair." For example, bulk traffic may simply require burstiness at the RTT level due to its flow-control mechanism, whereas real-time traffic such as video streaming via RTP may require burstiness based on its codec behavior happening at the so-called Group of Picture (GoP) intervals [19] that could be in the order of hundreds of msec to seconds independent of RTT.

L4S assumes that there are only three classes of traffic that need to be distinguished: scalable CC traffic, classic CC traffic with classic CC, and traffic without ECN, and L4S proposes to encode these differences into the 2 bit of ECN in the IP TOS field [20] together with the indication of whether or not CC was raised. While the proposal may be the best option under the assumption that the network header cannot be improved/extended, it also shows the extent of complexity and limitations if future networks are assumed to have to operate under this (no header change) limitation.

Real-time media, typically using the "Real-Time Protocol" (RTP) transport protocol [21], does not work well with CC algorithms designed for TCP or L4S. Their requirements against CC are described in [22] and include aspects such as the need for low jitter, the ability to maintain stable congestion control level when a flow can temporarily only send at lower than possible rate, but are also allowed to quickly utilize the maximum available rate for potentially large bursts of data such as the so-called I-frame as well as maintaining longer-term bitrates even under shorter-term bursting competing traffic.

Network-Assisted Dynamic Adaptation (NADA) [23] is one CC algorithm built in support of such real-time media requirements. One of its unique values beyond the requirements is the ability for different flows to automatically allocate different relative bandwidths under congestion. This is called the "weight" of a flow and allows to create better experience-based congestion control, for example, rate-adaptive "same-visual-quality-per-pixel" across video streams of different resolutions by setting the flows weight proportional to their resolution or intended display screen size. These novel benefits are largely unexplored in today's deployments.

11.2.4 Integrated Services

Integrated Services (Intsrv) was the first architecture developed by the IETF in the 1990s to distinguish hop-by-hop processing of traffic requiring differentiation. It defines two services: the Guaranteed Service (GS) [24] and Controlled Load Service [25], with GS being the more important service of the two. GS is often also used interchangeably with the architecture and called "IntServ." GS provides per-flow fixed bandwidth and latency guarantees. It is based on the concept of reserving bandwidth and buffer resources in advance for each flow. These resources are for exclusive use by packets of that flow and are not shared with other flows. To maintain bandwidth guarantees, GS traffic is shaped and policed at the ingress network edge as necessary so that the flow does not consume more resources than have been reserved for it. To support latency guarantees, flows need to be reshaped on every hop. Without shaping, collisions and resource contention between packets could occur, which would lead to the possibility of loss and unpredictable variations in latency. Because of the need for upfront resource reservations, IntServ solutions are also referred to as "admission controlled" (AC).

IntServ is a precursor for the IEEE L2 Time Sensitive Networking (TSN) solution (https://1.ieee802.org/tsn/), and recent IETF Deterministic Network (DetNet) solutions (https://datatracker.ietf.org/wg/detnet/about/), which are described separately below. Both TSN and DetNet are based on different variations of the reservation principle but both support fundamentally the same type of services with respect to bandwidth and latency guarantees.

IntServ with GS has seen little adoption. The cost of per-hop, per-flow processing with weighted fair scheduling as necessary for GS was and is considerably expensive for scalable, high-speed forwarding planes. TSN/DetNet improves on this challenge. More importantly though, applications initially considered to require IntServ guarantees where shown to be sufficiently well served in deployments which did not have explicit reservations, but instead relied solely on sufficient network wide capacity provisioning and scaling of quality under congestion/contention. This is particularly true for the majority of today's Internet traffic—non-latency bound streaming of entertainment content. With the focus of network development to support Internet best effort traffic, this resulted in a neglect of IntServ and GS up until DetNet today.

11.2.5   Differentiated Services

DiffServ [26] was designed shortly after IntServ in the 1990s, when it became apparent that the forwarding and control-plane cost for hop-by-hop IntServ processing was infeasible for most networks and that per-flow bandwidth guaranteed (but not latency guaranteed) service could also be achieved without per-hop/per-flow processing. In DiffServ, per-flow processing is only performed on the edge nodes of each DiffServ domain and on the "interior" nodes, which includes classifying packets of a flow into one of multiple classes by marking it with an appropriate DSCP in the IP header. The interior forwarders of a DiffServ domain then only perform class-based per-hop forwarding QoS based on that DSCP (and potentially the ECN marking of the packet). Per-flow bandwidth guarantees across interior hops of a DiffServ domain only require control-plane level tracking. It does not enforcement/per-flow processing in the forwarding plane as long as the ingress edge nodes ensure that traffic in the forwarding plane is restricted to the bandwidth reserved by the control.

The current framework for DiffServ processing on interior hops in the forwarding plane typically consists of one queue per set of classes. The classes sharing a queue are typically distinguished by a so-called drop-profile, which decides at which level of queue utilization packets for that class have to be discarded, or remarked or trigger CC marking (ECN; see below). Scheduling between the different queues is typically a mix between strict priority for network signalling and real-time traffic classes, and weighted-fair-queuing, or approximations thereof, for other traffic classes.

DiffServ is currently widely used in limited domains [27], also called controlled networks, such as enterprises, manufacturing/industrial, critical infrastructure management such as Road/Rail/River "Transportation," various other Internet of Things (IoT) verticals, as well as defense/military and federations of any such networks. It is not used in the Internet, but it is widely used in Internet service provider core, distribution, and access networks to multiplex different services, where Internet is one such service mapped into the best effort traffic class in hop-by-hop forwarding.

DiffServ's primary issue is the complexity in managing DiffServ configuration, and adjusting it to changing requirements: differentiated drop profiles for classes sharing a single queue are very difficult to optimize, and the scheduling weights for different queues need to adjust to the relative aggregate bandwidth of applications in those classes. Often, these weights may change rather frequently such as within the period of a day, in which case DiffServ is a rather bad fit. In addition, classifying application traffic into classes also becomes more and more difficult with most applications not supporting the setting of the DSCP markings themselves, and network ingress node-based classification being limited by the end-to-end encryption of payloads thereby prohibiting the so-called "Deep Packet Inspection." Most classifications done today are therefore usually based on attempts to identify the DNS name of the server side of a connection. This is difficult when cloud services are used where multiple services (DNS-names) use the same IP address.

For example, classically business critical applications were put into a different traffic-class/queue in enterprises from the traffic-class/queue used by best effort traffic. However, with more and more applications becoming HTTP/Web/cloud-based, this has resulted in the inability of the current forwarding plane framework to dynamically adjust to change weights between the different traffic classes for those traffic classes.

For congestion-controlled traffic, no resource reservations are made in advance, which leads to the possibility of network congestion. This congestion can be mitigated in several ways; the corresponding techniques are referred to as congestion control (CC) and described further below. Congestion control leads to the dynamic adjustment of flow bitrates based on the available bandwidth resources in the network. In the worst case, congestion can lead to loss which CC cannot avoid. In that case, retransmission is normally used for recovery. This becomes an issue for low-latency services, which often cannot afford retransmissions because this would result in the target latency being exceeded. While some applications that require low latency may be able to deal with low probability random packet loss resulting from transmission media bit error rates (BER), very few can cope with typical 50 ms interruptions resulting from equipment or link failures when reactive protection such as Fast ReRoute (FRR) mechanisms are used.

11.2.6   Gaps for Future Networks

The current Internet QoS architecture is insufficient to meet the needs of future networks for a number of reasons, including the following:

- The need for per-flow admission control makes IntServ expensive to support and scale, even if performed out-of-band via SDN.
- The inability to dynamically adjust the bitrate under varying network utilization makes this model too inflexible for both current and for future networks. Originally built for non-IP voice/video applications that required fixed bandwidth and support for constant bit rates (CBR), future applications require support for variable bit rates and elastic bandwidth together with differentiated loss, latency, and relative throughput guarantees. This is not adequately supported by the existing Internet QoS architecture.
- No mechanisms exist to support application-defined upper and lower bounds for the desired latency independent of the path round trip time (RTT).
- There are no mechanisms to slow down packets based on the desired earliest delivery time.
- Queuing cannot prioritize packets based on their desired end-to-end latency.

The same reasons fundamentally apply also to TSN and DetNet, which are described next.

11.2.7   Time-Sensitive Networking (TSN)

The Time-Sensitive Networking (TSN) [28] is a set of updates to the IEEE Ethernet standard that aims to empower standard Ethernet with time synchronization and deterministic network communication capabilities. For the purpose of discussing gaps for latency control, TSN can be best understood as an Ethernet layer 2 variation of the IntServ service, but with two important enhancements:

- With 802.1QCH (cyclic queuing), TSN supports a model for deterministic shaping that does not require per-flow state on transit nodes. It does require strict time synchronization and its throughput deteriorates with increasing network size. With 802.1QCR, TSN also introduces "Asynchronous Traffic Shaping" (ATS) in the style of IETF IntServ to avoid the need for time synchronization.
- With 802.1CB (Frame Replication and Elimination for Reliability—FRER), TSN introduces 1:n (n typically 1) path protection where packets are replicated n+1 times on ingress and sent across failure-disjoint paths and then the replicas are eliminated on egress. This so-called pro-active path protection supports close-to-zero loss in the face of link or equipment (node or linecard) failure. In contrast, reactive mechanisms such as L2 or L3 fast reconvergence or fast-reroute typically require up to 50 ms to patch the failure caused interruption—which is too long for low-latency traffic.

As a collection of layer 2 Ethernet services, TSN aims to provide deterministic service inside a LAN over a short distance and is thus not routing capable. TSN does not aim to provide on-time guaranteed service over large-scale networks and over longer distances. Like IntServ, TSN is geared towards CBR traffic, not VBR traffic, and does not support the slowing down of packets based on the required earliest delivery time.

11.2.8   Deterministic Networking Architecture (DetNet)

The Deterministic Networking Architecture (DetNet) [29] is an architecture that has been proposed by the IETF DetNet Working Group in order to ensure a bounded latency and low data loss rates within a single network domain.

The DetNet architecture intends to provide per-flow service guarantees in terms of (1) the maximum end-to-end latency (called bounded delay in DetNet) and bounded jitter, (2) packet loss ratio, and (3) an upper bound on out-of-order packet delivery. Some options considered in DetNet may in the future also be able to provide bounded delay-variation between packets of a flows. These service guarantees are ensured, thanks to three techniques used by DetNet. The first of these techniques involves resource reservations (to avoid the possibility for resource contention) as well as per-hop reshaping of traffic to avoid accumulation of bursts further downstream. The second technique involves protection against loss caused by random media errors and equipment failures. It is based on the PREOF mechanism (packet replication, elimination, and ordering functions). PREOF is similar to the TSN Frame Replication and Elimination for Reliability (FRER) mechanism and is based on duplicating single flows into multiple flows that traverse disjoint paths, then recombining them and dropping any duplicates near the egress point. The third technique concerns the use of explicit routing to take advantage of engineered paths with specific bandwidth/buffering properties and that are disjoint to other paths required for PREOF.

Although DetNet provides efficient techniques to ensure deterministic latency, scalability remains a challenge. In particular, implementing the DetNet techniques requires the data plane to keep track of per flow state and to implement advanced traffic shaping and packet scheduling schemes at every hop. This is not scalable because core routers can route millions of flows simultaneously. In the control plane, if resource reservation protocol (RSVP) is used, every hop needs to maintain per-flow resource reservation state, which is also not scalable.

The most fundamental limitation of DetNet, similar to IntServ, is in its targeted scope of constant bitrate (CBR) reservations. Future applications may have highly variable bitrates (VBR). Lower latency bounds, as required for on-time services, are also not directly supported in DetNet. While arguable bounded jitter in effect also imposes a lower bound in the case of CBR traffic, the same is not true for VBR. There is still much work to be done in order to design a solution that is both effective in ensuring deterministic latency for all types of traffic (not just at constant bitrates) and scalable to support a large number of simultaneous flows.

Additional limitations apply regarding the combination of PREOF and admission control, as applicable by IntServ as well as DetNet: fully distributed solutions such as distributed Maximum Redundant Trees cannot calculate the optimum paths and do not well work together with admission control. At the same time, centralized solutions have no scalable method to instantiate their desired paths in the network forwarding plane, and existing forwarding plane mechanism based on loose path steering can cause unexpected path traffic under failure.

11.3   Current QoS Scopes

The following sections discuss key aspects of important type of current network ("scopes") with respect to their QoS aspects and highlights evolving trends and gaps.

11.3.1   The Internet vs. Limited Domains

According to Wikipedia (https://en.wikipedia.org/wiki/Internet), "The Internet (or *internet*) is the global system of interconnected *computer networks* that uses the *Internet protocol suite* (TCP/IP) to communicate between networks and devices."

Because the IP protocol is called the "Internet Protocol," this often leads to the confusion that all use of the IP protocol is for the Internet. The desire to best support the Internet has led to the IETF standards body limiting support for QoS to only that which can be operated on the Internet regardless of the expected deployment case.

This expectation is based on the assumption that applications even if not designed for use on the Internet could be deployed such that their traffic intentional or unintentional could pass across (parts of) the Internet. The concern is that this might cause problems for other Internet traffic, if this new traffic is not compatible with the congestion prevention methods used in the Internet. Because the Internet today operates only for congestion controlled traffic, this means that any network/transport functionality is only Internet compatible when it supports congestion control. For example, Section 3.1 of [30], requires applications using UDP to limit their transmission rate. Otherwise such applications require the use of the so-called circuit breakers [31] which stops them sending traffic, when they are losing packets as that could indicate they may be causing Internet congestion control even though they are themselves not operating congestion control.

Whenever IP-based networks require and use additional functions not found on the Internet at large, they are often called "controlled networks" because provisioning and operations beyond what is required for the Internet is required to support such additional functions. For example, resource reservation-based QoS traffic requires such resource reservation mechanisms, which are not found on the Internet. Recently, the term "limited domain" was also phrased for such networks due to [32].

11.3.2 Home, Access, Service Provider

11.3.2.1 Home Networking

Home networks typically "connect to the Internet" via a single service provider and hence they utilize primarily best effort QoS. Nevertheless, they also often carry traffic that is not congestion controlled, for example, in-home streaming of non-congestion-controlled video from cameras, SAT-IP sources SES S.A., British Sky Broadcasting Ltd, Craftworks ApS, SAT>IP Protocol Specification, SES S.A., 2015, (https://www.satip.info/resources/specification/), or even lower bitrate traffic such as from audio or home automation systems. Service provider often also offers these non-congestion-controlled services (IPTV unicast/multicast video, camera streaming to the cloud, etc.). To the subscribers of these services, they look as if they are part of their "Internet" service, but in fact they are merely a service multiplex provided by their service provider of the subscriber but cannot extend beyond the realm of the service provider.

In all cases where non-congestion controlled traffic or other than reliable transport protocol protected services are used within homes, their reliable operation is based on an assumed overprovisioning of bandwidth. Typically, 1 Gbps Ethernet is assumed to be larger than the sum of any resource reservation-based traffic in the home network and loss in the home network is low enough that it does not impair the service experience. Bandwidth and non-congestion loss has therefore been an ongoing issue for these type of services whenever networking technologies other than Gbps Ethernet are used in the home network, especially PowerLAN or Wireless LAN. Because of the complexities of reliable and loss free home networks, service providers typically only "support" subscriber issues in well-defined setups, such as a service provider-controlled home gateway and Ethernet connection to the subscriber device.

The most significant improvements for better QoS in home networks in the past two decades have been improvements in performance of L2 technologies, especially the emergence of Gigabit Ethernet and the ongoing evolution of faster Wireless LAN technologies. The problems described above have led to a faster evolution of "Internet" compatible applications that can adjust their bandwidth based on available bandwidth. These applications operate with sufficiently large playout delays that they can overcome temporary loss or congestion such as commonly found in wireless LANs. Network support for real-time applications including low-latency TV streaming or video conferencing has largely stayed unchanged. Real-time applications including low-latency live event streaming or video conferencing is thus an ongoing challenge.

11.3.2.2 Access Networking

The use of better than best-effort QoS in broadband access network technologies is quite varied, but for residential services is uncommon. Service providers may set up DiffServ QoS for their own services such as (SIP or other) Telephony or IPTV (unicast/multicast), and this may extend all the way through the access network to include prioritization of such traffic on the subscriber access link. There are no established TCP/IP standards for how non-SP provided services traffic should be assigned to different traffic classes, triggered by subscriber demand/applications. Attempts to establish such standards, such as (https://datatracker.ietf.org/doc/draft-boucadair-pcp-sfc-classifier-control/), were not completed.

DOCSIS cable networks provide a rich L2 traffic class-based QoS framework, but no dynamic application triggered mapping to TCP/IP was defined; hence its adoption is commonly limited to service provider provisioned services, such as the aforementioned telephony/video/multicast services.

In digital subscriber loop (DSL), service provider-controlled QoS is sometimes used/provisioned via TR.69 [33] by the service provider onto the home access gateway (HAG), but no standards exist for any such QoS setups when the HAG is customer provided and not TR.69 controlled. Efficient QoS over DSL is further hampered by the fact that the DSL training rates can vary the up- and downstream bitrates available on the link, but those link rates are only available through proprietary methods to a HAG when it has the xDSL modem built-in. Likewise, if LAG/LNS are used to build PPPoE over xDSL, this information is also not always available, making the auto-adjustment of any form of QoS to actually available bandwidth on the DSL connection problematic.

In summary, QoS beyond support for best effort in residential "Internet"-type service offerings is quite sketchy and currently receives little investment, which in turn severely limits the ability to deploy low-latency, low-loss, high-reliability applications. This is even more so when those are not part of an explicit service bundle of the access service provider but instead are provided by a value-added provider (often called "over-the-top providers").

11.3.2.3 Service Provider Networks

In commercial services, such as L2VPN and L3VPN, DiffServ QoS is widely used and offered as part of the service package. As a result, service provider networks transporting such services including core, distribution and access networks themselves are most commonly built on multiservice designs to encompass support for not only best effort Internet service, but also those service-rich commercial offerings. Whereas access networks often utilize L2 Ethernet switching and hence IEEE 802.1p QoS (COS, similar to DiffServ), service provider distribution and core networks most often rely on MPLS [6] and SR-MPLS [34] with an even more limited DiffServ derivation, a 3-bit traffic class field. SRv6 [35] which relies on IPv6 and can hence use standard IP DiffServ is another evolving option in service provider networks.

The actual QoS designs used in service providers vary widely. Because DiffServ has no self-automation, it is cost intensive to set up and operate, and therefore the design principles for DiffServ in service providers are most often to determine the most lightweight approach to offer such services. For example, resource reservation-based services in most service providers add up to less than the minimum path capacities in the network, thereby alleviating the need for explicitly signalled on or off path admission control and allowing the operation such services purely through account system which in turn tracks the service capacities sold and causes the upgrade of capacity whenever needed. This is one of the major benefits of the economy of scale for service providers, where an overwhelming amount of best effort traffic results in very fast networks into which the addition of more advanced QoS service offerings can then be a lot less operationally complex and hence less expensive than it would be if dedicated networks for such services are built. Nevertheless, there is a wide range of open issues both with this current traffic models and when the amount of traffic requiring better than best effort traffic would increase further.

11.3.2.4 Service Provider QoS Gaps

Business class traffic often expects lowest available path latency. In high-speed service provider networks, best effort traffic paths are not optimized only for maximum utilization of the network capacity, often resulting in paths with higher than minimum available latency. Hence other than best effort traffic requires different path policies. This issue is still currently (as of 2021) being actively worked on in IETF routing protocols.

Deterministic services require per-hop regulation of traffic for non-hard calcula-tion of bounded latency goals. This regulation also needs to be per-flow when rely-ing on today's most widely recognized solutions such as IEEE 802.1Qcr Asynchronous Traffic Shaper (TSN-ATS).

TSN-ATS. This per-flow processing is not covered by todays DiffServ designs of service provider networks in which only edge nodes can optionally perform per-flow operations. Novel mechanisms such as [14] would allow the use of DiffServ designs in service provider networks and still provide guaranteed latency. Likewise, protection against non-congestion loss such as Bit-Error-Rate (BER) requires PREOF for which the most cost-effective operational models of diverse path routing have not been well explored.

The desire for mission-specific network services by large customers such as (vir-tual) mobile network operators, industrial uses cases, and infrastructure services for maintenance of power grids, roads/cities and other infrastructures has resulted in the concept of "network slices" whose desires for better QoS services can easily exceed what can be provided today via established L2VPN/L3VPN services. Defining architectures, protocols and operational principles for QoS in such slices is a wide-open field. Today's approach only goes so far as to introduce self-management mod-els for customers to easier instantiate available services in networks, but those too do not support virtualization concepts that extends much beyond what is utilized for L2VPN/L3VPN services.

11.3.3   Enterprise, IoT, Industrial

11.3.3.1   Enterprise

The range of non-SP centric networks utilizing TCP/IP protocols and some form of QoS is wide and it is not possible to describe them in detail in this chapter. We pick a few QoS relevant examples:

Most use of IETF defined IntServ/DiffServ QoS components happens in enter-prise networks ranging from financial institutions such as banks, brokers, and mar-ket exchanges, over Défense, government or public utility over to IT networks of any type of enterprise networks. A variety of applications from financial market data over live voice, video, telemetry, and distributed applications required better than best effort traffic classes support, and this has resulted in the wide use of some sub-set of DiffServ in many of these networks. These networks ultimately drove the need for the QoS support in L2VPN and L3VPN services of service provider net-works that are used by many of these enterprise type networks as WAN services.

The cost of operation of DiffServ, and the need for more bandwidth at lower costs, has in the last decade caused many enterprises to look for more lower-cost WAN offerings than those often premium priced L3VPN services. This has resulted in a higher growth of L2VPN services, and even more so, the growth of the so-called SD-WAN solutions, where only Internet connectivity is used to provide WAN services. This move was always easy when the applications, especially voice and (conferencing) video evolved to be able to operate better on best effort than they did 20 years ago, but it also goes along with an often observed reduction in service quality of such services such as reduction in interactivity because of higher RTT of such services over best effort transport.

To the authors of this chapter, these trends indicate a need to design better than best effort QoS services for service provider infrastructures that can be implemented and operated at scale with price points competitive to SD-WAN solutions, or better yet, in support of SD-WAN solutions.

11.3.3.2 Industrial

Industrial networks still use non-IP networks in many mission critical cases such as industrial control loops controlled by programmable logic controllers (PLC). The need for higher flexibility in industrial operations is leading to more and more move towards Ethernet and TCP/IP though, raising the need for the deterministic QoS services offering high reliability, predictable bounded latency, and low loss. Pre-IP network were typically synchronous and so were control loops. When moving to IP-based networks, the range between the best- and worst-case end-to-end latency (jitter) becomes a new challenge for these applications. Deterministic networking solutions that can provide low jitter can allow a network to provide a QoS service experience that mimics most closely that of those pre-IP networks.

Internet-of-Things

The first wave of the so-called IoT services was built around the attractive new business model of deploying variety of 'things" especially sensors/data collection on customer premises where connectivity to the Internet is available and then controlling them from a "cloud" location to provide a service. Preventive maintenance of equipment, monitoring, home control, and any other form of remote control are typical examples of this. All the successful instances of such services were built around requirements that could be met with best effort Internet service: no need for high reliability, controlled latency, or low packet loss. Surveillance cameras are likely the most widely used type of IoT appliances that best utilize the prime benefit of the Internet and its QoS model: best effort at ever-decreasing cost/bandwidth but otherwise no guarantees. As a result, cloud-based surveillance system is also an ever more successful type of IoT service.

Whenever IoT use cases require better than such best effort QoS, they currently do not get it unless they are deployed in networks solely designed for such QoS services, which can be a significant design/operations cost factor. Most of the critical IoT use cases can be subsumed under the variety of industrial networking use cases that are built using "things," but also new use cases are only evolving now.

A prime example of an evolving use case is vehicle networking. In-vehicle networking, especially for autonomous vehicles, transfers gigabits of telemetry, especially video. This needs to be processed at highest speeds for fast reaction. Real-time, low-latency human remote control of vehicles extends this model across metropolitan networks and could easily become one of the most sought-after low-est-possible network latency use cases, although this depends on the regulation of legal responsibility which often still outlaws fully autonomous cars. Networks supporting safety of life and other safety critical use cases would need to be built on much higher resilience and guaranteed latency control than today's approaches.

11.3.4   Mobile Networks

This section summarizes the evolution of mobile networks from 4G/LTE to 5G and puts it in perspective to transport network QoS.

11.3.4.1   LTE ("4G") Network Layer QoS

Figure 11.1 shows the LTE architecture and is included to provide the context for the QoS discussion. The QoS and user profile of the user equipment (UE) is enforced in the service end points E-UTRAN, and serving gateway/PDN gateway. Though various bearer types with different QoS requirements were defined in LTE specifications, only default bearer and voice bearer are normally deployed. The transport network is between E-UTRAN and S-Gateway on the S1-U interface and between S-Gateway and PDN Gateway on the S5 interface. Here, user data traffic is encapsulated in a GTP-U overlay which is normally carried using IP and MPLS as undelay technologies.

The default bearer is used to carry data traffic for Internet access and would be treated in a best effort manner in the transport network. To provide the QoS for the UE packets at E-UTRAN in the down link direction and S/P-GW in uplink direction, IP packet DSCP fields are copied into the outer IP header after GTP-U encapsulation. If MPLS is used in the transport network, then IP DSCP to MPLS TC bit mapping is done. The need for QoS in the transport network itself is primitive and basic prioritization of the voice packets is generally deployed to mitigate the congestion in the transport network.

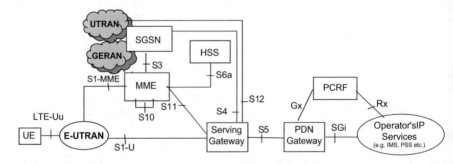

Fig. 11.1 LTE architecture (from [36])

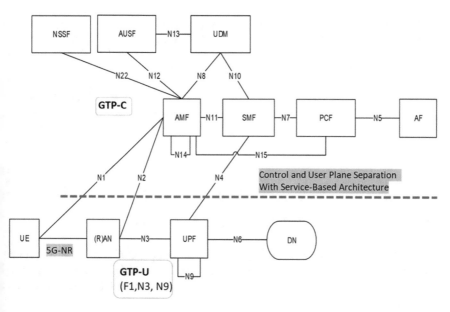

Fig. 11.2 5G architecture (from [37])

11.3.4.2 5G Network QoS Analysis and New Requirements

Figure 11.2 describes the system level architecture of the 5G network and will be discussed in the context of QOS.

While there are lot of changes in the (R)AN and the 5G control plane from LTE, significant difference with respect to QoS from the service level perspective is required because of the inclusion of slicing in 5G networks. An end-to-end slice is defined as a slice in the (R)AN, transport network, and core network. To provide the slice QoS, resiliency and hard separation characteristics, the characteristics of the transport network, need to be taken into account and tailored to meet the required service level.

One of the transport networks can be defined between (R)AN and UPF on the N3 interface. This is conceptually similar to the S1-U and S5 interface in LTE. Another transport network is the N9 interface, i.e., the network between multiple UPFs. N9 is a new architectural interface and has been designed to address multiple requirements in 5G, for example, URLLC, session offloading, and the new session and service continuity (SSC) modes as defined in [37]. The QoS characteristics for each slice need to be provisioned, applied, and honored for the data traffic passing in the transport network in the respective segments whether on the N3 or N9 interface. The QFI parameter in the GTP header describes the PDU session QoS requirements. Based on the SST in S-NSSAI in the 5G control plane, the QFI value would be set in the 5G service nodes, i.e., (R)AN and UPF in uplink and downlink directions, respectively.

An example mapping for QFI, SST, and transport path is shown below:

GTP/UDP SRC PORT (mapped to 5G dynamic QFI)	SST in S-NSSAI	Transport Path Info	Transport Path Characteristics
Range Xx - Xy X1, X2 (discrete Values)	MIOT (massive IOT)	PW-ID/VPN Info, TE PATH-X	GBR (Guaranteed Bit Rate) Bandwidth: Bx Delay: Dx
Range Yx - Yy Y1, Y2 (discrete Values)	URLLC (ultra-low latency)	PW-ID/VPN Info, TE PATH-Y	GBR with Delay Requirements Bandwidth: By Delay: Dy Jitter: Jy
Range Zx - Zy Z1, Z2 (discrete Values)	EMBB (Broadband)	PW-ID/VPN Info, TE PATH-Z	Non-GBR Bandwidth: Bx

Fig. 11.3 Mapping table for 5G slices to underlying transport paths

The delay and jitter defined in Fig. 11.3 are part of the QoS profile for that slice. The QoS state required is per traffic engineered path. The UE PDU sessions need to be mapped to the transport providing the traffic engineered path based on slice specific criteria of these QoS paths.

A difficulty in addressing the QoS requirements of 5G network slices (5GS) is that some of these requirements belong purely to the transport domain and are not governed by 3GPP. Transport aware mobility for 5G [38] discusses how a standardized mapping from the 3GPP domain to the transport domain can be done and the gaps in the available technologies from transport side with respect to QoS.

11.3.4.3 Beyond 5G (B5G) QoS Requirements

There have been some initial discussions in various forums regarding mobile network services "Beyond 5G" (B5G). Transport network characteristics for these networks need to be understood and be considered upfront for mission critical applications requiring future network support.

11.3.4.4 Mapping 5G/B5G to the Underlying Future Networks Infrastructure

In our reference case, the 5G or B5G network is mapped onto the future network as its transport underlay solely as an "overlay" network, where all 5G/B5G control-plane and user-plane functions are run in edge-data centers as virtual machines, containers, or lambda (https://en.wikipedia.org/wiki/Lambda). Even if some functions still require specialized hardware, such as NPU processing, they would still likely be positioned solely in edge-DC.

Figure 11.4 outlines how the 5G/B5G functions in fronthaul and backhaul would then integrate into future networks. The Future Networks Edgehaul access network

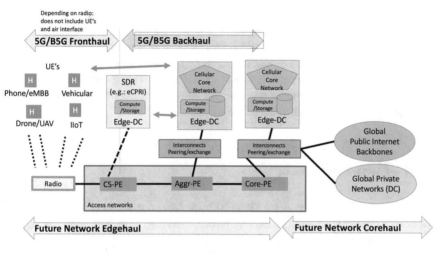

Fig. 11.4 5G/B5G fronthaul and backhaul

serves as the 5G/B5G transport network representing N3, N9, and N6 in the 5G architecture. When 5G/B5G functional blocks are distributed across multiple edge-DCs in the Future Networks Edgehaul, other Nx could also run across the Edgehaul; otherwise they would solely run between compute units providing VM/container/lambda to the 5G/B5G solution. In Sect. 11.3.4, we discuss the two key areas of future networks, Edgehaul and Corehaul in more detail. The current assumption is that Nx interfaces that pass through the Edgehaul would need to be provided with the QoS services described in this document. Nx interfaces that solely pass within the same edge-DC may be considered to always have negligible latency and no congestion relevant to the service provided. These assumptions may not hold if, for example, deterministic services or high-precision services are required across 5G/B5G, in which case burst collisions even within a data center (DC) could be detrimental to the required service, and QoS services would also need to extend into the edge-DC internal networks.

The use of slices in 5G/B5G can be independent or coupled with similar isolation mechanisms in the Future Networks Edgehaul. In addition to the 5G/B5G case, there is also the need to support softwareized radios in which the RRUs are connected to the baseband central processing unit over a eCPRI connection [39] and future techniques that evolve from this approach. It is highly desirable to enable the support of this functionality, but like other (software) services, it has strict latency and jitter requirements such as a one-way path delay of less than 25 us. With the evolution of access network switching speeds from 100 Gbps to beyond 1Tbps or more, the latency of the actual network equipment is unlikely to be a key impediment to this goal, but the speed of light will likely limit the access to at most one or two active switching components between the RRU device and the compute element. Hence, the above picture shows this option as one requiring a compute component that logically needs to be closest to the subscriber/radio edge which is considered to be part of the 5G/B5G fronthaul.

The specifics of the required QoS service attributes for eCPRI (or successor) traffic are the subject of further studies, but it seems clear that even a single switch that is connecting multiple radios with a single compute node would potentially have to deal with the problem that traffic arriving from those multiple radios (each from a different interface) could create undesirable FIFO burst collision delay when queuing towards the compute node and that fully synchronous solutions are likely to unacceptably increase the cost of the solution.

11.4   Future Networks

This section abstracts the networking infrastructure for Internet and private networks towards the conceptual building blocks described here. These will be used as references in the chapter for the feasible/required functionality.

QoS for future networks as discussed in this chapter is based on leveraging the reality of the evolution of the Internet architecture, such as the "The Death of Transit" (https://hknog.net/wp-content/uploads/2018/03/01_GeoffHuston_TheDeath_of_Transit_and_Beyond.pdf).

In its original form, as shown in Fig. 11.5, the Internet service is concerned with traffic between subscribers and servers that are interconnected by end-to-end network layer transit paths. In these paths, traffic is passed through the networks without any business relationship between the network and the subscriber or server. This is one of the core reasons why in the traditional Internet service model, only "best effort" (BE) traffic is supported.

For future networks, which we envision will to be structured as shown in Fig. 11.6, QoS evolution is primarily important for the subscriber edge where latency and services better than best effort will be required by applications. This will be in the region marked "Future Network Edgehaul" and reaches up to the edge compute/data centers. A metropolitan region is a typical instance of an Edgehaul.

Focusing on this part of the network also allows a reduction in the business and architectural complexity of providing differentiated QoS offering, because it can eliminate pure-transit network issues.

Corehaul networks will have specific QoS requirements/opportunities, but to the extent that these go beyond a subset of those QoS functions required in the Edgehaul, these will separately be considered for such specific type of Corehaul networks.

Fig. 11.5 Traditional Internet service model: worldwide end-to-end network paths with transit

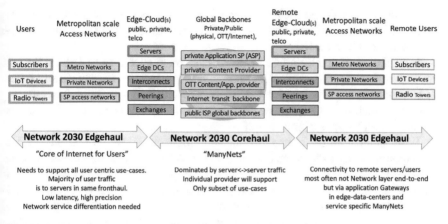

Fig. 11.6 Expected evolution of future network architecture

11.4.1 Edgehaul and Corehaul

Physical network infrastructure in future networks can roughly be divided into a "Edgehaul" and a "Corehaul." Edgehaul and Corehaul are interconnected by edge-data centers, exchanges/interconnects, and private peerings. This is called the Edgehaul/Corehaul edge. In the case of a classical Internet service provider, the Edgehaul/Corehaul edge could be the central office (CO) as long as these are sufficiently close to the network subscribers so as to permit latency constrained services with required RTT. For example, this is likely the case if the data centers are co-located in a metropolitan area with the subscribers they serve.

The Edgehaul of the network infrastructure consists of metropolitan size physical networks owned/operated by classical Internet/network service providers, application service providers (ASP, e.g., Facebook, Apple, Amazon, Netflix, Google (FAANG)), cities and other public operators, and other private networks (such as large manufacturers, transportation companies, and the like). These networks physically connect a set of users and/or (IoT) devices among each other via wired/wireless access and towards the Edgehaul/Corehaul edge.

The Corehaul of the network infrastructure consists both of the multi-AS hop "classical Internet" and a variety of private networks owned by a variety of institutions, network service providers, application service providers, public operators, and more specialized network operators.

11.4.2 QoS in the Edgehaul

The majority of application traffic flows that involve subscribers will stay within the Edgehaul because it is the part of the network connected to the subscribers. This includes, for example, consumer entertainment traffic from Edge-DC to consumers, and where it is desirable for lower latency also directly between subscribers, for example, with interactive virtual reality (VR), augmented reality (AR), or holography between subscribers.

New classes of applications such as Car2X communications, and the still very much evolving machine-to-machine communications in industrial solutions or other command and control within the city, will evolve with their own traffic flow characteristics, which may be more or less centered in the Edge-DC than the current widely deployed types of application.

While content/traffic for applications will ultimately extend far beyond a single metropolitan area, it should be expected that it will not flow end-to-end at the network layer, but instead it will be segmented at the application level at Edge-DC. This trend is already very strong in today's evolution of applications via distributed cloud-based application instances. One of the key reasons for this is that to an increasing extent the Corehaul network infrastructures is privately owned by ASPs, and access and resource management to their Corehaul is managed and only possible to applications running on those ASPs Edge-DC.

Included in these Edge-DC applications are virtual overlay network services that link and interconnect network layer access in the Edgehaul with Corehaul tunneling/transport of network layer traffic across private Corehauls. Today this is most often part of solutions called software-defined wide area networks (SD-WAN). Likewise, 4G/5G core networks can be considered to be intra-internet-Edgehaul overlay applications consisting of 4G/5G user and control planes typically implemented in VMs running on Edge-DC systems, and radio towers acting as another type of subscriber to the Internet Edgehaul.

11.4.3   QoS for the Corehaul

Many Corehaul networks will be built around the needs of specific use cases only, so they will not be necessarily general purpose. Many of latency and resilience-related service aspects will differ widely based on the use cases against which the Corehaul network is design. Corehaul networks may even have even larger more complex requirements such as those of low Earth orbit and geostationary satellite networks, or networks providing on-demand capacity. As a result of these considerations, this section does not address specific Corehaul QoS considerations. Instead, Corehauls could adopt a subset of the QoS functions derived from the Edgehaul considerations described here.

Note that Corehaul networks may reach all the way to subscribers, such as is planned for LEO satellite networks with direct subscriber terminals. How such "direct-to-subscriber" Corehaul networks integrate into the geographic Edgehaul network of the region where the subscriber is located is subject to the QoS design of these specific Corehaul networks.

11.4.4   Benefits

The key benefits and simplifications of this expected evolution of network services to QoS are as follows:

11.4.4.1   Simplified QoS on Paths Across the Edgehaul

Within the Edgehaul, the physical (speed of light) caused latency is low enough to allow traffic flows between any two points with very low RTT latency, for example, 7 ms including switching latency in equipment. This allows support for many foreseeable applications such as those described in the ITU-T Focus Group on Technologies for Network 2030, Representative use cases and key network requirements for Network 2030, ITU-T, 2020, http://handle.itu.int/11.1002/pub/815125f5-en.

Within the Edgehaul, the number of operators involved in end-to-end paths will in most cases be limited to one provider for each endpoint of a network layer traffic flow, directly connected only via exchanges/peerings—whether an endpoint is a server in an Edge-DC, a wired/wireless user or (IoT) device. This is an important shift and simplification for QoS from today's traditional Internet paths, where traffic typically passes not only through those two "endpoint access providers," but also through one or more additional "transit service providers" without any explicit business relationship to either of the endpoints.

To a large extent, the lack of support for better than best effort QoS in the Internet is caused by the inability to develop working business and technical solutions to support such QoS across such multi-AS (autonomous system) paths. In the Future Network Edgehaul, the simplification of these paths should enable the easier design of appropriate technical and business models to support the variety of QoS services desired, for example, by FGNET 2030 documents (https://www.itu.int/pub/T-FG-NET2030-2020-SUB.G1; https://www.itu.int/pub/T-FG-NET2030-2020-1).

11.4.4.2   Flat network QoS Design in Edgehaul ("Hop-by-Hop PE")

For the purpose of this document, we consider that Edgehaul networks should support the required per-hop QoS functions on every hop. This is different from the current practice in provider Core (P) and provider edge (PE) designs where QoS and other functions are organized hierarchically, in which the PE nodes and external out-of-band systems take on the responsibility for QoS and resource management is done such that no QoS service "impacting" congestion/contention can happen on P nodes. The existing hierarchical approach is used, because it is cheaper and easier to scale the P nodes.

"Impacting" in the previous paragraph does not mean that there cannot be any congestion/contention. For example, best effort traffic may still suffer delay and congestion on P nodes under higher loads when competing with guaranteed bandwidth services, but there are, for example, no expectations set regarding guaranteed service latency management on P nodes. For example, best effort traffic may cause delay on P nodes under higher loads when competing with guaranteed bandwidth services, as there are no expectations set regarding guaranteed service latency management on P nodes.

The reasons for not expecting P/PE differentiation in the Edgehaul is as follows:

1. Edgehaul networks should be able to support arbitrary, cost optimized topologies. Extrapolating from the past, this means that it could be complex topologies of subtended rings, which are the lowest cost (capital expenditure (CAPEX)) redundant topologies and are based on opportunistically available fiber trails. In these topologies, the probability of a node acting as a PE node is very high. In this case the benefit of optimizing the network architecture to support the reduced functionality of P nodes may therefore be insignificant compared to the additional system complexity that P node support may introduce to the overall system design.
2. The P/PE distinction is an optimization that evolved at least 10 years after the required services were understood and deployed in flat topologies. As of today, mechanisms to support P node equivalents of all future QoS services discusses are still evolving research topics and hence may take longer to become available.

11.5   New QoS Services

11.5.1   Experience-Based Resource Management

The evolution of widely adopted audio and video solutions over the last 30 years has shown that media can be made elastic, e.g., it can adjust to changes in available network bandwidth. For RTP real-time communications, this reaches as far back as [40] from 1996. In a simple model, each media flow may have a minimum acceptable bandwidth (resulting in minimum acceptable quality of experience) and a maximum desirable bandwidth (resulting best experience quality). This is not only true for today's media but also can safely be assumed to be true for at least some of the future media such as holography.

As of today, there is no standardized model for how these expectations should map to the allocation of network resources. When best effort flows compete through congestion control in the Internet, all flows are expected to roughly utilize the same amount of bandwidth. Under peak utilization, video streams with the lowest bandwidth for the best quality experience, such as small tablet displays, will get the best quality, while the most expensive display devices (requiring higher bandwidth for the same experience quality) suffer most. Worse still, traffic flows with arbitrary bandwidth requirements such as downloads will consume random, high amount of bandwidth, reducing the quality experience for all, more throughput critical applications.

Multimedia applications [41, 42] were, from an early stage, inherently multiuser and often operated in an environment where various participants are located on systems and communication links with different capacities and resource capabilities. Therefore, mechanisms were proposed to ensure appropriate quality media for different users. In order to achieve this, QoS filters were proposed as a way to adapt QoS to the user-specified level by changing the structure of a media stream in a well-defined way [41]. These filters are located along the data path (in contrast to the adaptation that happens at the server side as is the case in Dynamic Adaptive Streaming over HTTP (DASH) (https://www.broadband-forum.org/download/TR-069_Amendment-2.pdf)). Another advantage is that this supports one-to-many communication. Using QoS filtering in conjunction with other QoS provisioning allows for an integrated and optimized quality of experience (QoE) for individual users while optimizing communication and system resources [42].

For network 2030, it was considered important to investigate better support for elastic resource management. While there have been early architecture proposals for dynamic QoS at a comprehensive architecture level, as far back as the 1990s, e.g., [40], this has not been designed into current networks. Nevertheless, core mechanisms such as per-flow weighted congestion control schemes (e.g., NADA [23]) or congestion-based bandwidth reservation adjustments (RSVP Multi-TSPEC [52]) are recommended starting points at the lowest protocol levels.

The main challenge is the creation and deployment of appropriate policy frameworks, where experience quality and not simply absolute bandwidth becomes accepted factors in resource allocation, especially under congestion/contention for resources.

11.5.2   Lightweight, Scalable In-Network Resource Guarantees

The more complex the network, the more complex the resource reservation for bandwidth and even more so latency becomes at least with existing network technologies.

Off-path reservations as described above suffer from the problem of correctness in the face of complex dynamic path selection. SDN coupling has recently attempted to overcome this issue, but this results in very complex and fragile, tightly coupled systems. Nevertheless, this is the only currently feasible option in the absence of innovation for on-path resource management. It was therefore important for network 2030 to consider innovation in this area.

On-path bandwidth reservations such as via the RSVP protocol suffer the problem of scalability through per-flow control-plane state operations, and the proposed successor to RSVP (NSIS), which was developed around 20 years ago at the time of writing, made the overhead of these control plane operations even worse through even more complexity.

Whereas forwarding plane performances grew by factors of 10,000 or more in the last two decades, the performance of the control plane barely rose a factor 10 or

100 in the same period. This means that on-path resource reservation via traditional approaches such as RSVP, NSIS, or similar evolving protocols in IEEE can only be adopted by investing into significantly faster control plane performance.

An even better solution is to design new, on-path resource reservation protocols that are lightweight enough to be processed not by the control plane, but the actual (hardware accelerated) forwarding plane in 2030 network devices. Prototypes of such approaches, for example, with TCP, exist and are documented, for example, draft-han-6man-in-band-signalling-for-transport-qos.

Any form of reservations of bandwidth resources for network 2030 should support the handling of not only fixed reservations but also those of elastic media as described in the previous subsection, by, for example, combining the previously mentioned approaches.

Whereas bandwidth reservations "only" require accounting of per-hop/per-flow allocated bandwidth, guarantee of maximum end-to-end latency requires both bandwidth reservations *and* per-hop per-flow state with today's widely accepted mechanism such as in IETF IntServ Guaranteed Services, TSN, or currently envisioned DetNet mechanisms. Note that per-path aggregation of flows is possible to increase scalability.

This per-flow state, whose complexity may range from a per-flow shaper to the use of per-flow interleaved regulators, is technically feasible in some small networks, however, support is unlikely to scale even to the size and scale of flows required in metropolitan aggregation networks, where latency control can be critical with future network 2030 applications.

Solutions to provide better aggregated per-hop traffic shaping are being researched and are promising. An example of this is cyclic queuing for IP networks as described in [43]. This has been combined with the use of in-band signalling to provide both bandwidth and latency guarantees.

11.5.3   High Precision QoS

11.5.3.1   Service Level Objective-Based QoS

"Latency-based forwarding" (LBF, [28]) researches a new paradigm for QoS in the data plane. It carries the end-to-end Service Level Objectives (SLO) for latency as parameters in the network packet header to enable per-packet stateless latency SLO-based forwarding. Packets indicate their minimum and maximum desired end-to-end latency. The forwarding plane also tracks how much latency a packet incurred during forwarding and can therefore on every hop provide high-precision differentiation on forwarding of packets with different latency SLO. This enables a wide range of benefits. Packets delayed in queuing on a prior hop will automatically be given less latency when competing against packets with less queuing delay incurred on prior hops. CC unfairness for flows with different path RTT can be reduced, and playout buffering in receivers can be reduced and more. This behavior does not require resource reservation and is meant to support lightweight fine-grained differentiation of traffic. By being per-flow stateless, it is also to support scaling to Future Networks Edgehaul dimensions. The integration with resource reservation mechanisms to support harder guarantees is currently being researched.

The circa 20-year-old research into "Dynamic Packet State" (DPS) [44] is based on the same goals of per-packet stateless forwarding and fine-grained differentiation of QoS service experience. In the case of DPS, a relative-throughput weight parameter is carried in every packet to direct CC on every hop to provide weighted CC feedback with the result that traffic flows indicating different weights in their packets will receive relative weighted bandwidths when competing with each other. This is the same result that per-flow weighted fair queuing (WFQ) provides, but because of its per-flow state and the need to provision and manage its per-flow parameter on every congestion hop, WFQ is not appropriate for larger-scale network deployment.

This concept of expressing more fine-grained than per-DiffServ-class and more scalable than per-flow-based QoS forwarding mechanism that ideally also require no or minimum policy administration should in the opinion of the authors be explored more in support of future network QoS.

11.5.3.2 Fine-Grained, Path Aware Latency Management

The previous sections summarized the recommended directions for the QoS architecture for the following gaps to which solutions are already evolving:

1. Per-flow differentiated, non-reserved but congestion-controlled bandwidth management, for example, by supporting differentiated (weighted) bandwidths per flow.
2. Simplifying and scaling bandwidth admission control, by moving it to the high-performance/scalable forwarding plane.
3. Scaling guaranteed maximum (end-to-end) latency through forwarding plane mechanisms with less than per-flow complexity.

What these points do not cover is the differentiation of traffic in the network in a more fine-grained fashion by its latency requirements. These latency aspects are investigated in the [2] document by considering the requirements from [1], especially support for in-time vs. on-time latency management as part of high-precision communications and coordinated communications.

The majority of network 2030 applications will operate elastically without explicit resource reservations, if the experience of the last 20 years is good indicator of future trends. The strong resource-reservation-based approaches with fixed bandwidth reservations in IntServ/TSN/DetNet are not required for these, and therefore their guaranteed maximum bandwidth guarantee mechanisms are also not applicable as that depends on known reserved bandwidths. Nevertheless, by 2030 a growing amount of traffic will require lower and often also differentiated latency.

The first steps towards addressing these needs are the efforts in the last decade to reduce "bufferbloat" in TCP congestion control to minimize best effort traffic latency, together with the evolution of "low-latency" transport protocols such as DCTCP [14] for lower-than-best-effort latency. More recently, in the past few years, proposals have been developed for mechanisms that also allow these different types of traffic to coexist without per-flow-forwarding plane state (e.g., "PI^2: A Linearized AQM for both Classic and Scalable TCP" [25]). This allows operators to build networks with, e.g., both TCP and DCTP without bandwidth reservation for the DCTP traffic.

Whereas mechanisms such as PI^2 are only able to improve the management of latency classes of traffic (e.g., TCP/DCTCP) under congestion, explicit management of end-to-end latency in the per-hop forwarding without per-flow state has the potential to deliver even finer-grained latency differentiation benefits:

The differential latency of paths is not compensated for by the network, which would lead to differences in congestion control managed throughput, problems with reordering, and endpoint buffering in multi-participant applications ("coordinated communications") and multi-path flows (MPTCP or dual-path resilience). Congestion-induced latency is not compensated for later on in the path, and in paths with multiple congestion hops (such as metropolitan aggregation ring networks), differential latency between packet statistically increases (lucky packet vs. "biggest looser" packets experiencing worst congestion on multiple hops).

Absolute min/max desired end-to-end latency Service Level Objectives as defined in FGNET 2030 SubG2 cannot be specified with QoS existing mechanisms and therefore also cannot be used to deal with the path issues described.

Recent research has proposed per-packet forwarding mechanisms to support the FGNET 2030 High-Precision Communications requirements. See also "High-Precision Latency Forwarding over Packet-Programmable Networks" [45] (abbreviated LBF).

11.5.3.3 Resilience and Near-Zero Loss Forwarding

Today's networks offer protection against packet loss primarily via two mechanisms, one at the link layer and the other at the network layer. In each case, the target QoS is maintained using proactive recovery (resilience) techniques.

At the link layer, proactive redundancy such as forward error correction (FEC) is used against link bit errors such as in ADSL/VDSL, directed radio links, or 100 Gbps Ethernet and beyond to achieve a desired low level of lost packets (typically $<10^{-12}$ or lower). On radio links including 5G/B5G, WiFi, or directed radio links, reactive redundancy such as retransmission is used to overcome less predictable loss such as temporary radio path impairment. This typically leads to negligible loss in most fiber-optic links, but often leads to an increase in latency and temporary throughput for other, especially radio links.

While it is possible to expose worse than perfect links to the network layer and take those link properties into account for the path selection of different types of traffic, this is not provided as a part of the services in today's networks, although the introduction of segment routing may enable this. Whether this is relevant in 2030 networks depends primarily on how prevalent non-perfect links, such as microwave connections, will be in the relevant 2030 networks. With the increased use of fiber-optic links, this may not be a relevant issue. However, there is also a trend for more transit links using radio technologies (not only for mobile access), and those links would be much more useable if the end-to-end network service QoS would support distinguishing routing or even just retransmission across them in order to meet or exceed the required loss characteristics. For example, TCP best effort traffic (without latency requirements) is able to deal with sub-percent packet loss and therefore leverage such non-perfect links much better than traffic with higher QoS requirements (primary lower latency).

At the network layer, today's approach to component failure and recovery (link, interface, linecard, node) is at best via reactive rerouting, which typically achieves in the order of <50 ms interruption and recovery through technologies typically called Fast ReRoute (FRR). This level of recovery was recognized to be detectable in voice transmission over TDM, but was also shown to be indistinguishable from even longer outages such as <1 s interruptions for real-time streaming of video with Group of Picture (GOP) sizes of 1 s. This is because, with a significant probability, a single packet loss can invalidate a complete GOP. As a result, one of the main design criteria in networks for real-time services is not primarily to minimize the time of loss and recovery but to minimize their occurrence through the choice of reliable components, internal redundancies in components, and resilient make-before-break network operation procedures. Often interruptions for example are caused by break-before-make reconfigurations.

To support at the network layer less than this sub 50ms loss without the addition of latency through retransmission or FEC, it is necessary to transmit data multiple times across network paths without common failure points. This is called path diversity and the approach of sending traffic multiple times is called, for example, live-live or seamless protection switching as in broadcast video solutions using SMPTE 2022-7 [46]. The basic principle is to send each packet twice across diverse paths and eliminate the duplicate packets (when there is no loss) based on sequence numbers.

While such live-live services exist today in a variety of private network or private network services (broadcast video industries, financial industries), there is no standardized framework/protocol/signalling to request such a service experience over two access interfaces into a network, and there are no easy to deploy and widely available routing solutions to support this zero-loss solution. For example, Maximum Redundant Trees (MRT) [47] is one available IETF standard that can support this service from the routing perspective, but its main goal was not to enable live-live service but instead just the sub 50 ms FRR. For that solution a wide variety of alternatives exists, so the key unique benefit of the MRT solution to enable live-live services was not widely recognized. Nevertheless, MRT being distributed, its results for path latency are not as good as central PCE controller calculated live-live path sets.

In summary, one key recommendation for (near) zero-loss QoS in 2030 networks is to build a comprehensive resilience architecture that enables turnkey use of multipath redundancy for critical, low-latency applications, alongside traditional link layer methods such as FEC at the link layer (see also Section 2.8 on Resilience in Principles, and Sect. 2.4 on Assuring QoS and Resilience in Management [25]).

11.6 A New QoS Toolkit

The introduction of advanced in-network functions is challenging because of the dependency between the customer need, the operator finding a path to monetization of their investment, and the vendor financing the hardware development cost.

Over the past two decades, these interdependencies were resolved when the customer was the operator, so that only two entities where involved: the network + application owner/operator, and the equipment vendor. Similarly, application providers have developed or sponsored the development of equipment with the required functionality and turned themselves into network operators. The fewer parties involved, the more likely network functionality is developed and deployed. Softwareization through VNF/NFV dramatically improves this situation, as can be seen with the significant deployment of overlay, VPN, SD-WAN, and CORD (central offices of service) network services that emerged in the last decade. The following subsections describe the key areas of dependencies and proposals for solutions.

11.6.1 Programmable Virtual Networks

Programmable virtual networks are a key technology that allows future network application owners and operators to deliver their required end-to-end solution without being dependent on physical network operators or equipment vendors. Programmability means that the required functionality can be delivered virtually through the use of a common, cost-effective underlying physical network infrastructure.

The initial deployment of this technology can be seen in the softwareized overlay network solutions prevalent in SD-WAN and is likely to be deployed in future metropolitan size networks where distributed edge-data centers are able to host the VNF/NFV forwarding planes of such application-specific virtualized networks.

When we consider QoS, with the need to control latency and throughput, however, the data center-based softwarization approach is unlikely to bring the required benefits particularly at the multi-Tbps forwarding rates needed to form the physical infrastructure of a metropolitan area network (MAN).

When it comes to programmable forwarding planes, some initial industry-wide available mechanisms exist, driven by the need for programmable data planes in data centers, for example, via the P4 programming language [48] that still today primarily targets that market segment.

11.6.2   Reusable, Extensible Forwarding Protocol Packet Formats

The major challenge in developing a future network strategy is that the current network forwarding plane hardware is unable to scale to support a sufficient number of separately programmed virtual network contexts to allow operating multiple independent virtual network contexts. If each virtual network was to re-implement a network forwarding protocol stack from scratch, the total required context would be too expensive. A simple comparison of this problem to general purpose CPUs is to consider the total L1/L2 cache size in general purpose CPUs, and note the drop in performance, which would be inacceptable for packet forwarding, if the code side exceed those cache sizes.

To solve this problem, virtualized networks will require a common network packet forwarding framework, where individual virtual networks would only need to pick and choose required subsets of widely adopted network packet features and to only add new forwarding code for functions/actions that are novel to this virtual network. One proposed framework for such extensible, reusable network packet formatting to support new services is called "Big Packet Protocol," as described in "Packet-Programmable Networks and BPP: A New Way to Program the Internet" in the tutorials of the IM2019 conference [49].

11.6.3   High-Speed, Programmable QoS Algorithms

The programmability challenges for QoS go beyond the aforementioned programmability, scalability, and efficiency challenges for other components of the forwarding plane in network devices.

QoS support in today's programmable forwarding planes is usually based on well-established fixed functionality building blocks with a range of configurable parameters, for example, hierarchical DiffServ QoS with per-class programmable assignment to queues and drop behavior in queues, assignment to per-flow queues with similar parameters to name the most common functions. This functionality is insufficient to allow the programming of any of the aforementioned scheduling disciplines, the AQM mechanisms, or the LBF high-precision communications. Even within proprietary programmable vendor-specific forwarding plane chips, QoS is more ossified than other parts of network packet forwarding because of the absence of well-established, flexible programming models than previously mentioned configurable "legacy-QoS" toolset.

Only in lower-end forwarding planes with, for example, FPGA, is it currently possible to implement flexible new scheduling disciplines. This capability has been more widely available in Ethernet switches attempting to support the wide range of competing (proprietary) time-sensitive Ethernet and resilience options (redundant L2 rings). Nevertheless, FPGAs are generally considered to be too expensive and to consume too much power in high-speed networking equipment. Solving this problem is therefore an active area of research. In recent years proposals have emerged such as Push-In-First-Out (PIFO) and Push-In-Extract-Out (PIEO) queuing disciplines which allow the programming of new QoS disciplines by using a combination of these queuing disciplines and per-packet programmed forwarding code on packet enqueue and dequeue. However, while these approaches look very promising in enabling a wide range of future proof programmable QoS, it still has to be seen if they can be economically implemented in Tbps hardware, especially when being scaled to support suitably large numbers of flows. The aforementioned LBF QoS discipline in support of future network requirements has also been validated based on these queuing disciplines.

11.6.4   Instrumentation

Support for instrumentation of QoS services in most TCP/IP stacks is very limited but crucial to the ability to deploy, troubleshoot, operate, and sell QoS services. This topic is too broad to be captured here comprehensively. Instead we will attempt to motivate one core long-term instrumentation option.

For best effort traffic, the network operator typically just needs to focus on maximizing total network goodput. This is primarily achieved by maximizing network throughput through lightweight capacity engineering. In addition to this, some AQM may be added to critical congestion points to increase goodput and reduce latency. A simple well-understood core framework is insufficient to assess the service quality of better than best effort services, but it is also insufficient to easily isolate any problems even for best effort traffic, such as any form of error-based packet loss along the path.

Real-time traffic using RTP [21] carries sequence number and (equivalents of) timestamps in the RTP header. These header elements allow for the best payload independent quality monitoring in some of today's high-speed router implementations and also significantly eases pinpointing problems in the network. Sequence numbers in RTP could also serve as parameters for network-based PREOF functions.

Likewise, TCP headers allow the operator to assess throughput vs. goodput by recognizing primary data vs. retransmissions in some of today's high-speed routers. Whenever it is possible to recognize and distinguish request/reply traffic, it is also possible to determine the network performance impact on the so-called flow-completion times, e.g., recognizing the time spent/wasted on the application/host side vs. the network side.

Taken together, the authors argue that one fundamental direction to consider in support of better supporting latency, throughput, and loss QoS services in networks is to consider developing unified network/transport packet headers that would provide in a more efficient and current (varied) transport protocol independent from a superset of the aforementioned monitoring as well as active network enhancement functions such as PREOF. By creating and deploying such a common header, more consistent QoS monitoring and lower-cost, higher-quality QoS operations across traffic would be possible. Initially, such headers could be added by network-based proxies close to the hosts. Ideally, they could be added by proxy functions in the actual host stack, but still without the need to update any application code. Application change would only be required where deemed to be beneficial to the application, such as when using such a header to replace all or parts of an actual RTP or TCP or other transport header.

11.6.5 Economic Incentives

Monetization of differentiated QoS for different traffic is currently limited to private networks, in applications such as potential different charging for different classes of traffic in L3VPN services. There are only a few and ad hoc price differences for different QoS services beside the ubiquitous "peak bitrate" charging for Internet services, and in less developed countries still the "volume charging." The exception within public networks is where statically charged overall lower latencies are provided over access technologies such as xDSL. Monetization is an important dependency for making future QoS services successful in networks, but is a subject outside the scope of this book.

11.7 Summary and Next Steps

This section gave an overview of the current state of QoS in TCP/IP networks, its past and present focus on congestion controlled best effort, recently with more focus on low latency, but also the revival of interest in better controlled latency, loss, and throughput for deterministic and more general high-precision use case scenarios. Support for several of these use cases is required to expand TCP/IP networks and the future Internet towards a network for real-time applications required for often critical infrastructures and applications not yet using TCP, such as in industrial scenarios.

When mapping these evolving QoS requirements against the extrapolation of the evolution of the Internet, we conclude that an improved QoS architecture will predominantly be required on the edge, which we call Edgehaul, spanning topologies from industrial campus all through metropolitan/regional areas, often based on permissible RTT of control loops of applications.

The authors believe that a core reason for the past limited success of better QoS services in networks is based on the limited functionality of the current IntServ and DiffServ QoS models of TCP/IP. They were designed 25 years ago and carry forward assumptions about feasible forwarding plane functionality, putting the majority of work on an operationally expensive control and management plane, therefore making QoS an expensive to sell service. Today's high-speed forwarding planes could provide a lot better scalable QoS functionalities with significantly reduced operational complexity.

Operationalizing new QoS services also depends on the model of how the involved business entities are interacting. Vertical networks in which a single interested entity can control the applications and the network services to support them have in the past shown to be the most successful model. Service providers can enable this model by working towards QoS centric slices across their networks so that slice owners have complete control over the design and operations of the QoS services their applications require.

Today's evolving slice services in 5G/B5G only allow the network operator to parameterize existing QoS services. Per-slice programmable QoS via forwarding plane programming evolving from network forwarding plane programming languages such as P4 and programmable QoS abstractions such as PIFO (and beyond) that can enable high-precision QoS services designed by and for the actual owners of the application use cases and their partners. The authors think that these directions should be a core target for network research funding focused to enable better industrial and critical infrastructures in the coming 10 years.

References

1. I. Stoica, H. Zhang, F. Baker, Y. Bernet, Per hop behaviors based on dynamic packet state, draft-stoica-diffserv-dps-02 (work in progress) (2002)
2. J. Postel, Internet protocol, STD 5, RFC 791, DOI 10.17487/RFC0791 (1981), https://www.rfc-editor.org/info/rfc791
3. B. Braden, D. Clark, J. Crowcroft, et al., Recommendations on queue management and congestion avoidance in the internet, RFC 2309 (1998), https://www.rfc-editor.org/info/rfc2309
4. E. Rosen, A. Viswanathan, R. Callon, Multiprotocol label switching architecture, RFC 3031 (2001), https://www.rfc-editor.org/info/rfc3031
5. R. Pan, P. Natarajan, F. Baker, G. White, Proportional integral controller enhanced (PIE): a lightweight control scheme to address the bufferbloat problem, RFC 8033 (2017), https://www.rfc-editor.org/info/rfc8033
6. D. Newman, Benchmarking terminology for firewall performance. RFC 2647 (1999), https://www.rfc-editor.org/info/rfc
7. S. Floyd, M. Handley, J. Padhye, J. Widmer, TCP friendly rate control (TFRC): protocol specification, RFC 5348 (2008), https://www.rfc-editor.org/info/rfc5348
8. S. Blake, D. Black, M. Carlson, E. Davies, Z. Wang, W. Weiss, An architecture for differentiated services, RFC 2475, DOI 10.17487/RFC2475 (1998), https://www.rfc-editor.org/info/rfc2475
9. Netflix and YouTube are slowing down in Europe to keep the Internet from breaking, https://edition.cnn.com/2020/03/19/tech/netflix-internet-overload-eu/index.html
10. S. Shenker, C. Partridge, R. Guerin, Specification of guaranteed quality of service, RFC 2212, DOI 10.17487/RFC2212 (1997), https://www.rfc-editor.org/info/rfc2212
11. L. Andersson, R. Asati, Multiprotocol label switching (MPLS) label stack entry: "EXP" field renamed to "traffic class" Field, RFC 5462, DOI 10.17487/RFC5462 (2009), https://www.rfc-editor.org/info/rfc5462
12. F. Baker, G. Fairhurst, IETF recommendations regarding active queue management, BCP 197, RFC 7567 (2015), https://www.rfc-editor.org/info/rfc7567
13. P. Eardley, Pre-congestion notification (PCN) Architecture, RFC 5559 (2009), https://www.rfc-editor.org/info/rfc5559

14. S. Deering, R. Hinden, Internet protocol, version 6 (IPv6) specification, STD 86, RFC 8200 (2017), https://www.rfc-editor.org/info/rfc8200
15. K. Schepper, B. Briscoe, Identifying modified explicit congestion notification (ECN) semantics for ultra-low queuing delay (L4S), draft-ietf-tsvwg-ecn-l4s-id-12 (work in progress) (2020)
16. O. Albisser, K. De Schepper, B. Briscoe, O. Tilmans, H. Steen, DUALPI2 - low latency, low loss and scalable (L4S) AQM. Proc. Linux Netdev 0x13 (2019) https://www.netdevconf. org/0x13/session.html?talk-DUALPI2-AQM
17. X. Zhu, R. Pan, N. Dukkipati, V. Subramanian, F. Bonomi, Layered internet video engineering (LIVE): network-assisted bandwidth sharing and transient loss protection for scalable video streaming. *2010 proceedings IEEE INFOCOM*
18. H. Schulzrinne, S. Casner, R. Frederick, V. Jacobson, RTP: a transport protocol for real-time applications, STD 64, RFC 3550 (2003), https://www.rfc-editor.org/info/rfc3550
19. M. Handley, Congestion control for real-time media: history and problems, 2012 IAB workshop, https://www.iab.org/wp-content/IAB-uploads/2012/07/2-iab-cc-workshop.pdf
20. L. Han, G. Li, B. Tu, T. Xuefei, F. Li, R. Li, J. Tantsura, K. Smith, IPv6 in-band signaling for the support of transport with QoS, draft-han-6man-in-band-signaling-for-transport-qos-00 (work in progress) (2017)
21. K. Ramakrishnan, S. Floyd, D. Black, The addition of explicit congestion notification (ECN) to IP, RFC 3168 (2001), https://www.rfc-editor.org/info/rfc3168
22. R. Jesup, Z. Sarker, Congestion control requirements for interactive real-time media, RFC 8836 (2021), https://www.rfc-editor.org/info/rfc8836
23. X. Zhu, R. Pan, M. Ramalho, S. Mena, Network-assisted dynamic adaptation (NADA): a unified congestion control scheme for real-time media, RFC8698 (2020)
24. J. Wroclawski, Specification of the controlled-load network element service, RFC 2211 (1997), https://www.rfc-editor.org/info/rfc2211
25. P. Almquist, Type of service in the internet protocol suite, RFC 1349 (1992), https://www.rfc--editor.org/info/rfc1349
26. K. Nichols, S. Blake, F. Baker, D. Black, Definition of the differentiated services field (DS Field) in the IPv4 and IPv6 headers, RFC 2474 (1998), https://www.rfc-editor.org/info/rfc2474
27. B. Carpenter, B. Liu, Limited domains and internet protocols, RFC 8799 (2020), https://www. rfc-editor.org/info/rfc8799
28. A. Clemm, T. Eckert, High-precision latency forwarding over packet-programmable networks. In *NOMS 2020 - 2020 IEEE/IFIP Network Operations and Management Symposium* (IEEE, 2020)
29. S. Bensley, D. Thaler, P. Balasubramanian, L. Eggert, G. Judd, Data center TCP (DCTCP): TCP congestion control for data centers, RFC 8257 (2017), https://www.rfc-editor.org/ info/rfc8257
30. L. Eggert, G. Fairhurst, et al., UDP Usage Guidelines. RFC8085 (IETF, 2017)
31. G. Fairhurst, Network Transport Circuit Breakers. RFC8084 (IETF, 2017)
32. B. Carpenter, B. Liu, Limited Domains and Internet Protocols", RFC8799 (IETF, 2020)
33. IEEE, Time-Sensitive Networking (TSN) Task Group, https://1.ieee802.org/tsn/
34. N. Finn, P. Thubert, B. Varga, J. Farkas, Deterministic networking architecture, RFC 8655 (2019), https://www.rfc-editor.org/info/rfc8655
35. C. Filsfils, P. Camarillo, J. Leddy, D. Voyer, S. Matsushima, Z. Li, Segment routing over IPv6 (SRv6) network programming, RFC 8986 (2021), https://www.rfc-editor.org/info/rfc8986
36. 3GPP TS 23.401 System architecture for the 4G System, figure 4.2.1-1, https://www.3gpp.org/ ftp/Specs/archive/23_series/23.401/23401-g90.zip
37. 3GPP TS 23.501 System architecture for the 5G System (5GS), figure 4.2.3-2, https:// www.3gpp.org/ftp/Specs/archive/23_series/23.501/23501-g70.zip
38. U. Chunduri, et al., Transport Aware Mobility for 5G (2020), https://tools.ietf.org/html/ draft-clt-dmm-tn-aware-mobility
39. Common Public Radio Interface: eCPRI Interface Specification, http://www.cpri.info/downloads/eCPRI_v_1_0_2017_08_22.pdf

40. I. Busse, B. Deffner, H. Schulzrinne, Dynamic QoS control of multimedia applications based on RTP. Comput Commun **19**(1), 49–58 (1996)
41. A. Campbell, G. Coulson, D. Hutchison, A quality of service architecture. SIGCOMM Comput Commun Rev **24**(2), 6–27 (1994). https://doi.org/10.1145/185595.185648
42. N. Yeadon, A. Mauthe, F. García, D. Hutchison, QoS filters: Addressing the heterogeneity gap, in *Interactive Distributed Multimedia Systems and Services*, Lecture Notes in Computer Science, ed. by B. Butscher, E. Moeller, H. Pusch, vol. 1045, (Springer, Berlin, 1996)
43. L. Qiang, B. Liu, T. Eckert, et al., Large-scale deterministic IP network (2019), https://tools.ietf.org/html/draft-qiang-detnet-large-scale-detnet
44. B. Briscoe, K. Schepper, M. Bagnulo, G. White, Low latency, low loss, scalable throughput (L4S) internet service: architecture, draft-ietf-tsvwg-l4s-arch-08 (work in progress) (2020)
45. A. Clemm, T. Eckert, High-precision latency forwarding over packet-programmable networks. IEEE/IFIP NOMS 2020 (2020)
46. SMPTE ST 2022-7-2019 seamless protection switching of RTP datagrams https://ieeexplore.ieee.org/stamp/stamp.jsp?tp=&arnumber=8716822
47. A. Atlas, C. Bowers, G. Eynedi, An architecture for IP/LDP fast-reroute using maximally redundant trees (MRT-FRR), RFC 7812 (2016), https://www.rfc-editor.org/info/rfc7812
48. P4 language and related specifications, http://p4.org/specs/
49. ISO/IEC 23009-1, https://www.iso.org/obp/ui/#iso:std:iso-iec:23009:-1:ed-4:v1:en
50. A. Mauthe, F. Garcia, D. Hutchison, N. Yeadon, QoS filtering and resource reservation in an internet environment. Multimedia Tools Appl. **13**, 285–306 (2001)
51. TS 24.401
52. J. Polk, S. Dhesikan, Integrated services (IntServ) extension to allow signaling of multiple traffic specifications and multiple flow specifications in RSVPv1
53. https://www.rfi.fr/en/science-and-technology/20200320-french-telecoms-struggle-user-data-surge-netflix-cuts-quality-europe-coronavirus-lockdown-work-from-home
54. B. Briscoe, K. De Schepper, Resolving tensions between congestion control scaling requirements (2017), https://arxiv.org/pdf/1904.07605.pdf
55. https://www.ietfjournal.org/bufferbloat-dark-buffers-in-the-internet/
56. A. Bashandy, C. Filsfils, S. Previdi, B. Decraene, et al., Segment routing with the MPLS data plane, RFC 8660 (2019), https://www.rfc-editor.org/info/rfc8660
57. X. Zhu, R. Pan, M. Ramalho, S. Mena, Network-assisted dynamic adaptation (NADA): a unified congestion control scheme for real-time media, RFC 8698 (2020), https://www.rfc-editor.org/info/rfc8698

Chapter 12
Burst Forwarding Network

Jingcheng Zhang

12.1 Introduction

A burst is the basic application data unit that can be processed by the application. For example, a burst can be a photo for the image processing system, or it is a video clip in the video streaming service. Instead of per packet forwarding, the basic transmission unit is the application data unit and the data source sends the entire burst using the line rate of the network interface card (NIC).

The burst forwarding aware network is based on cut-through forwarding paradigm and a congestion-free virtual circuit is established between the source and sink. The assumption here is that the receiver (or sink) application must receive the entire burst before it can start processing the received data. In burst forwarding network, the bursts are received in sequence; therefore, no other packet processing such as ordering is necessary. The application in the receiver node can begin processing the data without any further data buffering. This mechanism helps in optimizing the compute resource utilization.

The use cases and the problem analysis and the category of the applications that can benefit most from burst forwarding technology are also described in Chap. 2. In this chapter we summarize the theoretical foundation and the results of the study.

The necessity of using burst forwarding in the future network is discussed, which includes the analysis results of the network throughput, the end host performance, the application data processing efficiency, and the router buffer requirement. Finally, we describe the architecture design of the burst forwarding network in detail.

J. Zhang (✉)
HuaWei Technologies Co. Ltd, Shenzhen, Guangdong, China
e-mail: zhang.jingcheng@huawei.com

© The Author(s), under exclusive license to Springer Nature   345
Switzerland AG 2021
M. Toy (ed.), *Future Networks, Services and Management*,
https://doi.org/10.1007/978-3-030-81961-3_12

12.2   Use Case Description

The current network is a packet forwarding network. The data generated by the applications are usually much larger than the packet MTU size. Before transmitting to the network, the application data is segmented and encapsulated into many MTU size packets. During the data forwarding, the packets from different flows are interleaved in the network congestion link. Congestion control algorithms are designed to equally share the congestion link bandwidth between different flows. In the receiver side, the application needs to retrieve the entire application data to start processing. In a congested network, the data transmission time in the network could be much longer than the data processing time in the receiver node. In this case, the compute resource utilization in the receiver node is very low. Additionally, uncorrelated data transmission in a bandwidth converged network usually has incast (drops due to router buffer overflow) problem and reduces the network utilization. As a result, it takes even longer time to finish the data transmission.

In burst forwarding technology, each application related data is transmitted to the destination node in sequence. The application in the receiver node can immediately start the data processing in pipeline. Therefore, the computation resource utilization in the receiver node is optimized. Moreover, by careful arrangement of each burst transmission, the network controls the ingress traffic always below the network egress capacity. In this case, the network can be congestion-free.

We describe two use cases of burst forwarding below: the metro gate control using face recognition system and the video surveillance system with real-time image processing. Simulation result of computation resource utilization and data transmission latency are presented while running TCP network and burst forwarding network.

12.2.1   Metro Gate Control Face Recognition System

Figure 12.1 illustrates the sample network architecture of the metro gate control face recognition system. In order to guarantee the high recognition accuracy, the metro gate camera takes high-resolution picture for each passenger. The average size of the image generated by a camera for each passenger is around 8 MB. The cameras connect with the cloud AI system using 10 Gbps leased lines. The recognition result should be sent back to the metro gate within 200 ms after the picture is taken. The timing details of the system are shown in Table 12.1.

The average service time for each passenger should remain below 1.5 s, of which 1.3 s are consumed when the door opens (0.3 s), the passenger passes through (0.7 s), and the door closes (0.3 s). Thus, only 200 ms are available for end-to-end network communication and data processing. The face recognition application consumes 7 ms to process a photo per network processor core. Therefore, the maximum

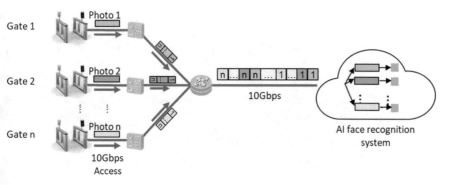

Fig. 12.1 Metro gate control face recognition system architecture

Table 12.1 Latency requirement of the metro gate control face recognition system

Total time	AI	Tx	Data size	BW per gate	Access BW	No. of lines
200 ms	7 ms	193 ms	8 Mb	332 Mbps	10G	30

end-to-end data transmission time is 193 ms. The available physical bandwidth for the cloud access is 10 Gbps, which can support 30 concurrent photo transmissions.

12.3   Problem Analysis

The AI face recognition application cannot process partially received photo. It needs to wait until the entire photo to be received. As shown in Fig. 12.2, if all cameras start sending photo at the same time, ideally, the 30 flows will be fully interleaved packet by packet. Thirty concurrent photo transmissions take 193 ms to deliver 8 Mb photo over a 10 Gbps link. In this case, the AI cloud service has only 7 ms to process 30 pictures. Therefore, the cloud service needs to reserve 30 network processor cores for the upcoming data processing. However, during the data transmission period, no data is received in the AI cloud, and the NP cores are left idle. The efficiency of AI computation resource utilization rate is only 3.5%.

If the burst forwarding technology is utilized, the network forwards each photo at a time. The photo can be received by the AI cloud service much faster. As shown in Fig. 12.3, every photo transmission occupies the entire bandwidth. For a 10 Gbps link, it only takes 6.4 ms to transmit one photo. Once the photo is received by the cloud service, it can be immediately processed. Since each core takes 7 ms to process one photo, it requires maximally two NP cores to process the data. The computation resource utilization in this case is 54.6%.

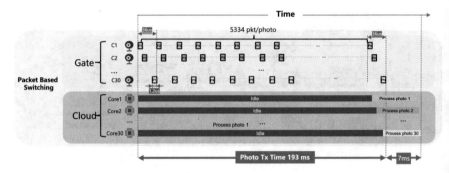

Fig. 12.2 Computation resource consumption of 30 concurrent photo transmissions

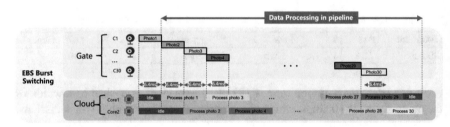

Fig. 12.3 Application-aware data forwarding

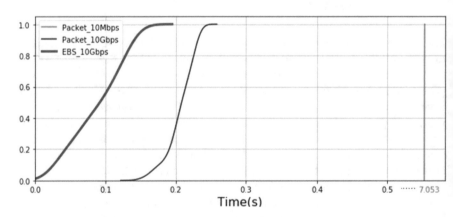

Fig. 12.4 CDF plot of the photo arrival time

The scenario described previously is the worst case which assumes that all the cameras send data at the same time. We have done simulations where the image arrival traffic pattern follows Poisson distribution. As shown in Fig. 12.4, in the packet forwarding network, more than 60% of the traffic failed to meet the 200 ms deadline. The latest photo was received at 260 ms. During this period, up to 5 NP cores are needed to process the concurrently received photos.

12.3.1 Video Surveillance System with Real-Time Image Processing

The video surveillance system uploads the video clips filed from different cameras to the remote server, where the received video streams are analysed in real time. The data generated from different cameras are required to be uploaded to the remote server within 1 s.

As shown in Fig. 12.5, cameras access the network using Fast Ethernet (100 Mbps) link. The average code rate for one camera is 8 Mbps. The egress port rate of the access switch is 1 Gbps. In theory, such a switch can support 125 camera connections. However, based on the field test result, the switch can only support 30 cameras without losing any packet. The equivalent bandwidth consumption is only 24%.

12.4 Problem Analysis

As shown in Fig. 12.6, the cameras access the network using Fast Ethernet port. The GE egress port can only support 10 concurrent camera data transmission. If there are more than 10 concurrent transmissions, the switch buffer starts to store the overloaded data. The access switches usually have shallow buffers and overflows lead to packet losses (incast problem). Although TCP will guarantee a reliable delivery, the retransmission mechanism causes extra time delay.

In burst forwarding technology, a dedicated virtual channel is created for each video clip transmission. If there is more data to be transmitted, the current burst needs to wait for the previous burst to finish data transmission. Burst forwarding network engineering limits the number of concurrent data transmission below 10 flows, to maintain a congestion-free network. The cumulative ingress speed never exceeds the egress speed. Therefore, packet losses due to buffer overflows are prevented.

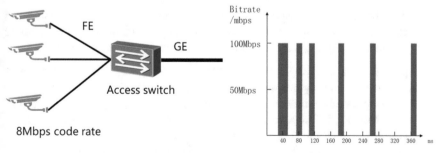

Fig. 12.5 Video surveillance system data uploading

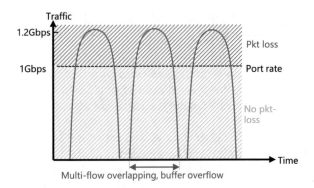

Fig. 12.6 Packet loss due to uncoordinated multi-flow overlapping

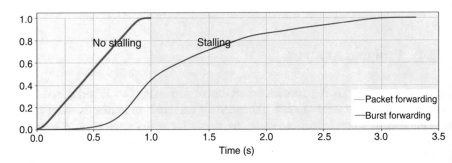

Fig. 12.7 CDF plot of video chunk uploading interval

Figure 12.7 shows the CDF of the data arrival rate with 110 camera connections. By using burst forwarding technology, all data can be delivered from the camera to the remote server within 1 second. However, when using TCP to transmit the same amount of data, more than 55% of the data failed to meet the deadline.

12.4.1 Scope of the Burst Forwarding Technology

Based on the previous description, we further generalized these use cases into a use case category, aka multisource convergence with large data transmission units under bounded latency. Such applications have the following common characteristics.

- Application data are originated from different data sources. However, the generated data are centrally processed, e.g. in a remote cloud service.
- The network architecture of the application is usually aggregation tree with converged bandwidth. The accumulated physical bandwidth of all the data source is much higher than the access bandwidth to the cloud. However, the equivalent code rate matches the cloud access bandwidth. The data sources use high bandwidth to access the network, but only transmit data sporadically.

- The data transmission needs to be finished within a bounded latency. Overdue data are either too late to be useful or it might break the pipeline of a closed loop control system.

Such network architectures are found in most of the IoT applications. A large number of sensors keep reporting measurement results to the remote server. The actuators are manipulated by the post-processing results from the remote server, for example, an indoor climate control system. However, processing such data becomes challenging as the volume of data being uploaded becomes larger and larger and the latency requirement gets more and more tight.

12.5 Theoretical Analysis of Burst Forwarding Mechanism

In this section, benefits of using burst forwarding in the current network are theoretically analysed. The results show an improvement in the network throughput, the end host performance, and the application data processing efficiency. We also analysed the router buffer requirement of the future large bandwidth application, e.g. holographic type of communication.

Our model shows that increasing burst size can significantly improve the network throughput as against the TCP. The end host performance analysis reveals the relationship between packet per second (PPS) and CPU resource allocation. A mathematical model is built to describe the packet transmission and packet receiving process. In the end host operating system, using small packet size triggers excessive packet tx/rx interrupts. In the worst case, the host CPU can be completely occupied by the interrupt service handling and leaves little resource for other applications. In the data transmission complete time study, we utilized the queueing theory on the burst level. It shows that the entire burst receiving time is minimized when the bursts are transmitted in sequence without any interleaving. In the router buffer requirement study, we present the relationship between the router buffer consumption and bandwidth requirement. Based on the current data transmission technology, the future ultra large bandwidth applications will require too much router buffer which is difficult to be fulfilled. A new congestion-free data forwarding method needs to be utilized for the near future ultra large bandwidth applications.

12.5.1 Network Throughput Study

According to [1], the TCP Reno network throughput can be calculated using Eq. (12.1). The MSS (maximum segment size) is the burst size, RTT is the round trip delay time, and ρ is the packet loss rate. At the first glance, the network throughput is proportional to the MSS in a fixed RTT network.

$$\text{Throughput} \approx \sqrt{\frac{3}{4}}\frac{\text{MSS}}{\text{RTT}\sqrt{\rho}} \tag{12.1}$$

However, the MSS also affects the RTT value and packet loss rate. For the store and forward network, the RTT time is increased since the router needs longer time to receive the whole burst before it can be processed and forwarded. According to [2, 3], the packet loss rate also increases when the MSS increases. By taking all these considerations, Eq. (12.1) can be further expanded as

$$\text{Throughput} \approx \sqrt{\frac{3}{4}}\frac{\text{MSS}}{\left(\sum_{i=1}^{N}\frac{\text{MSS}}{R_i}+T_p+T_c+T_q\right)\sqrt{\sum_{i=1}^{N}\frac{\text{MSS}}{B_i}+\text{MSS}\cdot\rho_{\text{bit}}}} \tag{12.2}$$

where R_l is the link rate, N is the hop number of the path, and T is the sum of the propagation delay (T_p), the computation processing delay (T_c), and the packet queuing delay (T_q). ρ is the link error rate and B is the router buffer size. According to Eq. (12.2), the network throughput reaches maximum when $\text{MSS} = \dfrac{T_p+T_c+T_q}{\sum_{i=1}^{N}\dfrac{1}{R_i}}$.

Figure 12.8 shows the relationship between the MSS and the network throughput. The path consists of 8 hop, the link rate is 1 Gbps, router buffer is 10 MB, $T_p =$ 0.5 ms, $T_c = 5$ ms, $T_q = 20$ ms, and bit error rate BER is 10^{-12}. The throughput reaches maximum 300 Mbps when MSS≈400 Kb. However, if 1.5 Kb MSS is used, the throughput is only around 10 Mbps.

From this analysis, we conclude that using large MSS as the basic data forwarding unit can increase the throughput of the current TCP network. However, there is an upper limit of the burst size. This problem is due to the TCP network dynamics. Larger burst result of a smaller number of packet per bandwidth delay product (BDP). In this case, a burst loss can easily trigger network retransmission timeout (RTO) which greatly reduces the network throughput.

12.5.2   Host Performance Study

The PPS (packet per second) value has a great impact on the host side performance. Packet sending and receiving are processed in the kernel space of the operating system. These operations have higher priority than the applications in the user space. Markov state machine to create a packet receiving model is used in [4]. Based on the same idea, we created a similar mathematic model on packet transmission, and the relations between PPS and CPU utilization are shown in Fig. 12.9.

As shown in Fig. 12.9a, when the MSS is small, the PPS is extremely high so that all the CPU resources are occupied by the packet receiving interrupt service routine (ISR). As the MSS increases, CPU resource is released. These resources are firstly

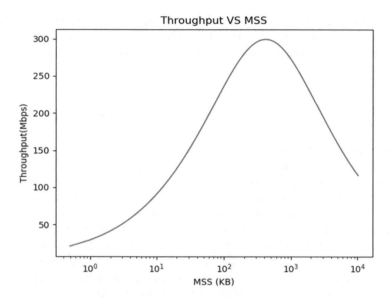

Fig. 12.8 Relationship between the network throughput and the MSS

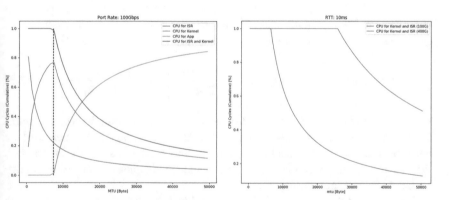

Fig. 12.9 Relation between MSS and CPU utilization. (**a**) Packet receiving analysis. (**b**) Packet transmission analysis

utilized by the kernel stack to process the received packets. Since both ISR and kernel stack have higher priority than the user space application, the MSS needs to be large enough so that the CPU can have extra resource for application data processing. Figure 12.9a shows the CPU utilization of a server with 3.3 GHz and 100 Gbps network interface card (NIC). The MSS needs to be larger than 7.5 kb so that the accumulated CPU usage of ISR handling and kernel logic is less than 100%. Similarly, as shown in Fig. 12.9b, the burst size needs to be larger than 6.4 kb at 100 Gbps link and 25 kb at 400 Gbps during the data transmission.

Based on this study, we conclude that using burst as the basic data forwarding unit can greatly reduce the PPS. In order to save CPU resource for other tasks, it is essential to use large burst size by the end host with high NIC bandwidth.

12.5.3 Data Transmission Completion Time Study

A bust contains the application related data. The application needs to receive the entire burst to begin data processing. In order to increase the data processing efficiency in the host side, the burst needs to be received in sequence. If different bursts are interleaved, the end host needs to buffer the data until the entire burst is received. Figure 12.10 shows the burst transmission complete time of different forwarding methods. As shown in Fig. 12.10a, the four bursts are transmitted in sequence. The total waiting time of the four bursts is minimized. If the bursts are forwarded with interleaving, as shown in Fig. 12.10b, a burst transmission is only completed when the last data block of that specific burst is received. As long as bursts are interleaved, the averaged burst transmission complete time is not optimized.

This observation can be explained using the M/M/1 queue theory. We assume that there are N bursts that need to be transmitted. The burst size is L and each burst contains small data blocks with size l. In this case, the waiting time of all the bursts is the accumulated queuing delay of the last data block of different bursts in the queue. The average waiting time of the burst can be expressed in Eq. (12.3)

$$t = \left(t_r + t_y + \cdots + t_b + t_g\right)/N = \frac{(N+1)L}{2\rho l} + \left[\frac{(N-1)L}{2\rho l} - \frac{(N-1)}{2\rho}\right]x \quad (12.3)$$

where ρ is the data block service ratio and x is the interleaving degree. The interleaving degree is a discrete distribution indicator ranged from 0 to 1.0 means all the burst are transmitted in sequence, while 1 means the burst are fully interleaved. When x is zero, the equation is the same as the classic queue theory which corresponds to the minimum waiting time.

Based on this study, we concluded that sending the entire burst to the destination node without any interleaving can optimize the average burst delivery time.

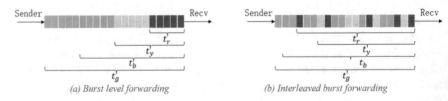

(a) Burst level forwarding   (b) Interleaved burst forwarding

Fig. 12.10 Burst forwarding with or without interleaving. (a) Burst level forwarding. (b) Interleaved burst forwarding

12.5.4   Router Buffer Requirement Study

The router buffers are needed to ensure the high network utilization. Congestion control algorithms such as Reno and Cubic rely on the packet loss to detect the network congestion. Due to the addictive increment multiplicative decrement (AIMD) algorithm, the data transmission speed is decreased after the packet loss. The buffered data are used to compensate the low network utilization which is caused by the temporary low transmission speed. The buffer should store enough data so that the sender can recover from the previous transmission speed decrement.

As shown in [5], the router buffer size which ensures high network throughput can be calculated using the following equation:

$$\text{BufferSize} = C * \text{RTT} / \sqrt{n} \qquad (12.4)$$

where C is the congestion link capacity, RTT is the round tripe delay time, and n is the number of uncorrelated flows/users. It is worth to note that the buffer requirement is inverse to the square root of the user number. As shown in Fig. 12.11, we have experienced massive user increment during the past 10 years. However, the bandwidth requirement per user only increased from 480P video to 1080P HD video. This situation will change for the next 10 years. The emerging media technologies consume significantly higher bandwidth. For example, basic VR consumes 50 Mbps bandwidth which is 8 times higher than HD video. Extreme VR consumes 15.2 Gbps bandwidth which is 2500 times higher than HD video. Such great bandwidth increment also requires proportional increment of the router buffer. It is believed that the current network processor architecture can only support up to good VR [6].

Figure 12.12 summarized the buffer requirement of different applications. We assume a dedicated router with 100 Tbps switch capability which serves different applications every time. Based on the bandwidth requirements shown in Fig. 12.11, the concurrently supported user can be calculated. Meanwhile, the required router buffer size can be calculated using Eq. (12.4). As shown in Fig. 12.12, in order to support HD video streaming, the router only consumes 31 Mb buffer. For Basic VR,

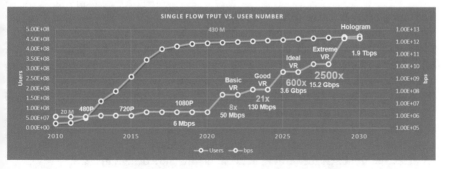

Fig. 12.11 Future trend of user number and the bandwidth requirement of different applications

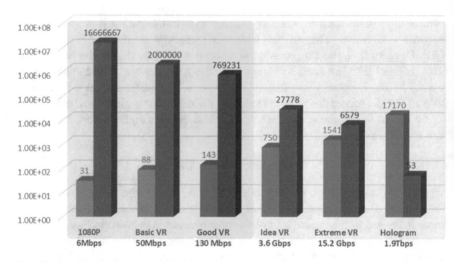

Fig. 12.12 Buffer requirement of different applications

88 Mb router buffer is needed. As the bandwidth increases per application, the con-current supported user number decreases. For good VR, ideal VR, and extreme VR, the buffer requirement is 143 Mb, 750 Mb, and 1.541 Gb. For the hologram, an astonishing 17.17 Gb router buffer is needed. According to [6], the practical NP cache size should below 256 Mb. In this case, the current NP technology can only support up to good VR application.

In order to decouple the buffer usage from network throughput, a new data for-warding flow control algorithm is needed.

12.6   Burst Forwarding Motivation

The current network architecture is originally designed for packet-oriented data for-warding. Numerous efforts have been done to smooth the data transmission, evenly share the congestion link bandwidth, and predict the available bandwidth and RTT. Uncoordinated burst transmission could cause a severe incast problem in the current network architecture and therefore reduce network performance. However, the concept of burst forwarding network is different from the mindset of the tradi-tional data forwarding. Instead of evenly share the bandwidth, each burst transmis-sion occupies all whole bandwidth of the link for a short period. The network should guarantee that the burst is sent to the destination without any congestion. This chap-ter describes the burst forwarding network architecture design in detail.

The burst forwarding network requires the collaboration between the network side and the host side. Both sides work together to provide a burst forwarding service infrastructure. This chapter begins with the general description of the burst forwarding network architecture. The network creates virtual channels for each burst transmission to guarantee cut-through forwarding. Secondly, the data plan design is presented. When being forwarded, a burst is split into multiple small data chunks, aka burstlet. On-demand local forwarding table entries are created for burstlet forwarding. The forwarding entry is deleted once the burst is successfully transmitted. Thirdly, the host architecture consideration is described. A new data interface is proposed for burst data sending. Moreover, the host also collaborate with the burst grant send algorithm. It blocks the application data transmission until the network is free. Finally, the burst grant send algorithm requirement for burst forwarding is presented. The goal of this algorithm is to guarantee that the burst transmission is congestion-free and consumes limited router buffer.

12.6.1 Architecture Overview

The store and forward mechanism require the routers to buffer the entire packet before forwarded to the next hop. In the burst forwarding network, a burst can be 10x Mb or even 100x Mb in size. Store and forward bursts consume huge amount of router buffer. An alternative method is the cut through forwarding. The cut through method starts forwarding a packet after the address fields were received. It is a good candidate for burst forwarding since it requires minimum router buffers. However, the limitation of cut through forwarding is that it requires the symmetric link speeds end to end. Burst forwarding leverages virtual channel technology to create path with same link speed on demand. Figure 12.13 shows a sample burst forwarding network architecture.

As shown in Fig. 12.13, three data sources access the network via 10 Gbps link. The access router connects to the cloud via 20 Gbps links. In this case, the 20 Gbps link is divided into two 10 Gbps virtual links. The links can be concurrently used by any two data sources. In burst forwarding network, if all three data sources send bursts at the same time, one of them will be blocked until the previous transmission finishes. By doing this, network guarantees that the burst can be forwarded in the path using cut through for the two selected flows. The data can be received by the destination as fast as possible.

12.6.2 Packet-Oriented Network Data Plane Limitations

Forwarding a burst in a traditional packet-oriented network has many challenges. One obvious problem is the head of line (HOL) blocking. High priority packets can be blocked by a long-lasting burst transmission. In the worst case, a small packet

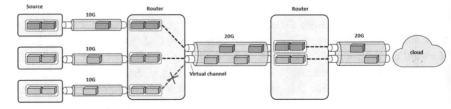

Fig. 12.13 Burst forwarding network architecture

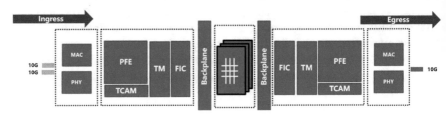

Fig. 12.14 HOL problem of router forwarding a non-splittable burst

could be blocked by the burst twice inside a switch. As shown in Fig. 12.14, if a small packet and a burst are received from two different ingress ports almost at the same time, the smaller packet could be blocked by the burst before being sent to the packet forwarding engine (PFE) for further processing. Moreover, if the two packets happen to be scheduled to the same egress port, the burst could block the small packet one more time. This problem could increase the service jitter (for small packet flow) and reduce the network determinacy.

As another practical problem, forwarding a burst also increases QoS scheduling interval inside the switch. It reduces the shaping effect from the switch traffic management system. For the switch traffic management, the minimum scheduling interval should be longer than the transmitting period of the biggest frame as shown in Fig. 12.15. Since a burst takes longer transmission time, it prolongs the scheduling interval of the router traffic management. For short frames, the shaping effect of traffic management is decreased by using long scheduling interval. If too many short frames are scheduled in the same interval, it could form a microburst.

12.7 Network Side Design

To solve the mentioned problem while largely maintaining the current router architecture, we need to decouple the router basic forwarding unit from IP packet size. This section provides the high level description of this mechanism. Instead of forwarding the entire burst at once, the burst is further split into smaller data blocks, aka burstlet. The burstlets are sent in a wormhole-switching-alike mechanism along the virtual channel. In this case, the high priority small packet transmission only

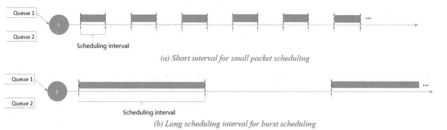

(a) Short interval for small packet scheduling

(b) Long scheduling interval for burst scheduling

Fig. 12.15 Packet scheduling interval composition between packet and burst. (**a**) Short interval for small packet scheduling. (**b**) Long scheduling interval for burst scheduling

Fig. 12.16 A burst consist of head burstlet, body burstlet, and tail burstlet

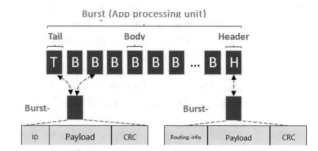

needs to wait for a burstlet instead of the entire burst. It also improve the accuracy of QoS since burstlet level scheduling providing finer granularity.

12.7.1 Burst Data Packetization

Depending on the data size, a burst is split into head burstlet, one or more body burstlets and a tail burstlet. As shown in Fig. 12.16, the header burstlet includes the routing information of the entire burst, e.g. source and destination IP addresses and port numbers. The body burstlet and the tail burstlet only contains the data of the burst. The burst ID uniquely identify a burst which links the head burstlet with the remaining body and tail burstlet. This is especially useful when multiple virtual channels share the same physical link where burstlet from different burst are interleaved.

Typically, a burstlet should include the following information in order to be correctly forwarded:

- Burstlet type: Flags indicating the type of the burstlet, i.e. head, body, or tail burstlet.
- Burst ID: Uniquely identify a burst from the same data source.
- SEQ: Burstlet sequence within a burst. Used by the burst receiver host for ordering reliability check.

- Port rate: Identify the sending speed of a specific burst. Carried in the head burst-let to dynamically create corresponding virtual channel.

12.7.1.1 Burst Forwarding Network Data Scheduling

The burst forwarding network scheduling function mainly serves two purposes, creation of on demand virtual channel and data forwarding over virtual channel. A typical data forwarding process includes the virtual channel creation, data forwarding, and virtual channel tear down.

As shown in Fig. 12.17, 5 data sources access the network using 10 Gbps NIC. The bandwidth of the backbone network is 40 Gbps. The egress port maintains a table (scheduler) which records the accumulated bandwidth allocated for the virtual channels. During the initial phase, no virtual channels are allocated at the egress port. When burst 1 data transmission starts, the network creates a virtual channel for the burst transmission. The egress port checks the port rate field of the header burst-let and allocates 10 Gbps resource for the burst. Virtual channels are also allocated for burst 2, burst 3 and burst 4. After this, the 40 Gbps backbone network is virtually divided into four 10 Gbps link segments. When the fifth burst arrives, the data transmission is blocked since the egress port cannot allocate any more bandwidth. The data transmission of burst 5 can only start when one of the previous 4 data transmissions finishes.

The router maintains the burst transmission speed. Since all data sources access the network with the same speed (10 Gbps), the scheduler uses round robin to forward each burstlet. In the 40 Gbps link, it seems like the burstlets from the four bursts are interleaved, but the forwarding speed of each burst is maintained at 10 Gbps. In the ingress port side, the burstlets are identified and categorized into different burstlet buffer using the burst ID. The burst ID management mechanism is described in the following part of this section.

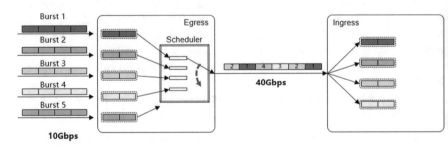

Fig. 12.17 Burst scheduling mechanism

12.7.1.2 Channel Allocation Process

Figure 12.18 depicts the detailed virtual channel creation process. A virtual channel is created on demand for each specific burst transmission. In the first step, once the head burstlet is received, the router first selects the egress port based on the routing information carried in the head burstlet. Based on the selected port, the router starts to allocate the bandwidth required for the specific burst transmission. As shown in the figure below, in the second step, an internal resource-ID is assigned. Each egress port maintains an ID resource list which records the previously allocated virtual channel. The ID number corresponds to the available bandwidth of the physical port. Each ID represents the greatest common divisor of the bandwidth in the network, e.g. FE port. Based on the port rate field carried in the head burstlet, one burst virtual channel might require multiple IDs in the port. As shown in Fig. 12.18, port 4 is selected as the egress port. Based on the head burstlet information, only one ID is required. By checking the ID resource list of P4, ID 3 is available. In the third step, ID 3 is marked as "occupied" in the ID resource list indicating this ID is allocated for the virtual channel being created. If the ID resource list is fully occupied, the burst forwarding is blocked. It is resumed once the ID is released by other burst transmission. Once the ID is allocated, as shown in step 4, an entry is added in the forwarding table. The following body burstlets and tail burstlet will be forwarded according to the records in the forwarding table.

The burstlet forwarding table is a port-based local forwarding table. It is created by the head burstlet and used by the body and tail burstlet for data forwarding. As shown in the burst forwarding table of Fig. 12.19, the burstlet with ID 1 received from port 1 is forwarded to egress port 3 with a new ID 3. The value of NewID is unique per port at a time, and it is a mechanism to guarantee that different outgoing

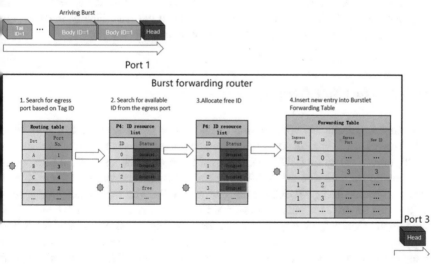

Fig. 12.18 Virtual channel allocation process

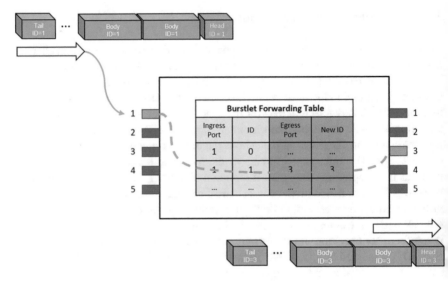

Fig. 12.19 Burstlet forwarding procedure

bursts from the same egress port have different burst ID. For example, if both ingress port 1 and ingress port 2 receives bursts with same ID and they are heading to the same egress port 3. Assumes that the port rate of port 3 is higher than port 1 plus port 2, two virtual channels are established, and the burstlets from port 1 and port 2 are interleaved. However, if the outgoing burst ID is not changed, it is impossible for the router in the next hop to identify the body and tail burstlet of these two bursts.

The virtual channel in the burst forwarding router is destroyed after the complete burst has been forwarded. As shown in Fig. 12.20, when the tail burstlet is received by the router, the egress port is checked in the ID forwarding table. In the second step, the previously allocated ID in the resource list is released. Finally, in step 3, the forwarding table entry is removed after the tail burstlet forwarding.

12.7.2 Host Side Design

The burst forwarding network requires an end host to send each burst using NIC line rate. However, the current socket interface only supports sending data as a stream (TCP) or as a datagram (UDP). TCP is used by most of the application because of the reliable transmission and self-tuning transmission rate control. Other popular transport protocol, e.g. QUIC, is built on top of UDP. The flow management, reliability, and security features are developed in the user space. Both TCP and QUIC send application data as data stream. As shown in Fig. 12.21, the end host OS that supports burst forwarding should provide a new socket function. The new socket interface should support the burst sending at NIC line rate. The transmission speed should not be limited by any flow control algorithm.

Port 1

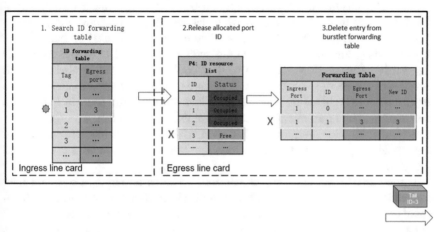

Fig. 12.20 Virtual channel tear down procedure

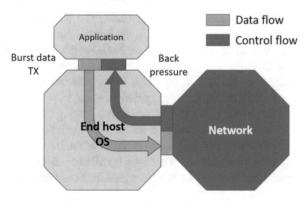

Fig. 12.21 Burst forwarding host data transmission and flow control interface

However, sending uncoordinated burst to the network is dangerous. It can easily create network congestion and packet loss. The burst forwarding host first asks for the permission to transmit. Once granted, the burst can be transmitted. Instead of implementing self-maintained congestion control algorithm, the host cooperates with the flow control function of the network to ensure congestion-free. It keeps monitoring the received traffic information. If the received data is too much to handle by the host, a back pressure message should be issued to the application and block the burst transmission.

12.8   Coordinated flow control function

The flow control function of the burst forwarding network mainly serves two purposes—(a) to ensure the network congestion-free and (b) to arrange the burst transmission in sequence. The burst forwarding network does not bind with any specific flow control functions. If the burst forwarding router has very shallow buffer or the application requires extremely low end-to-end latency, global TDMA-like scheduling can be utilized. However, such method could sacrifice the bandwidth which depends on the network scale and time synchronization accuracy. Another possible approach could be based on transmission token. Only the data sources with tokens may start the burst transmission. The total number of tokens depends on the egress port bandwidth. However, this method usually works best in the application with aggregation tree topology where the message destination is centralized, e.g. cloud access service.

If the burst forwarding network can tolerate some buffering, a quantum flow control (QFC) mechanism may also be utilized. Different from traditional host-based congestion control algorithm, QFC is a distributed port-based credit flow control algorithm. By using QFC, the amount of packet that can be sent from the egress port to the next hop ingress port is explicitly calculated. In order to accommodate burst forwarding, the algorithm is updated to support virtual channel creation.

The QFC mechanism is described in Fig. 12.22. The ID marked in the connection link is the created virtual channel ID. In the ingress port side, a burst buffer is allocated for each virtual channel. During the initialization phase, the capacity of the burst buffer of the ingress port is sent to the egress port via BSL-I message. This message is confirmed by the BSL-C message. During the run time, the burst buffer utilization can be calculated by subtracting fwd_counter from rx_counter value. In order to avoid buffer overflow, the ingress port keeps posting the fwd_counter value using the BSU message to the egress port device. On receiving the BSU message, the egress port device calculates the available buffer of the ingress port using Buffe rLimit $-$ (TxCounter $-$ FwdCounter). The result is called credit balance. Meanwhile, the egress port periodically sends the BSC message to correct the possible mismatch

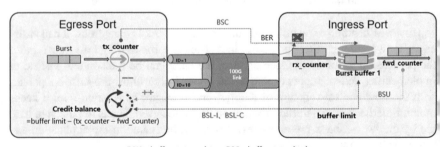

BSU: buffer state update; **BSC**: buffer state check;
BSL-I: buffer state limit indicate; **BSL-C**: buffer state limit confirmation;

Fig. 12.22 QFC flow control algorithm for burst forwarding network

between tx_counter and rx_counter due to packet transmission error. Since the egress port only sends data which can be stored in the ingress port buffer, the link is lossless from buffer overflow.

12.9 Reliability with Burst Losses

Today's transmission media is quite reliable, but errors may still occur due to hardware failures. In such cases, a retransmission strategy has an adverse impact on the network. In QFC, acknowledging each burstlet in the network can slow down the transmission. As mentioned earlier, the reliability is a function of the receiving side, which may discard the entire burst and request the retransmission. Using this approach, only one burst flow is impacted, and the rest of the congestion-free network continues to operate normally. Alternately, instead of the entire burst, an end-to-end protocol may request only lost burstlet with its sequence number which may be delivered as normal packet.

12.10 Summary

This chapter describes burst forwarding technology, an application oriented data forwarding mechanism. A burst is a basic application process unit. The size of the burst depends on the application type. The host sends its burst at line rate of the NIC. The burst forwarding network forward the burst with the same rate as it is injected into the network. If multiple data transmissions exceed the network capacity, the extra transmission is blocked until the previous burst transmission finishes. Since an entire burst is forward by the network from the data source to the destination, the average application data transmission time is much shorter. The application in the destination node can immediately start processing the data once the burst is received; thus the utilization efficiency of the compute resource is optimized.

References

1. N. Cardwell, S. Savage, T. Anderson, Modelling TCP latency. INFOCOM 2000. Proc. IEEE **3**, 1742–1751 (2000)
2. K. Chan, F. Tong, C.K. Chan, et al., An all-optical packet header recognition scheme for self-routing packet networks, in *Optical Fiber Communication Conference*, (Optical Society of America, Washington, DC, 2002), p. WO4
3. R.S. Prasad, C. Dovrolis, M. Thottan, Router buffer sizing for TCP traffic and the role of the output/input capacity ratio. IEEE/ACM Trans. Netw. **17**(5), 1645–1658 (2009)
4. K. Salah, On the accuracy of two analytical models for evaluating the performance of Gigabit Ethernet hosts. Inf. Sci. **176**(24), 3735–3756 (2006)

5. G. Appenzeller, I. Keslassy, N. McKeown, Sizing router buffers. ACM SIGCOMM Comput. Commun. Rev. **34**, 4 (2004)
6. J. Zhang, M. Zha, L. Niu, EBS, electric burst scheduling system that supports future large bandwidth applications in scale, in *Proceedings of the 15th International Conference on Emerging Networking EXperiments and Technologies, CoNEXT 2019*, Companion Volume, Orlando, FL, USA, December 9–12, 2019

Chapter 13
Security, Anonymity, Privacy, and Trust

Simon Scherrer and Adrian Perrig

13.1 Introduction

Given the broad scale of security, anonymity, privacy, and trust, we need to properly scope these notions and to define them. First off, we are going to consider these notions mainly in the context of inter-domain network infrastructures—as the challenges are much reduced in an intra-domain context, which is typically under administrative control of a single entity. We consider the security and trust of end hosts and the privacy of data stored on end hosts to be out of scope for this chapter. We do consider the security of network infrastructure devices, however, as their compromise can result in threats to network security. We will focus on network properties and not on individual services, unless the services are directly relevant to achieve the properties we seek.

We pursue security in terms of these network properties: a network is considered secure if it can achieve the desired properties even in the presence of an active adversary. One prominent property is availability, i.e., the control, data, management, and configuration planes should be protected such that an adversary cannot disrupt connectivity. Another important property is trust, which we understand here as the ability of network nodes to verify origin and content authenticity of messages passed through the network. Furthermore, desirable, but difficult to achieve properties are privacy and anonymity, treated here as the ability of nodes to communicate without other network entities being able to identify the communication parties. (Privacy typically refers to the secrecy of personal information, whereas anonymity is a more specific property that refers to the identity of the user or end point. Since personal information is usually carried within the communicated data, we focus on

S. Scherrer (✉) · A. Perrig
Department of Computer Science, ETH Zurich, Zurich, Switzerland
e-mail: simon.scherrer@inf.ethz.ch; adrian.perrig@inf.ethz.ch

achieving anonymity in the network-focused context of this chapter. However, we consider the privacy of network metadata to be outside the scope of this chapter.)

In order to concretize the notions of security, anonymity, privacy, and trust, we first state the goals of a secure inter-domain network infrastructure in Sect. 13.2. While pursuing these goals, a number of requirements have to be respected, which are listed in Sect. 13.3. Finally, Sect. 13.4 sketches possible pathways for achieving security and trust under the mentioned requirements.

13.2 Goals

Concerning the security, anonymity, privacy, and trust of a next-generation Internet, we consider the following aspects as the most critical to consider:

- Improved trust model: A new network trust model should be deployed to provide decentralized verifiability. Based on the new model, important network information, such as BGP, DNS, and RPKI information can be verified in a more trustworthy way to prevent any single point of failure. The network trust model should also provide trust transparency, i.e., for any piece of information, a verifier should be able to identify all entities that have to be relied upon for the information to be trusted.
- Transparency and control for forwarding paths: Network paths in today's Internet lack transparency. In a first step, it would be useful to know as a sender which entities are traversed by a packet. In a second step, it would be useful for a receiver to achieve ingress path control for incoming traffic. Finally, in a third step, end hosts could benefit from controlling the packet's forwarding path. These are important properties to prevent eavesdropping and man-in-the-middle attacks of intermediate entities, as well as to increase availability in case of maliciously congested paths that can be circumvented with path control. An important aspect of this property is path correctness: the sender should be able to verify path information and the receiver should be able to verify for each packet that the selected path was correctly followed. As a result, an off-path adversary should not be able to alter a packet's path.
- Efficient and scalable authentication mechanisms for AS and host-level information: Such properties will prevent IP source address spoofing attacks, for instance. Such a service would, for instance, enable a receiver to verify the origin of error packets.
- Availability in the presence of an active adversary: Communication between two end points should be possible, as long as a functional and connected sequence of intermediate network devices and links exists. This is the foremost goal of the network to provide utility to demanding use cases. A particular challenge is to ensure a Service-Level Objective (SLO) or Service-Level Agreement (SLA) network contract even in adversarial contexts.

- Pseudonymous sender/receiver anonymity: Untrusted nodes (i.e., nodes under control of an adversary) in the network cannot identify the sender and/or receiver of communication without resorting to timing analysis (contrast with *perfect* sender/receiver anonymity below). This property is typically achieved by identifier-translation services. Note that there exists an inherent tension between the goals of anonymity and source accountability.
- Algorithm agility: Cryptographic algorithms need to be replaced in case of breakthroughs in cryptanalysis or computation technology such as quantum computers. Thus, it is necessary that the network architecture and infrastructure are prepared to replace cryptographic mechanisms. A challenge is if algorithms are implemented in hardware, which requires a hardware replacement cycle to upgrade. Consequently, techniques need to be devised to retain secure operation through a potentially multiyear algorithm replacement cycle.
- Class of security level: Not all applications or processes need the same level of security. Security schemes typically require additional resources or time which may not be necessary nor available in some scenarios. A class of security level should be considered to support different requirements.

There are other network properties, which, albeit desirable, should not be provided by the network infrastructure itself, either because the properties can be achieved without network support or because the properties are too costly to achieve as basic network primitives. We thus consider the following goals to be out of scope:

- Communication secrecy: Achieve secrecy for communicated data. This property is typically well understood and can be achieved with encryption between the end points, for instance, using a VPN.
- Perfect sender/receiver anonymity, anonymous communication: Untrusted nodes in the network cannot identify the sender and/or receiver of communication, even when performing timing analysis. Although sender and receiver identities can be concealed by services providing name-to-address translation, perfect anonymity can only be achieved by thwarting timing attacks, which requires an expensive traffic-mixing infrastructure.

13.3 Requirements and Challenges

The nature of inter-domain networks constrains the set of security solutions that are practically feasible. To achieve meaningful progress for the broad challenge of security, anonymity, privacy, and trust in networks, we provide a list of requirements that need to be considered by any proposed system:

- Heterogeneous trust relationships: Given the difficulty to establish globally accepted trust roots, allowing for choice among decentralized, diverse trust roots (sovereignty) is advantageous.

- DoS and DDoS attacks at all levels (e.g., also against services, infrastructure, etc.): The diversity of different types of (D)DoS attacks are substantial, for instance, algorithmic complexity attacks on the implementation, or resource exhaustion on a network link (bandwidth) or service (computation).
- Difficulty of latency guarantees: Due to complexity of inter-domain networks and interactions between high numbers of flows, latency guarantees are very challenging to achieve even in non-adversarial contexts. When considering an adversary, they become exceedingly challenging.
- Protocol complexity requires formal verification: Modern distributed systems reach a scale that eludes people's mental capacities for considering all possible states and interactions, thus necessitating automated protocol verification techniques. Such formal verification achieves a high level of assurance. Protocol flaws can be avoided through formal verification tools, such as the Coq, ProVerif, and Tamarin systems. However, verification tools encounter scalability challenges with increasing protocol complexity.
- Large network-technology diversity: Ensuring security properties across ManyNets, a wide diversity of different network technologies, is a challenge. For instance, resource-constrained network environments may not provide sufficient resources to carry needed cryptographic information in each packet.
- Software vulnerabilities throughout infrastructure and applications: Although not directly connected to network security, the fact that some network infrastructure devices and end points will be under the control of an adversary needs to be considered. Implementation security can be achieved through formal code verification, which unfortunately is still quite costly and does not scale well beyond tens of thousands of lines of code. Current state-of-the-art tools for code verification include Dafny and Viper. Examples for large-scale verification efforts include the seL4 secure microkernel, the project Everest verified HTTPS stack, or the VerifiedSCION project. API-level attacks can be prevented through the combination of protocol and implementation verification techniques.

In addition to the nature of inter-domain networks, the adversary model constrains possible security solutions. A general adversary model should consider the following types of attackers:

- Nation-state adversary: Well-funded, large numbers of trained personnel and vast infrastructure resources can exploit vulnerabilities in devices, set up malicious entities/infrastructure, or control a large number of devices for DDoS attacks. Among the main motivations are industrial espionage, critical infrastructure attacks at the network level, and preventing network availability in general.
- Criminal organization: Significant resources can control a smaller amount of infrastructure resources than the nation state adversary.   Main motivation is to profit through contracted attack services, to a lesser extent espionage.
- Independent hacker groups: Individuals or small political and ideological targets, smaller-scale attacks.

Ideally, even for nation-state adversaries, the security properties shall be achieved assuming the existence of a network path that is not controlled by the adversary.

13.4 Design and Method

In this section, we present proposals for achieving the goals laid out in Sect. 13.2, where each of the following subsections corresponds to a security goal. It is important to note that there exist dependencies between individual proposals. For example, the decentralized trust model introduced in Sect. 13.4.1 enables the source-authentication architecture presented in Sect. 13.4.3.

13.4.1 Improved Trust Model

The currently existing public-key infrastructures, e.g., the DNSSEC PKI, the TLS PKI, and the RPKI used in BGP, are based on a centralized system architecture or a centralized trust model. Such centralized architectures suffer from the problem of trust-anchor compromise. In the centralized model, since descendants need to rely on some common ancestors or authorities as trust anchors, a central authority node has privilege over all descendants. Central authorities can unilaterally perform malicious actions like revoking certificates, issuing fraudulent certificates, or providing fake information. Since all these infrastructures are widely used across the world, malicious actions of central authorities may thus adversely affect the Internet. Trust anchor failures may happen for many reasons. A central authority may be hacked or compromised to perform malicious actions unintentionally. In other cases, an authority may not be fully neutral and perform malicious actions for economic gains or political reasons.

For next-generation networks, a decentralized trust model should be provided, which can be achieved with the SCION secure network architecture [1]. In SCION, the Isolation Domain (ISD) comprises a group of autonomous systems (AS) and enables setting localized trust roots defined in a trust root configuration (TRC). The ISD can operate independently of any external network entity and thus achieve sovereignty and address global heterogeneous trust relationships. The TRC of each ISD serves as the root of trust for the local control-plane PKI [2], which provides AS-level certificates. These AS-level certificates can be used to establish control-plane functionality in a secure manner. Thanks to the structure of SCION's control-plane PKI which explicitly enumerates all trust roots, trust transparency is achieved.

13.4.2 Transparency and Control for Forwarding Paths

A promising development over the past decade are path-aware network (PAN) architectures, where senders embed the network path into the packet header. This seemingly simple concept results in exciting security opportunities for next-generation networks. Packet-level path information enables delivery as long as the path is functional, independent of actions by the routing protocol. Path information also enables predictability of which ASes need to be relied upon for the packet to arrive at the destination. Given topological path information, the reliance on any single AS can be minimized by using multipath transmissions over maximally dis-joint paths. Moreover, topological path information enables exclusion of some routes altogether, e.g., for the purpose of surveillance resistance. Stable paths are also a necessary precondition for future QoS mechanisms that are based on band-width reservation along paths (cf. Sect. 13.4.4). Even without QoS systems in place, transparency and control over forwarding paths provide some protection against DDoS attacks, as path control allows the circumvention of maliciously congested paths (given that alternative paths exist).

However, path awareness requires dissemination of path information, which is confronted with the following four challenges. First, path information must be disseminated in an *authenticated* fashion such that the information can be verified. Second, path information must be disseminated in a *scalable* fashion, i.e., the dissemination complexity in terms of messages should not become overwhelming in large topologies. Third, path information, in particular dynamic path properties such as load on the path, should be disseminated in a *timely* fashion in order to be useful. Fourth, path information should be disseminated in a *policy-compliant* fashion such that only paths that are explicitly allowed by ASes are usable.

In order to solve these challenges, the key idea in the SCION network architecture is to use a form of network partition, i.e., to split the network into Isolation Domains (ISD), each containing multiple ASes (cf. Fig. 13.1). A subset of ASes in each ISD forms the ISD core, which both initiate intra-ISD path discovery and provide inter-ISD connectivity. For intra-ISD path discovery, an ISD-core AS sends a beacon to each of its customer ASes, where the beacon contains information about

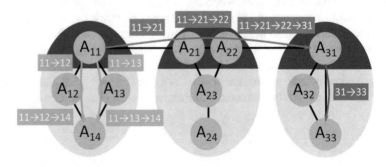

Fig. 13.1 Path-information dissemination across isolation domains

the link to the respective customer AS. In turn, each customer AS forwards the beacon to its own customer ASes after updating the beacon with the necessary link information, and so on. The same path-segment construction process takes place between core ASes of different ISDs. The resulting path segments can be combined to connect any AS to any other AS. For this purpose, the core ASes maintain a destination-based database of active path segments and respond to path-segment queries of other ASes.

The segmentation of paths allows the path-discovery process to remain *scalable* while preserving universal connectivity. In comparison to pure source routing, segmentation is much more scalable while only marginally reducing the space of possible paths, as business-logic constraints on possible paths are practically identical with the constraints enforced during segmentation. Since the number of individual path-dissemination messages is reduced, their frequency can be increased, leading to a more up-to-date view of the network (which achieves *timeliness*). Moreover, isolation is a security feature, as intra-ISD forwarding is completely independent of the less trusted exterior ISD network.

In order to provide *authenticity* of constructed path segments, the beacon-forwarding AS always has to include the AS to which the beacon is forwarded, as well as sign all the information added to the beacon (similar to BGPsec). Since paths are only constructed from segments which are authorized by all involved ASes (forwarding implies authorization) and the path construction itself cannot result in unauthorized paths, the policy compliance of offered paths can be ensured. Moreover, each beacon-forwarding AS also associates a hop-authenticator value with the forwarded path segment, which cryptographically encodes that the segment was received from the preceding AS; it suffices that this encoding is later verifiable by the forwarding AS alone. If these hop authenticators are then later inserted into the packet header by the end-host using the path (in a manner described in the EPIC system [3]), it is also ensured that no paths can be maliciously assembled from parts of other paths. Hence, not only all offered paths are guaranteed to be policy-compliant but also all usable paths.

Finally, in order to guarantee that a packet in fact follows the path selected by its sender (path validation), every packet carries a series of hop-validator fields in its header, where each field corresponds to an AS hop and is cryptographically linked to the packet source, the packet timestamp and the hop authenticator mentioned above. Such packet-carried forwarding state allows any AS on the path to verify that the sender intended to send the packet through the AS. Moreover, when forwarding the packet, each AS replaces its corresponding hop-validator field in the packet by a cryptographic proof that it saw the packet (also described in the EPIC system [3]), in which the destination then reflects to the source. Since forwarding misbehavior can be detected and deterred using this technique, the end-hosts gain full control over the forwarding paths that their packets follow.

Finally, while the mechanisms described above provide end-hosts with control over forwarding paths, they do not guarantee that this path control is exercised in a manner favorable to network efficiency. If the distributed path-selection decisions led to heavily suboptimal traffic distributions or persistent load oscillation, the resulting congestion and jitter might endanger availability. However, recent research has shown that end-host path control can be expected to lead to nearly optimal traffic distribution [4], even if end-hosts obtain only limited information about the network, and the end-hosts can be incentivized to select paths in a non-oscillatory fashion with appropriate mechanisms [5].

13.4.3 Efficient Authentication Mechanisms for AS and Host-Level Information

In addition to path authorization (packets only follow policy-compliant paths) and path validation (packets follow the path selected by the source), source authentication is an extremely valuable property of a secure network architecture. For instance, denial-of-service attacks are possible nowadays by spoofing an error message, supposed to originate from an on-path router R in AS A_1, and sending it to a host H in AS A_2, avoiding the path including router R as a result. For the sake of efficiency, this authentication needs to be performed on the basis of a symmetric key, as usage of asymmetric cryptography by the router R would make it vulnerable to denial-of-service attacks based on resource exhaustion. However, storing a symmetric key for every relevant host in the Internet is not viable.

This problem can be addressed by the DRKey (*dynamically recreatable keys*) system [6], which could be employed as follows. The egress border router of the source AS could compute the authentication tag of the packet on the basis of a dynamically recreatable key specific to the destination host H. The source egress border router R could derive $K(A_1{:}R, A_2{:}H)$ by means of an efficient pseudo-random function (PRF) computation with key $K(A_1, A_2)$ (which has been previously securely exchanged between A_1 and A_2) and arguments R and H. Standard hardware allows such a derivation to be highly efficient, even more efficient than a memory lookup for stored keys. On the other end of the communication, host H can obtain $K(A_1{:}R, A_2{:}H)$ from its local key server, which can perform the same computation as router R in AS A_1. As a result, host H could verify all packets from router R on its own after only one request to its local key server, which is necessary to learn $K(A_1{:}R, A_2{:}H)$.

Using this lightweight approach to source authentication, a next-generation network can inhibit IP spoofing and attacks that make use of IP spoofing, such as session hijacking, man-in-the-middle attacks, and many forms of DDoS attacks. In particular, instead of using the dynamically recreatable key to $K(A_1{:}R, A_2{:}H)$ to verify packets from router R at host H, key $K(A_2{:}H, A_1{:}R)$ can be used to perform source authentication at router R as described in the EPIC system [3]. As a strong notion of authenticity also rules out replaying authentic packets, replay-suppression system [7] can complement DRKey-based packet authentication at each router. Such a multiple-verification design stops malicious traffic early in the network and prevents malicious traffic from converging on the victim host, thereby limiting the effectiveness of DDoS attacks. However, DDoS attacks are still possible if the source domain is malicious, e.g., if the source domain does not restrict flows that misbehave despite blocking requests from the destination domain. Even if the destination domain identifies the source domain as malicious, the border router of the destination domain as well as the paths leading to the destination domain could be overpowered by a sufficiently powerful malicious source domain. If a DDoS attack is carried out by an attacker with AS-level capabilities, the QoS systems described in the next section are required.

13.4.4 Availability in the Presence of an Active Adversary

DDoS attacks are still a stubborn problem that undermines network availability. In 2020, DDoS attack traffic has exceeded 2.3 Tbps in a single attack. As more vulnerable IoT devices are deployed across the Internet, DDoS attack threats will continue to intensify and break the existing firewall-based security defense baseline. 5G network technology will support millions of connections per square kilometer, so DDoS attack traffic from the same administrative domain should not be underestimated.

As explained in the previous section, pervasive packet-origin verification on the basis of EPIC can prevent DDoS attacks in some cases. However, if the attacker has AS-level capabilities, this line of defense fails, as a malicious source domain can continue to overload targets along a certain path while ignoring the shutoff requests from the destination domain. For such attacks, quality-of-service (QoS) systems based on bandwidth reservation are an effective mitigation tool.

The rationale of bandwidth-reservation systems is as follows. In return for a payment (which could be of a monetary nature or of a virtual resource), end-hosts obtain a share of the available bandwidth along a certain path. The reserved bandwidth amount is the assured minimum amount of bandwidth usable in any case, i.e., even in case of a link overload along a path. In case of a link overload, flows on the link without a reservation might be dropped, while flows with a reservation can continue using the link to the extent of their reservation. The bandwidth not used by flows with reservations is available to flows without reservations on a best-effort basis. With a bandwidth-reservation system in place, predictable quality of service can thus be ensured even in the presence of AS-level attackers.

In order to obtain a reservation, an end-host would need to send a reservation request along the desired path, where the request would contain the desired amount of guaranteed bandwidth. When passing the request in the initial direction, every AS on the path incorporates into the packet the amount that the AS is willing to allocate for the reservation. After reflection at the destination, the ASes along the path could then allocate the actual available bandwidth, given by the minimum amount of bandwidth that has been appended to the reservation request. Figure 13.2 illustrates the reservation process. Developing a scalable, fair, and efficient method of bandwidth allocation is a subject of ongoing research, which has resulted in the

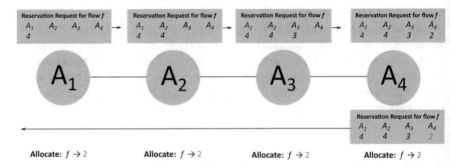

Fig. 13.2 Distributed management of reservation requests in bandwidth-reservation architectures

lightweight, Pareto-optimal GMA algorithm [8] that even works without explicit reservation requests.

When passing back the reservation request, every AS also inserts a reservation tag into the packet, which cryptographically protects the AS-specific reservation information. This reservation tag is a message authentication code (MAC), based on a local secret known only to the AS. An end-host with a reservation must include all the reservation tags for a path into its packets. When checking packets that include a reservation tag, each AS can efficiently verify that a flow indeed corresponds to a reservation, without keeping reservation state on the border routers.

13.4.5 Pseudonymous Sender/Receiver Anonymity

As mentioned in Sect. 13.2, strong anonymity guarantees in the sense of complete anonymity can only be given by resistance to timing attacks, which so far cannot be achieved in an inter-domain context without inacceptable degradation of network performance. Therefore, this section describes how to achieve anonymity under the assumption that attackers cannot perform timing analysis. Moreover, anonymity should be preserved in a way that does not undermine the ability of network entities to perform source authentication (cf. Sect. 13.4.3). In particular, the anonymity mechanism should still provide source accountability. We consider the APNA system [9] as a promising approach under these requirements.

To reconcile the seemingly conflicting requirements of source accountability and anonymity, domains (ISPs) act as privacy brokers in APNA. In short, the fundamental idea of APNA is that domains equip end-hosts with an ephemeral ID that has time-limited validity and does not allow an external observer to identify the real origin within the AS. More precisely, an end-host H registers at its domain with credentials that have previously been provided out-of-band, securely sends a symmetric key k_H to the AS (which will later be used), and obtains an ephemeral ID that is cryptographically linked with an AS secret k_A to its real IP. When sending a packet outside the AS, the host H replaces the source IP with its ephemeral ID and extends the packet with a MAC computed with the self-generated symmetric key. The egress border router in the AS then derives the real host IP from the ephemeral ID (knowing k_A) and checks if the MAC is in line with the symmetric key k_H that the host sent to the AS beforehand. If these checks succeed, the egress border router forwards the packet. If a remote domain detects misbehavior of a flow, it can send a shutoff request to the source domain, containing the packets that demonstrate the misbehavior.

This construction rules out abuse of the pseudonymization system. Ephemeral IDs cannot be spoofed because each packet must contain a MAC that is computed with the secret, previously registered host key k_H. Moreover, the registration process makes sure that no end-host can register such a host key under the ephemeral ID of another host. Shutoff requests cannot be misused for denial-of-service attacks because malicious packets, including the MACs that are only computable with knowledge of k_H, as proof of misbehavior.

13.4.6 Algorithm Agility

Algorithm agility is a property that enables migration from one algorithm to another one. It is especially important in the context of cryptographic algorithms, which become weaker over time. Since it is impossible to predict advances in cryptanalysis techniques, every future-proof architecture that employs cryptographic algorithms should provide a mechanism for algorithm agility.

In particular, achieving algorithm agility is a challenge if the exchangeable cryptographic algorithm has to be harmonized network-wide. In the following, we point out the elements of the proposed security architecture for a next-generation network where cryptographic algorithms are needed and explain how to provide algorithm agility in these settings:

- Signatures for path information (Sect. 13.4.2): In the path-discovery process, the path-segment construction beacons are extended by ASes with path information, which needs to be protected with a signature in order to be universally verifiable. In order to obtain algorithm agility for the signature algorithm, we envision that an AS can protect its added information by multiple signatures using different algorithms, while always explicitly naming the used signing algorithm. A consumer of created path segments can thus always check whether a trusted signature algorithm was used in the creation of the path segment. Algorithm diversity may also give rise to varying security properties across path segments in a transparent manner, enabling end-hosts to take account of the desired security level in their path selection.
- Hop authenticator in EPIC (Sect. 13.4.2): The hop authenticators embedded in path segments are mainly required to enable path authorization. In EPIC, the hop authenticator corresponding to an AS is computed as a MAC, which is based on a secret of the AS and needs to be later verified only by the AS itself. Hence, the

MAC algorithm can be chosen without coordination by each individual AS, providing perfect algorithm agility.

- Hop validation field in EPIC (Sects. 13.4.2 and 13.4.3): The hop-validation fields carried by packets allow to both authenticate the packet source and to validate that the path directive was followed during forwarding. Cryptographically, these hop-validation fields correspond to a MAC, which is computed with a DRKey shared between the hop AS and the source host. This MAC needs to be computable both by the source host and the hop AS, which requires negotiation of a MAC algorithm between these two entities. However, as no further parties are involved, a high degree of algorithm agility is still provided.
- Key derivation in DRKey (Sect. 13.4.3): The dynamically recreatable keys that are required in EPIC (see above) are derived with a pseudo-random function (e.g., a MAC), which must be known to both involved end-point ASes. Hence, negotiation of the PRF algorithm is necessary in order to use dynamically recreatable keys, but this negotiation only takes place between two parties, which yields a high degree of algorithm agility.
- Computation of reservation tags in QoS system (Sect. 13.4.4): In the reservation process, the authenticity of AS-specific reservation information is protected by a MAC, resulting in a reservation tag. Since this MAC is only intended for the AS itself to verify, the MAC algorithm can be chosen at the discretion of the respective AS, without requiring any coordination.
- Computations in APNA (Sect. 13.4.5): The main use of cryptography in the anonymity-enhancing APNA system is (1) in the computation of the ephemeral ID by the host AS and (2) in the computation of the MAC appended to packets to authenticate usage of the ephemeral ID in the packet. While the ephemeral ID computation must only be verified by the AS itself and the underlying cryptographic operations can thus be arbitrarily determined by any AS, the packet-appended MAC must be computable by both hosts and AS entities. Therefore, the currently usable MAC algorithm must always be communicated by the AS to the hosts residing in it. However, as this coordination effort is limited to an intra-AS context, algorithm agility is still given.

To obtain algorithm agility in certificates used in the PKI, provisions for allowing multiple signatures need to be made. Thus, a new signature algorithm can be introduced in the certificate, requiring an additional signature. A verifier who is only aware of one signature algorithm can validate that algorithm, but a verifier who is aware of both signature algorithms can verify both signatures. Such an approach enables switching to a new signature algorithm, and later phasing out the old algorithm.

With the emergence of quantum computers, preparation for quantum-safe cryptographic algorithms is appropriate. Although quantum computers capable of endangering current cryptosystems are still expected to be decades away, preparing a next-generation network architecture for algorithm agility is advisable.

13.4.7   Class of Security Level

Although security is desirable for almost any use case in an inter-domain network, security often comes at the price of additional processing, latency, or complexity, reducing the efficiency of communication. For some use cases, it may thus be desirable to trade security for efficiency. An end-host should thus be able to employ security functions depending on the desired security properties. The proposed security architecture for a next-generation network allows an end-host to adapt its guarantees to its demand for security in manifold ways:

- Path awareness (Sect. 13.4.2): Having path awareness allows an end-host to strike the optimal balance between security and performance in a multitude of ways. For example, an end-host can leverage path information to balance the degree of multipath transmissions with the overhead of managing multiple connections. Moreover, an end-host can choose paths according to performance properties (bandwidth, latency, loss, etc.) or according to security properties (location, confidence in path-information authenticity, disjointness, etc.).
- Bandwidth reservation for a QoS system (Sect. 13.4.4): By design, bandwidth reservation is an on-demand service. An end-host can purchase a bandwidth reservation for critical communication or rely upon best-effort transmission for less critical communication. By adapting the reservation amount, an end-host can obtain the optimal degree of insurance against link overload.
- Pseudonymous anonymity on request (Sect. 13.4.5): The APNA system allows an end-host to perform its inter-AS communication under a pseudonym (given by an ephemeral ID), but this anonymity comes at the cost of an additional check at the AS egress. If an end-host values the latency savings higher than the additional anonymity given by APNA, it is perfectly possible to avoid the check at the AS egress by using the real IP in packets instead of the ephemeral ID.

13.5   New Roles and Features

In this section, we aim at listing the new devices, services, and processes that are needed in the security architecture proposed in Sect. 13.4.

13.5.1   ISD-Specific Trust Roots

In order to provide a less vulnerable PKI for control-plane operations, Sect. 13.4.1 proposes to adopt the trust-root model of the SCION architecture. This model consists of grouping ASes in Isolation Domains (ISDs), which are spanned by a corporation or a jurisdictional region. Each of these ISDs has one or more core ASes that collaboratively determine the allowed roots of trust for control-plane operations in a so-called trust-root configuration (TRC). Hence, the trust infrastructure for a next-generation network requires (1) forming ISDs, (2) determining core ASes of each ISD, and (3) negotiating a trust-root configuration.

13.5.2   Path-Aware Network Architecture

In order to enrich a next-generation network with inter-domain path awareness, additional services are needed. For instance, in the SCION architecture, every AS deploys the following two additional services:

- Beacon service: Required for managing the path-segment construction beacons. The beacon service adds the relevant information to beacons and forwards the beacons to downstream ASes according to the domain's policy.
- Path service: Required for enabling lookups of paths for a given destination. The path servers cache path segments, providing end-hosts with the necessary information to reach destinations. In case there are no cached path segments for a given destination, the path service of a domain requests corresponding path segments from another path service, usually from an ISD core path server.

In order to grant path control to end-hosts, border routers must be extended such that the data-plane processing can check the path representation in the packet header. In particular, the border routers should be able to verify that the packet in fact follows the intended path (path validation), that this intended path is valid (path authorization), and that the packet is not spoofed (source authentication).

13.5.3   Key Servers for Enabling Source Authentication

For source authentication, the security architecture presented in this chapter builds on dynamically recreatable keys that have to be accessible by both the source AS and the authenticating AS, as well as both the source host and the authenticating host. In order for all entities to access the relevant keys, each AS requires a key server that can derive keys according to the DRKey rationale, and exchange keys with key servers in other ASes.

13.5.4   Bandwidth-Reservation System for Inter-Domain QoS

For bandwidth-reservation systems such as the system proposed in Sect. 13.4.4, every AS requires a reservation accounting server that manages the reservation requests arriving at the border routers as well as keeps track of available bandwidth that can be reserved. The border routers need to be extended with MAC computation functionality such that the data-plane processing can verify the reservation tag in packets.

13.5.5   Services and Border-Router Features for Pseudonymization

For the anonymity-enhancing APNA system presented in Sect. 13.4.5, ASes need two new services. First, they need a service that equips end-hosts with ephemeral IDs and communicates to other AS entities what key is associated with each ephemeral ID. Second, border-router functionality needs to be extended to retrieve the key associated with each ephemeral ID and check the MAC in packets that proves legitimate use of the ephemeral ID.

References

1. A. Perrig, P. Szalachowski, R.M. Reischuk, L. Chuat, *SCION: A Secure Internet Architecture* (Springer, Cham, 2017)
2. SCION Control-Plane PKI, https://github.com/scionproto/scion/blob/master/doc/ControlPlanePKI.md
3. M. Legner, T. Klenze, M. Wyss, C. Sprenger, A. Perrig, EPIC: Every packet is checked in the data plane of a path-aware internet. *29th USENIX Security Symposium (USENIX Security 20)* (2020)
4. S. Scherrer, A. Perrig, S. Schmid, The value of information in selfish routing, in *International Colloquium on Structural Information and Communication Complexity*, (Springer, Cham, 2020)
5. S. Scherrer, M. Legner, A. Perrig, S. Schmid, Incentivizing stable path selection in future Internet architectures. *Perf. Eval.* 144, 102137 (2020)
6. B. Rothenberger, D. Roos, M. Legner, A. Perrig, PISKES: Pragmatic internet-scale key establishment system, in *Proceedings of the ACM ASIA Conference on Computer and Communications Security (ASIACCS)*, (ACM, New York, 2020)
7. T. Lee, C. Pappas, A. Perrig, V. Gligor, Y.-C. Hu, The case for in-network replay suppression, in *Proceedings of the ACM on Asia Conference on Computer and Communications Security*, (ACM, New York, 2017)
8. G. Giuliari, M. Wyss, M. Legner, A. Perrig, GMA: A pareto optimal distributed resource-allocation algorithm. In *Proceedings of the International Colloquium on Structural Information and Communication Complexity (SIROCCO)* (2020)
9. T. Lee, C. Pappas, D. Barrera, P. Szalachowski, A. Perrig, Source accountability with domain-brokered privacy. In *Proceedings of the 12th International on Conference on emerging Networking EXperiments and Technologies* (2016)

Chapter 14
Intent-Based Network Management

Alexander Clemm

14.1 Introduction

As outlined in earlier chapters of this book, the networking landscape is expected to undergo profound changes over the coming years. New networking services are expected to emerge that will enable new applications, such as the tactile internet, holographic-type communications, or tele-driving. Many of these services will bring about new management challenges, such as the need to provide service assurance for unprecedented service level guarantees, given many of services are much less forgiving of network performance glitches than services in the past. At the same time, existing services will continue to not only evolve, but the volume of network traffic, the number of connected devices, and the number of service instances continue to explode. This creates significant challenges for the management of those networks and services, which are of course expected to operate smoothly, securely, and efficiently. Among the biggest challenges are the requirement to keep up with exploding scale and, from a business perspective, the need to minimize cost.

Clearly, networks are a long way past being able to rely on heroics of individual network administrators as a viable business strategy. Instead, they have to rely heavily on automation. This trend has been at work for a long time. Functions such as automated provisioning and service fulfillment workflows for new customers and automated diagnosis and correlation of alarms have been standard operator practice for decades. Since then, automation has accelerated to include functions such as automated performance trend analysis, machine learning to no longer have to rely on human expertise to detect operationally relevant network telemetry patterns, or real-time analysis of traffic matrices to optimize dynamic allocation and placement of networking resources.

A. Clemm (✉)
Futurewei Technologies, Inc., 2550 Central Expressway, Santa Clara, CA 95050, USA
e-mail: alex@futurewei.com

© The Author(s), under exclusive license to Springer Nature
Switzerland AG 2021
M. Toy (ed.), *Future Networks, Services and Management*,
https://doi.org/10.1007/978-3-030-81961-3_14

The goal for network automation has culminated in the vision of networks that no longer need management involvement by human users because they have been fully automated. This vision has had many names: self-CHOP (configuring, healing, optimizing, protecting) [1], autonomic networking [2], and self-driving networks [3], among the more prominent ones. Regardless of differences in nuances, all these visions have in common the concept of automated control loops: networking devices generate data that is then collected and analyzed by intelligent systems, which automatically derive conclusions and automatically adjust configurations and parameter settings of networking infrastructure as needed, subsequently observing the outcomes as the loop closes and the next cycle begins. The other common requirement is that no human must be in the loop for a long list of reasons, including the lack of ability to scale operations, the inability to operate at very short time scales, the possibility for human error, and cost.

Network management has made continuous advances towards this vision. While arguably networks are still not fully automated, a lot of routine management tasks have been absorbed into networking functions while at the same time the role of traditional network administration has evolved into DevOps, i.e., the integration of continuous operations and development cycles in which the role of the traditional networking engineer evolves to include continuous development of automation.

However, regardless of the level of network automation, networks should neither be free-willed nor will they be clairvoyant. Instead, they will still need to accommodate human input – not to conduct routine operational tasks, but to allow humans to give direction and guidance for how the network should ultimately be used, what services (and to whom) need to be provided, what operational goals to prioritize, and what other aspects to take into consideration that should affect the way the network operates. This guidance and direction is what is now commonly referred to as "intent."

Intent is defined as the ability to allow users to define management outcomes, as opposed to having to specify precise rules or algorithms that will lead to those outcomes [4]. This requires an intent-based system to possess the necessary intelligence to identify the required steps on its own. Networks that are supported by intent-based systems that allow them to be managed using intent are referred to as "intent-based networks" (IBN).

In the remainder of this chapter, an overview of intent-based network management and IBN is given. Section 14.2 provides a more detailed look at the concept of "intent" and what it entails. While "intent" is a fairly new term, there have been related concepts in the past, such as policy-based management and service management, which will be discussed in Sect. 14.3. Section 14.4 lays out various functions of an intent-based system. How these function and interrelate and how they can be combined into a reference architecture for intent-based networking is subsequently described in Sect. 14.5. It should be noted that intent-based networking is largely still an emerging topic, its more advanced promises in many cases still more of a vision than an actual reality, clever marketing of some products that are commercially available today notwithstanding. Hence, Sect. 14.6 points out a number of problems and research challenges that require further work. Section 14.7 provides brief conclusions.

It should be noted that intent is still an emerging topic that is subject to active discussion in research and standardization fora, notably the IRTF. Accordingly, the contents in this section draws heavily on material from those fora, specifically on [4, 7] for Sects. 14.2, 14.3, 14.4, and 14.5, which we hereby reference and acknowledge.

14.2 Intent Concept Overview

Intent is a declaration of operational goals that a network should meet and outcomes that the network is supposed to deliver, without specifying how to achieve them. Those goals and outcomes are defined in a manner that is purely declarative. This means that they specify what to accomplish, not how to achieve it. They reflect what is on the operator's mind, not a specific plan or procedure for specific steps to take.

"Intent" thus applies several important concepts simultaneously. For one, it provides data abstraction: users and operators do not need to be concerned with low-level device configuration and nerd knobs. Instead, they are allowed to think in terms of higher-level concepts. Of course, the concept of data abstraction by itself is not new and is used in other contexts. However, in addition, it also provides functional abstraction from particular management and control logic: users and operators do not need to be concerned even with how a given Intent might be achieved. What is specified instead is a desired outcome, with the intent-based system automatically figuring out a course of action for how to achieve the outcome. Determining a course of action could involve applying an algorithm, applying a set of rules derived from the intent, even machine-learning applications that assess which actions are having the desired effect, in extreme cases using feedback loops based on trial-and-error and observation. This goes well beyond the type of event-condition-action logic ("if event e happens and condition c holds, then perform the following action") familiar from other systems in the past. Those systems still rely on an administrator to enumerate rules that define what to do under any given circumstance, rather than focusing on the outcome that should be achieved.

The following are some examples of intent:

- "Steer networking traffic originating from endpoints in one geography away from a second geography, unless the destination lies in that second geography." This simply states what the network should achieve without saying how.
- "Avoid routing networking traffic originating from a given set of endpoints (or associated with a given customer) through a particular vendor's equipment, even if this occurs at the expense of reduced service levels." Again, this simply states what to achieve, not how. In addition, guidance is given for how the system should trade off between different goals when necessary.
- "Maximize network utilization even if it means trading off service levels (such as latency, loss), unless service levels have deteriorated at least 25% from their

historic mean." This clearly defines a desired outcome. It also specifies a set of constraints to provide additional guidance, without specifying how to achieve any of this.

- "VPN service must have path protection at all times for all paths." Again, a desired outcome. How to precisely accommodate it is not specified.
- "Generate in situ OAM data and network telemetry across that will be useful for later offline analysis whenever significant fluctuations in latency across a path are observed." This goes well beyond traditional event-condition-action rules because it is not specific about what constitutes "significant" or what specific data items need to be to collected.

In contrast, the following are examples of what would not constitute intent:

- "Configure a given interface with an IP address." This would be considered device configuration and fiddling with configuration knobs, not intent.
- "When interface utilization exceeds a specific threshold, emit an alert." This amounts to a rule that can help support network automation, but a simple rule is not an intent.
- "Configure a VPN with a tunnel from A to B over path P." This would be considered as configuration of a service.
- "Deny traffic to prefix P1 unless it is traffic from prefix P2." This would be an example of an access policy or a firewall rule, not intent.

All of the above examples are expressed in natural language for sake of clarity. Indeed, ideally an operator conveying intent to the network would be able to use natural language and have the network infer intent from it, asking for clarification where required. This way, the network would speak the operator's language, as opposed to imposing the hurdle of forcing the operator to learn the network's language. However, in practice a special intent language or, for machine-to-machine communications, intent API will be used.

The possibility of an intent language leading to greater operator convenience should not distract where intent's real significance lies. While convenience and ease of use are nice, the real significance of intent lies in its ability to scale operations. As networks continue to grow increasingly complex and the number of devices and services explodes, keeping up with that growth from a management perspective becomes a daunting challenge. By not requiring network operators to define individual steps or rules to manage devices, even when those steps can be automated, and by not requiring them (for example) to come up with plans for how to optimize deployments, a problem which can be NP-hard, intent-based networks put network operations in a position where scalability is no longer the issue that it used to be in the past.

An intent-based network (IBN) is a network that can be managed using intent. This means that the network is able to recognize and ingest intent of an operator, or user, and to configure and adapt itself autonomously according to the user intent, achieving an intended outcome (i.e., a desired state or behavior) without requiring the user to specify the detailed technical steps for how to achieve the outcome.

Similarly, an intent-based system is a system that allows users to manage a network using intent. Such a system will serve as a point of interaction with users and implement the functionality that is necessary to achieve the intended outcomes, interacting for that purpose with the network as required.

In an ideal world, an intent-based network would not even need an intent-based system to achieve intent. Instead, network devices themselves would be able to achieve intent using a combination of distributed algorithms and local device abstractions. In this idealized vision, because intent holds for the network as a whole, intent would ideally be automatically disseminated across all devices in the network as needed, which would themselves decide whether they needed to act on it and how to coordinate with other devices where required. However, such decentralization will not be practical in most cases and certain functions will need to be at least conceptually centralized. For example, users may require a single conceptual point of interaction with the network. Likewise, the vast majority of network devices will themselves be intent-agnostic and focus only (for example) on the actual forwarding of packets. This implies that intent functionality needs to be provided functions that are specialized for that purpose. Depending on the scenario, those functions may be hosted on dedicated systems or cohosted with other networking functions. For example, functionality to translate intent into courses of actions and algorithms to achieve desired outcomes may need to be provided by such specialized functions. Performing those functions may require significant processing power, visibility of large parts of the network, and large datasets. Of course, to avoid single points of failure, the implementation and hosting of those functions may still be distributed, even if the functions themselves are conceptually centralized.

It should be noted that other definitions of intent exist, such as the one used in [6]. Intent there is simply defined as a declarative interface that is typically provided by a controller. It implies the presence of a centralized function that renders the intent into lower-level policies or instructions and orchestrates them across the network. While this is certainly one way of implementation, this definition is fairly narrow and does not emphasize what it is that makes intent-based network management truly unique: namely, the ability to manage the network by specifying desired outcomes without the specific steps to be taken or algorithms to be performed in order to achieve the outcome. According to this, a controller API that simply provides a network level of abstraction would not necessarily qualify as intent (even if it is marketed as such). Likewise, ingestion and recognition of intent by the network may not necessarily occur via traditional APIs, but may involve other types of human-machine interactions.

14.3   Related Concepts

The concept of automatically breaking down management requests from higher levels of abstraction into low-level management actions has been applied by other technologies in the past. Examples include service order provisioning systems that break down requests for user services, or policy-based management, which allows operators to specify policies, often conditioned around rules that express what set of actions to take under which circumstances (often a combination of conditions and event triggers). However, in each case, the rules to apply or the mapping steps to take need to still be specified by a network administrator.

In contrast, intent is about letting users specify desired outcomes without having to specify the specific set of steps to get there or spelling out which actions to take under which condition. The set of actions to take or even the set of policies or algorithms to apply in order to achieve the outcomes may not even be predetermined but could be learned automatically by an intent-based (management) system over time. Of course, in simple cases, a simple mapping or translation step similar to what a policy-based system would perform may be enough also in the case of an intent-based system. In some cases, the translation steps may themselves result in network policies. However, more advanced and sophisticated systems may be able to apply artificial intelligence techniques to identify courses of action, dynamically moderate in real-time competing demands from millions of service instances, and apply learning techniques to optimize outcomes over time. Likewise, the specification of intent by users may follow unconventional interfaces, not necessarily based on a traditional command syntax or request pattern, but allowing for human-machine dialog that allows for iterative refinement and includes explanation components. These aspects set intent-based networking apart from other technologies.

Of course, intent-based networking would not be conceivable without the technologies that came before it. Over time, they have evolved along two dimensions, towards increasing levels of abstraction as well as greater degrees of associated automation, as depicted in Fig. 14.1.

With this, let us take a brief look at some of the closest "relatives" of intent-based networks in order to explain their differences.

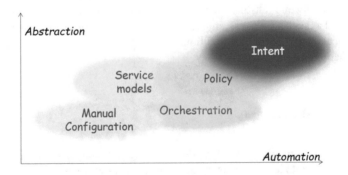

Fig. 14.1 Evolution of intent (figure adopted from [7])

14.3.1 *Service Models and Service Management*

A service model is a model that represents a service that is provided by a network to a user. Per [8], a service model describes a service and its parameters in a portable, implementation-agnostic way that can be used independently of the equipment and operating environment on which the service is realized. Similarly, [9] introduced a management reference model that includes a service management layer on top of a network management layer.

There are two aspects of a service model: the model of the actual service is used to describe instances of a service as provided to a customer, possibly associated with a service order. This is sometimes also referred to as "customer service model." Examples would be residential Internet access, a layer 3 VPN service, or a network slice. In addition, there is a model that describes how a service is instantiated over existing networking infrastructure. Sometimes this is referred to as "service delivery model." This includes the allocation of network resources such as ports, IP addresses, or bandwidth, as well as configuration steps that need to be taken, often in an orchestrated manner that involves particular sequences of steps.

What service management has in common with IBN is the fact that it provides an abstraction of the network that allows operators to focus on the ultimate purpose of the network, which is to provide services. This abstraction shields operators from low-level configuration knobs and from the need to be aware of details of how to configure service instances across the network. This allows operators to take a holistic, end-to-end perspective. However, service management is confined to services, not management of other aspects of a network that still need to be addressed, such as, say, the management of device and software upgrades, planning of paths or link capacities, or management of managing security for the network infrastructure as a whole. Unlike intent, service management does not allow to define a desired "outcome" that would be automatically maintained by the intent system. Instead, the management of service models serves a much more limited purpose that still requires the development of sophisticated algorithms and control logic by network providers or system integrators.

14.3.2 *Policy-Based Management*

Policy-based network management (PBNM) is a popular management paradigm that separates the rules that govern the behavior of a system from the functionality of the system. It is also the subject of a rich set of literature [5].

At the heart of policy-based management is the concept of a policy. Multiple definitions of policy exist: "Policies are rules governing the choices in the behavior of a system" [10]. "Policy is a set of rules that are used to manage and control the changing and/or maintaining of the state of one or more managed objects" [11]. Common to most definitions is the definition of a policy as a "rule." Typically, the definition of a rule consists of an event (whose occurrence triggers the rule), a set of conditions (which get assessed when trigger event occurs and which must evaluate to true before any actions are actually "fired"), and finally a set of one or more actions that are carried out when the conditions hold. The rules allow to automate the dynamic behavior of systems in a smart way and allow it to react to dynamic occurrences and changes in context.

Like intent, policies provide a higher layer of abstraction. In general, the events, conditions, and actions that are defined as part of policies are defined in a device-independent manner based on abstract models that are "rendered" to and from device-specific representations as needed. However, unlike intent, the definition of those rules (and of the courses of action that they imply) still needs to be articulated by users. Since the intent behind the policies is unknown, conflicts between policies which contradict one another cannot be easily detected or resolved. Instead, it requires invention by the user or by some kind of logic that resides outside of PBNM. In that sense, policy constitutes a lower level of abstraction than intent. That said, it is conceivable for intent-based systems to generate policies that are subsequently deployed by a policy-based management system, allowing PBNM to support and complement intent-based networking.

A good analogy that captures the difference between policy and intent systems is that of expert systems and learning systems in the field of artificial intelligence. Expert systems operate on knowledge bases with rules that are supplied by experts, analogous to policy systems whose polies are supplied by users. They are able to make automatic inferences based on those rules and explain how they arrived at their conclusions, but are not able to "learn" new rules on their own. Learning systems (popularized by deep learning and neural networks), on the other hand, are able to learn without depending on user programming or articulation of rules. However, they do require a learning or training phase, and providing explanations for actions that the system actually takes may sometimes prove challenging. Analogous to intent-based systems, learning systems allow users to focus on what they would like the system to accomplish, not how to do it.

14.3.3 Autonomic Networking

As mentioned earlier, autonomic networking deals with the vision of networks that do not require any management at all. Over the years, different names have been used to define this vision, including "self-managing networks," "self-CHOP networks" (i.e., self-configuring, self-healing, self-optimizing, and self-protecting), and, most recently, "self-driving networks."

However, even if a network were fully autonomic, it would still not be clairvoyant. In other words, it would not be able to read the operator's mind as to what services it should provide and to whom, which requests are legitimate and which not, or how to resolve conflicting goals and decide between tradeoffs. This is where intent comes in, allowing an operator to specify which outcomes are desired, in the process giving the direction and guidance that is needed even by an autonomic network. Translating these outcomes into what corresponding steps to take, communicating with the operator the tradeoffs that may need to be made when certain outcomes are only partially achievable are likewise part of the IBN, which may of course leverage autonomic network functions in order to make sure the intent is being followed.

In that sense, autonomic networking and IBN naturally complement one another. Both are ultimately about the same things:

- Improving ROI (return on investment) by reducing cost (opex, e.g., by simplifying operations, as well as capex, e.g., by making more effective use of networking resources than would be "manually" achieved) and increasing revenue (e.g., enabling faster time to service or providing the ability to provide better and hence more valuable and expensive service level guarantees).
- Scaling of operations, allowing network providers to keep up with the growth of their networks and of the amount of complexity involved. The ability to scale operations is just as important as the ability to scale the network itself.

14.4   IBN Functionality

With the background from the previous subsections, let us dive into the functionality that needs to be provided by an intent-based network. This functionality has been thoroughly described also in [4], from which the following description therefore heavily borrows.

It turns out that intent-based networking involves a wide variety of functions which can be roughly divided into two categories:

- Intent fulfillment provides functions and interfaces that allow users to communicate intent to the network and that carry out the necessary actions to ensure that intent is achieved. This includes algorithms to determine proper courses of action and functions that learn to optimize outcomes over time. In addition, it also includes more traditional management functions such as any required orchestration of coordinated configuration operations across the network and rendering of higher-level abstractions into lower-level parameters and control knobs.
- Intent assurance provides functions and interfaces that allow users to validate and monitor that the network is indeed adhering to and complying with intent. This is necessary to assess the effectiveness of actions taken as part of fulfillment, providing important feedback that allows those functions to be trained or tuned over time to optimize outcomes. In addition, intent assurance is necessary to address "intent drift." Intent drift occurs when a system originally meets the intent, but over time gradually allows its behavior to change or be affected until it no longer does or does so in a less effective manner.

We will describe IBN functionality along those two categories in the following subsections.

14.4.1 Intent Fulfillment

Intent fulfillment is concerned with the functions that take intent from its origination by a user (generally, a network administrator or the responsible organization, not an end user of a communications service) to its realization in the network.

The first set of functions is concerned with "ingesting" intent, i.e., with extracting intent from users and communicating it to the IBN. The functions involve the ability to recognize intent from interactions with the user, including functionality that allows users to refine their intent and articulate it in such ways so that it becomes actionable by an intent-based system. Typically, those functions go beyond a traditional API, although they may include APIs provided for interactions with other machines. However, they may also support unconventional human-machine interactions, in which a human will not simply give simple commands, but which may involve a human-machine dialog to provide clarifications, to explain ramifications and trade-offs, to avoid ambiguities and contradictions, and to facilitate refinements. The goal of those functions is to make intent-based systems as easy and natural to use as possible. This enables the user to interact with the intent-based system in ways that does not involve a steep learning curve forcing the user to learn the "language" of the system, but that simply makes the user as effective as possible.

A second set of functions is needed to be able to translate user intent into courses of action that need to be performed against the network, for example, to determine which requests need to be directed at network configuration and provisioning systems. Intent translation lies at the core of intent-based systems. It bridges the gap between interaction with users on one hand and, on the other hand, the traditional management and operations infrastructure that orchestrates provisioning and configuration activity across the network. Beyond merely breaking down a higher layer of abstraction (viz., intent) into a lower layer of abstraction (such as sets of policies or individual device configurations), intent translation functions can be complemented with functions and algorithms that perform optimizations and that are able to learn and improve over time in order to result in the best outcomes, specifically in cases where multiple ways of achieving those outcomes are conceivable. For example, satisfying an intent may involve computation of paths and other parameters that need to be configured across the network and that can be optimized along multiple criteria. Heuristics and algorithms to do so may evolve over time to optimize outcomes which may depend on a myriad of dynamic network conditions and context.

Finally, a third set of functions deals with the actual configuration and provisioning steps that need to be orchestrated across the network and that were determined by the previous intent translation step.

14.4.2   Intent Assurance

Intent assurance is concerned with the functions that are necessary to ensure that the network indeed complies with the desired intent after it has been fulfilled. This includes a set of functions that simply monitor and observe the network and its behavior. This includes (but is not limited to) functions that monitor the network for events and alarms, that perform measurements to assess service levels that are being delivered, that detect performance outliers and analyze bottlenecks, and that generate and collect telemetry data. These are all the usual assurance functions that have traditionally been part of operating a network for a long time; they are not specific to IBN. However, the data and insights from monitoring and observing the network are required as basis for the next set of functions that assess whether the observed behavior is in fact in compliance with the behavior that is actually "intended," i.e., the expected outcome per the intent.

The functions that assess whether the network is indeed complying with the stated intent are at the core of intent assurance. These functions compare the actual network behavior that is being observed with the intended outcomes and behavior that is expected per the intent. These functions continuously assess and validate whether the observation indicates compliance with intent. This includes assessing the effectiveness of intent fulfillment actions, including verifying that the actions had the desired effect and assessing the magnitude of the effect as applicable. It can also include functions that analyze and aggregate other raw data that was collected from monitoring, measurement, and observation activities. The results of the observations and assessment, i.e., data on how close the network is adhering to intent and how well it is performing, can be fed to learning functions that help intent fulfillment functions to optimize overall outcomes.

Intent compliance assessment also includes assessing whether intent drift occurs over time. Intent drift can be caused in many ways, for example, by control plane or lower-level management operations that cause behavioral changes which inadvertently conflict with intent which was orchestrated earlier. Intent-based systems and networks need to be able to detect when such drift occurs or, better yet in order to be able to react in time, when is about to occur. When intent drift occurs or network behavior is inconsistent with desired intent, functions that are able to trigger corrective actions are needed. This includes actions that are needed to resolve intent drift and to bring the network back into compliance.

As an alternative and where necessary, reporting functions can be triggered that alert operators and provide them with information and tools that empower them to react appropriately, for example, by helping them articulate modifications to the original intent in order to moderate between conflicting concerns.

The outcome of intent assurance needs to be reported back to the user in ways that allows the user to relate the outcomes to their intent and assess the effectiveness both of the IBN and of the intent. This requires a set of functions that are able to analyze, aggregate, and abstract the results of the observations accordingly. In many cases, lower-level concepts (such as detailed performance statistics related to low-level settings) need to be "up-leveled" to concepts the user can relate to and take action on. In addition, the associated aggregation and analysis functionality should be complemented with functions that report intent compliance status and that provide adequate summarization and, ideally, visualization to the user.

14.5   IBN Control Loops and Reference Architecture

The functions that make up an IBN do not exist side by side but complement one another, forming broader control loops.

For one, there is a continuous control loop that takes place within the IBN itself: intent is fulfilled, and then compliance of the network with the intent is assessed, which in turn may result in further fulfillment actions in order to make adjustments.

In addition, there is a larger control loop that involves the user: the user expresses intent which is subsequently fulfilled and assured. In addition to assessing compliance of the network with the intent and making continuous adjustments as needed by the "inner" control loop, as part of the larger control loop, the outcomes and the effectiveness of the intent are reported back to the user. This information is then taken into account by the user in order to refine or modify intent as needed. Accordingly, intent is subjected to a life cycle: it comes into being, may undergo changes over the course of time, and may at some point be retracted.

The way in which the various functions complement each other and where they are positioned on the intent control loops is depicted in the Fig. 14.2. The figure also depicts the flow of control and data between the various functions. This way, it really also reflects an IBN reference architecture.

Intent functionality is arranged into two functional (horizontal) planes that reflect the distinction between functions that are related to fulfillment and functions that are related to assurance. In addition, there are three (vertical) spaces into which

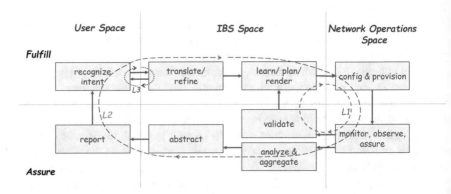

Fig. 14.2 Intent life cycle and IBN reference architecture (figure adopted from [4, 7])

functions are arranged. The spaces indicate the different perspectives and interactions with different roles that are involved in addressing the respective functions:

- The user space involves the functions that interface the network and intent-based system with the human user. This includes the functions that allow users to articulate and the intent-based system to recognize that intent. It also includes the functions that report back the status of the network relative to the intent and that allow users to assess whether their intent has the desired effect.
- The translation or intent-based system (IBS) space involves the functions that bridge the gap between intent users and network operations. This includes the functions used to translate an intent into a course of action, the algorithms used to plan and optimize those courses of action also in consideration of feedback, and the functions to analyze and abstract observations in order to validate compliance with intent and take corrective actions as necessary.
- The network operations space, finally, involves the traditional orchestration, configuration, monitoring, and measurement functions which are used to carry out the various actions, e.g., perform necessary configuration steps, and to observe the effects of those actions on the network.

Also depicted in the figure are the mentioned control loops. The "inner" loop, L1 (depicted in with a blue dashed line), is completely autonomic and does not involve any humans or operators. It involves rendering of intent by performing configuration and provisioning operations, followed by monitoring and observing the network and feeding those observations into functions that analyze those observations and validate that the network is indeed conforming with the intent, and then feeding the results of that into analysis back the into function that plans the rendering of networking intent and thus closing the loop. This allows the function that renders the intent to make adjustments as needed to the configuration of the network. It also provides important feedback that allow for further optimization as well as for learning.

In addition, the "outer" intent control loop, L2 (depicted in green with a wider dashed line), involves the user space. It reflects the fact that the user may take actions and adjust intent based on feedback from the IBS. Finally, a third and very small control loop, L3 (depicted with a dotted purple line), reflects the fact that the communication of intent may involve unconventional interfaces and user interaction patterns. For example, the articulation of intent by users may involve feedback from the IBS to help clarify intent, refine its articulation to provide additional details as required, and potentially inform users of ramifications of intent they articulate.

14.6 Research Challenges

Intent-based network management is at this point still more a vision than a reality. While great strides are being made and commercial offerings are beginning to appear, plenty of work remains to be done and many fundamental problems still need to be solved. This creates significant opportunities for research. In the following, some of these challenges are described. While by no means a complete list, advances in any of those areas will have the potential to not only satisfy academic curiosity and interest, but to also "move the needle" for the networking and network services industry quite significantly.

One set of research challenges concerns the area of human/machine interaction, where simple command-based interfaces may need to suffice for now but will not be sufficient in the longer term. Interfaces will need to support interactions that go beyond simple request/response patterns to help users articulate actionable intent in ways that are effective and easy. This goes well beyond capabilities that are offered by digital voice assistants.

For example, interfaces should incorporate abilities to refine intent when it is needed, possibly involving a dialogue, negotiation, or interview rather than a simple request. Likewise, interfaces should anticipate the possibility of intent ambiguities and conflicts and allow for their resolution. For example, optimizing utilization of a certain resource may conflict with the intent of optimizing performance of a particular service, as it increases the likelihood of collisions and resource contentions. Also, they should allow to explain unintended consequences and ramifications of certain intent choices to users. For example, the decision to minimize energy use may result in loss of elasticity when a sudden surge in demand for networking resources arises. In order to conserve energy, some resources might be simply shut down instead of being let run idle, but starting them up again and bringing them back online may take time.

Interfaces should also allow a system to acknowledge a user's intent while managing user expectations in cases where it may not be possible to fully meet the intent, or only meet it to a certain degree. An example here might be the ability to protect every service instance with an alternate path, which may have topology-based limitations. There are also synergies to explore with Promise Theory [12], with the commitment of a system to fulfill an "intent" amounting in essence to a promise, which may not necessarily and under all circumstances be possible to keep, or which may only be given to a degree and with certain caveats and limitations.

A second set of research challenges involves the ability for intent-based systems to explain the causes of actions that are taken as well as the reasoning behind it. One concern for network providers is to lose control over their network and no longer understand what it is doing. As intent allows users to merely specify desired outcomes, with no need (and quite possibly, no idea) to specify how to achieve that outcome, this is a real possibility. As networks are very large and distributed systems, a particular fear concerns that network behavior might no longer consistently converge and essentially spin out of control, with no ability to counteract. This particular set of challenges is closely related to explainable AI [13]. Indeed, as far as learning and AI techniques are made use of to achieve intended outcomes, IBN may be a prime use case here.

A third set of research challenges concerns assessing intent compliance. Inferring from a network whether and to what degree it complies with intent, in which ways it might deviate, and how to best bring it into full compliance from its current state presents interesting challenges. Doing this is a significant challenge, as it requires the ability to relate low-level state in the network and networking devices to higher-level abstractions and understanding the impact and contribution to those abstraction. Additional challenges relate to automatic determination and planning of steps that will move the network to a state of better compliance.

Addressing the problem of intent compliance is all the more important due to the related problem of intent drift, i.e., the fact that over time, a network may "drift away" from earlier intent as the network undergoes changes over time and other intent is accommodated. While traditional commands and requests are one-time type of affairs of a transactional nature, intent typically persists over time and hence also needs to be maintained over time.

There are many other challenges beyond these. In no particular order:

- Automated learning techniques to help improve outcomes, particularly in ways that can be deployed as part of a production system without negatively affecting it. In particular, advances are needed that allow for learning in real-time and during production deployment, as training phases with advance training sets may in many cases not be an option.
- Prediction techniques to predict intent compliance from network telemetry and state data, to predict the effectiveness of courses of actions taken, to predict intent drift. Similarly, forecasting techniques (that make a forecast in time).
- Automated planning techniques that can be applied to NP-hard network optimization problems related to important outcomes, such as network resource assignment or path configuration.
- Maintaining visibility in the face of encrypted and tunneled traffic which may complicate functions related to intent assurance.

This is by no means an exhaustive list. However, it should convince the reader that IBN is a fertile area for research. Several of those problems intersect with the exploding fields of artificial intelligence and machine learning and should also benefit from progress made there. For all these reasons, we should expect exciting progress in IBN over the coming years.

14.7   Conclusions

Intent-based networking represents the latest stage in the long-running quest for smarter networks that are able to simply "run on their own," minimizing or, better yet, eliminating the need for management invention entirely. In many ways, IBN complements and indeed completes the vision of autonomic networks, which, while being largely "self-managed," do not have clairvoyant abilities and still depend on guidance from users. At the same time, IBN represent a significant step forward from, for example, the policy-based network management paradigm that preceded it, allowing to define expected outcomes without needing to specify a particular course of actions or even rules of what to do under which circumstance.

While first intent-based products are beginning to appear, many challenges remain until the full IBN vision can become a reality. These include the need for advances in human/machine interaction, the ability to add intent assurance functions such as validation of intent compliance and detection of intent drift to the arsenal of tools, as well as advances in automated planning and learning capable of adapting in real time. In that last area, IBN has plenty of natural synergies with artificial intelligence and machine learning and stands to benefit from the rapid advances made in those fields. However, what will propel IBN to further advances and increasing relevance in the future will simply come down to economics: namely, IBN's potential to significantly improve network provider's return on investment due to operational efficiencies, and the necessity to be able to scale the ability to operate networks as the complexity of networks and interdependencies, the demands placed on future networking services, and scale of networks continue to explode.

References

1. N. Agoulmine, Introduction to autonomic concepts applied to future self-managed networks, in *Autonomic Network Management Principles: From Concepts to Applications*, (ACM, New York, 2010)
2. M. Behringer, M. Pritikin, S. Bjarnason, A. Clemm, B. Carpenter, S. Jiang, L. Ciavaglia, Autonomic networking: definitions and design goals. IETF RFC 7575 (2015)
3. K. Kompella, Self-driving networks, in *Emerging Automation Techniques for the Future Internet*, (IGI Global, Hershey, 2019), pp. 21–44
4. A. Clemm, L. Ciavaglia, L. Granville, J. Tantsura, Intent-based networking – concepts and definitions. IETF I-D draft-irtf-nmrg-ibn-concepts-definitions (2021)
5. R. Boutaba, I. Aib, Policy-Based Management: A Historical perspective. JNSM 15, 4 (2007)
6. Open Networking Foundation, Intent NBI – definition and principles. ONF TR-523 (2016)
7. A. Clemm, M.F. Zhani, R. Boutaba, Network management 2030: Operations and control of network 2030 services. JNSM 28(4), 721–750 (2020)
8. C. Wu, W. Liu, A. Farrel, Service models explained. IETF RFC 8309 (2018)
9. ITU-T, Principles for a telecommunications management network. ITU-T M.3010 (2000)
10. M. Sloman, Policy driven management for distributed systems. JNSM 2, 4 (1994)
11. J. Strassner, *Policy-Based Network Management* (Elsevier, London, 2003)
12. J. Bergstra, M. Burgess, *Promise Theory: Principles and Applications* (CreateSpace, Scotts Valle, 2014)
13. A. Adadi, M. Berrada, Peeking inside the black-box: a survey on explainable artificial intelligence (XAI). IEEE Access 6, 52138–52160 (2018)

Chapter 15
AI-Based Network and Service Management

John Strassner

15.1 Introduction

Current network and service management provisioning and monitoring functions are increasing in complexity. The proliferation of different technologies, as well as different implementations from different vendors, demands human-in-the-loop processing, which is time-consuming and error-prone. In addition, users are demanding more complex services (e.g., context-aware, personalized services).

However, these problems pale in comparison to being able to provide network services that are offered according to the current business needs of the organization. This problem was first conceptualized in 2002 [1], called business-driven device management (BDDM). Network management architectures suffer from the inability to define and use business processes to drive the configuration and management of network resources and hence, network services. BDDM is a paradigm that enables business rules to manage the construction of configuration files and commands for a device as well as enforce how the configuration of a device is created, verified, approved, and deployed. BDDM uses different types of policies to manage the different aspects of providing network services. These policies form a continuum that represents the complete life cycle (from order to creation to teardown) of network services, bridging the automation gap between the service and element layers, and controlling which network services and resources are allocated to which users. More importantly, a continuum of policies is critical for representing the needs of different constituencies [2]. For example, there is no command line interface command that corresponds to a business concept like "Gold Service," nor is their understanding at the network configuration level of differences between service levels

J. Strassner (✉)
Futurewei Technologies, Inc., Santa Clara, CA, USA
e-mail: john.sc.strassner@futurewei.com

M. Toy (ed.), *Future Networks, Services and Management*,
https://doi.org/10.1007/978-3-030-81961-3_15

399

(e.g., Gold vs. Platinum). This makes it very difficult to translate Service Level Agreements (SLAs) into commands that network devices can understand.

This problem is exacerbated as the level of business abstraction increases. For example, suppose a network operator wants to optimize the set of services offered to maximize revenue while minimizing customer churn by ensuring that features such as security and availability are not comprised for more important customers. This can be formulated as a multi-objective optimization problem, where optimal decisions need to be made, even though all objectives may not be able to be simultaneously optimized. If there is some amount of uncertainty as to whether attributes of an option are guaranteed, the problem can also be defined as a multi-attribute utility function, where uncertainty and risk for each attribute of an objective are modeled; this could then be optimized by a number of methods, such as multi-objective integer linear programming. The problem with these approaches is not the mathematics, but rather, how to use the results to properly configure and manage network services.

Another factor is taking into account the monetary cost associated with over-provisioning or under-provisioning of networking capacity, computational power, and other capabilities. This can be captured in a number of ways, such as constraints on realizing an objective or additional attributes of choosing a particular approach.

Hence, operators are concerned about the increasing complexity of integration of different platforms in their network and operational environment. These human-machine interaction challenges increase the time to market of innovative and advanced services. Moreover, there is no efficient and extensible standards-based mechanism to provide contextually aware services (e.g., services that adapt to changes in user needs, business goals, or environmental conditions). These and other factors contribute to a very high OPerational EXpenditure (OPEX) for network operation and management. Operators need to optimize the use of networked resources (e.g., through the automation of their network configuration and monitoring processes to reduce this OPEX). More importantly, operators need to improve the use and maintenance of their networks.

The above examples are exacerbated by the frequent changing of user needs, business goals, and environmental conditions. This requires improved automation and real-time closed control loops. Thus, network intelligence is needed to detect these contextual changes, determine which groups of devices and services affect each other, and manage the resulting services while maintaining SLAs.

The above problems are examples of the *static configuration* of current networks and services. One solution is to realize a *cognitive* network, where offered services can be more easily related to business needs. In this approach, intelligence is imbued into the governance of the network system and its services by making use of three important design principles: situation awareness, experiential learning, and decision-making using adaptive closed control loops.

15.2   Terminology

This chapter defines key terminology as follows (Table 15.1):

Table 15.1 Terminology

Term	Definition
Closed control loop	A control loop whose controlling action is dependent on feedback from the object or process being controlled to achieve desired behavior
Adaptive closed control loop	A closed control loop whose controlling function adapts to the object or process being controlled using parameters that are either unknown and/or vary over time
Cognitive closed control loop	A closed control loop that selects data and behaviors to monitor that can help assess the status of achieving a set of goals and produce new data, information, and knowledge to facilitate the attainment of those goals
Cognition	The process of acquiring and understanding data and information and producing new data, information, and knowledge
Cognitive network	A network that uses contextual and situational awareness to understand new data and behavior, compare those new inputs to its current goals, and then formulate actions to protect and achieve those goals, while learning from the consequences of its actions
Context	The collection of measured and inferred knowledge that describe the environment in which an entity exists or has existed
Contextual awareness	The gathering of information about itself and its environment to provide personalized and customized services and resources corresponding to that context
Experiential	Learning through experience (both from actions taken by the system and actions taken outside the system that affect it)
Policy	A set of rules that is used to manage and control the changing and/or maintaining of the state of one or more managed objects
Reference point	A conceptual point at the conjunction of two non-overlapping functions that can be used to identify the type of information passing between these functions
Reference point, external	A reference point between two different systems
Reference point, internal	A reference point within different functions of a system that is not visible to external systems
Situational awareness	The perception of data and behavior that pertains to the relevant circumstances and/or conditions of a system or process ("the situation"), the comprehension of the meaning and significance of these data and behaviors, and how processes, actions, and new situations inferred from these data and processes are likely to evolve in the near future to enable more accurate and fruitful decision-making

15.3 Problems to Be Solved

This section provides an overview of current problems in network and service management. It emphasizes the needs to use business goals to determine the set of network services offered at any given point in time. Finally, it discusses why using AI algorithms to solve part of the network management problem, such as improving telemetry information, is not sufficient to meet existing user needs.

15.3.1 Current Network Management Problems

Most current business support systems (BSSs) and operational support systems (OSSs) are designed in a stovepipe fashion that consists of best-of-breed systems to perform specific tasks [3]. For example, it is common to have multiple inventory systems, each designed to support a specific set of network devices and systems. However, this impedes interoperability, since each such stovepiped system uses its own view of the managed environment. This of course impedes the fusion of information from different systems. An example of such an OSS is shown in Fig. 15.1. This creates a number of problems, including the following:

- Best-of-breed systems exhibit high coupling and low cohesion. This means that a component may depend on many other components, so when it is changed, that change impacts other components.
- There is no easy way for this OSS to interoperate with the BSS, as well as with lower-level management entities (e.g., an SDN controller, or an element manager, or an orchestrator). This turns the OSS into a system-level stovepipe.
- The lack of commonly defined data prohibits different components from sharing and reusing common data, both within its own components, but also between other systems (e.g., the BSS).

Of these three problems, the most common and detrimental is the lack of a unifying information architecture. This causes a number of problems that prevent information from different sources to be used together to form a more complete picture of the environment. For example, suppose that data referring to the same person has different names (e.g., JohnS vs. Strassner.John vs. jstrassn). While a human may be able to equate these, it is very difficult for machine to do so. As another example, consider the same person which, when represented in three different systems, has three different forms of IDs (e.g., employeeID of 123456, EmpID of "SJ033ab," and ID of "123456"). These three different IDs each have different names and datatypes, making it almost impossible for a device to realize that these IDs identify the same object.

Technical incompatibilities abound. For example, there is no information or data model, let alone standard, to help translate SNMP commands to either command line interface commands or newer models, such as YANG. This is primarily because

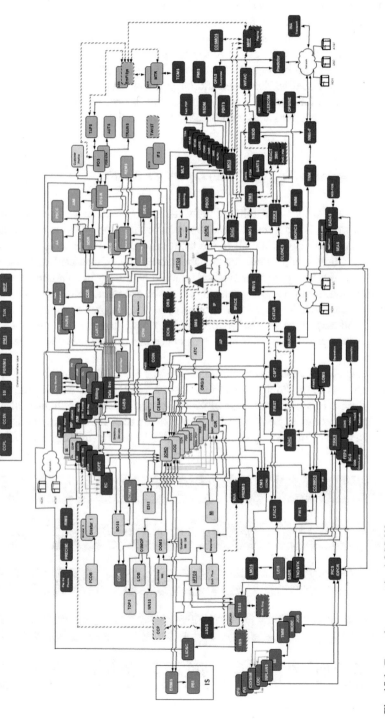

Fig. 15.1 Exemplary stovepiped OSS [3]

the syntax and semantics of each variant of these three approaches used by each vendor is different. In addition, there are hundreds of versions of a vendor's operating system. For some vendors, two devices that run the exact same version of the operating system, but have the ability to use different line cards, can display different responses to the same command. Note that there are other incompatibilities, such as protocols and APIs. These and similar problems can be addressed by, first, using the concept of Reference Points and, second, by adding formal semantics (e.g., by using formal logics and/or ontologies) to the models.

15.3.2   Understanding User and Operator Needs

The above technical incompatibilities imply the need for a single Esperanto-like language. However, this assumes that all users have the same goals. For example, business users rarely understand all of the technical details of a service, and similarly, network administrators rarely understand concepts like Customer Relationship Management and why a particular customer should be treated in a particular manner. For example, a business user may think of the economic implications of an SLA, while a network administrator may think of how to program services specified by the SLA. Ideally, there should be different languages that can be used by each constituency to express their needs using concepts and terminologies that are familiar to them.

This was the idea that motivated the concept of the Policy Continuum [4, 5]. The original depiction of the Policy Continuum is shown in Fig. 15.2. Each of the five views is optimized for a different type of user that needs and/or uses slightly different information. For example, the business user wants SLA information and is not interested in the type of queuing or routing that will be used in the implementation

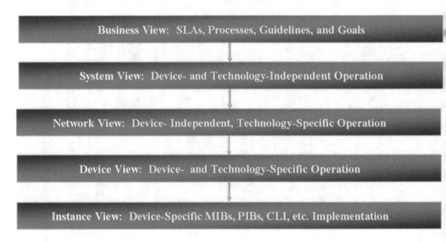

Fig. 15.2   The policy continuum

of the service. Conversely, the network administrator may want to develop CLI commands to program the device and may need to have a completely different representation of the policy in order to develop the queuing and routing CLI commands. More importantly, consider a Service Level Agreement that specifies a number of Service Level Objectives. This needs to be translated into a form that the network administrator can use to program. This is a seriously underestimated task! For example, a customer may have purchased a single service (e.g., "Gold Service") that applies to any application they run. Each application may consist of multiple services. Each service needs to be programmed according to its own needs in order to properly interact with other services, not only of that same customer but also with other customers of the service provider. Hence, the single business policy of assigning a customer a particular service will be translated into a set of policies at a lower level of the Policy Continuum; this process will continue until the service can be properly instantiated.

Thus, the requirement is for policy to be treated as a continuum, where different policies take different forms and address the needs of different users. However, it was previously noted that this is actually a very difficult proposition. Not only do business terms not translate into network commands, but also the network itself is highly heterogeneous and uses different data and commands for different vendor devices.

The approach being used in both ETSI ENI [6] and the MEF Policy Driven Orchestration (PDO) [7] project is similar. Both of these approaches use the same starting information models (the MEF Core Model [8] and the MEF PDO model) to represent different types of policies used in the system. The Policy Continuum is used to define different constituencies that can build policies. The PDO is a novel information model that can define different types of policies (e.g., imperative, declarative, and intent), which facilitates their interaction. Note that this strategy is unique in the MEF and ETSI; other standards bodies typically address different types of policies in a non-unified, one-off approach.

15.3.3   Translating Business Needs to Network Services

Business agreements state the conditions for provided services and any associated penalties and incentives for their use. One or more business agreements can be combined making a Service Level Agreement (SLA). Each SLA specifies what set of services is to be provided when and where, along with its costs, performance, and other metrics (e.g., reliability and availability). SLAs are typically written using business language and terminology that is not amenable to direct programming of network flows. A Service Level Agreement is typically a business contract stating the consequences of failing to achieve each SLO in the SLA (e.g., "If 99% of the Customer system requests aren't completed in 10 ms, the Customer receives a refund.")

SLAs can be written for any service, and multiple SLAs can be used to define the characteristics and behavior of a single service. The details of service quality and performance are defined by one or more Service Level Objectives (SLOs) for each SLA based on context (e.g., 90 and 120 ms maximum round-trip time for domestic and international packets, respectively). Services themselves are varied. Examples of different services are a haptic service whose latency must be less than 10 ms or the service fails (i.e., it has not met its contractual obligations and it is unusable) versus a tiered service (e.g., Bronze, Silver, Gold, and Platinum each define different SLOs for the applications included), where each tier defines different functionality based on cost and class of service. Note that in the second example, if a tiered service has a violation, depending on its frequency and severity, the customer is owed penalty fees, but the service is not considered "failed."

Services are typically ordered using a product. Continuing the above-tiered example, two customers could order the same product using different SLAs that provide different behavior (e.g., Gold provides extra applications, greater speed, and better quality than Silver or Bronze). Similarly, the haptic service could provide the choice between visual, audio, and/or touch feedback.

In addition, a product could also be made up of one or more SLAs, where each SLA is designed to cover a specific set of resources and/or services provided by the product. For example, a product could include two line items, one for a service and a separate line item for coverage and warranty information. As another example, a product could be made up of several components, where the characteristics and behavior of each component is specified by its own SLA (e.g., a mobile phone could be offered using different components (e.g., codecs or cameras) that each have associated SLAs).

An SLA may have one or more SLOs. Some SLOs are context-dependent (e.g., geo-location, time of day, and national vs. international traffic) and may be constrained by business rules of the provider (e.g., maximize revenue for some classes of customers, but maximize security for other classes of customers). An SLO may itself be made up of different performance levels and associated metrics. For example, the following three statements could be considered as three parts of the same SLO:

- 99.99% of customer system requests will complete in less than 15 ms.
- 99.9% of customer system requests will complete in less than 5 ms.
- 90% of customer system requests will complete in less than 1 ms.

Each SLO has two service level values (SLVs): (1) a percentage of customer service requests (e.g., 99.99%, 99.9%, and 90%) and (2) a completion time (i.e., 15, 5, and 1 ms). Each SLV has a corresponding metric to enable its measurement. Hence, a generic relationship between an SLA, its SLOs, and associated SLVs is:

An SLA has 1..n SLOs, and each SLO has 1..n SLVs; each SLV is measured by a metric.

In general, there is no standard for translating business documents, let alone SLAs and SLOs, into a form that network engineers can use to program services. However, understanding this business language is key, since it defines contractually what the service is, what the responsibilities of the provider and the customer are, and the specific characteristics and behavior of the service.

There are two obvious solutions to this dilemma. The first is what most businesses are doing: having one or more people manually translate the business documents into a form that network engineers can use. The problem with this approach is twofold. First, it is manual and hence delays the official testing and offering of a service. This type of translation needs to be done for each pair-wise set of constituencies that need to understand contractual information, or regulatory policies, or business rules of the organization, or in general anything that can affect the programming and management of offered services.

The second obvious solution is to introduce parsers or compilers to help automate the process. The problems are again twofold. First, parsing natural language is a very difficult and computationally intensive task. Second, there is no standard to define the terms that can appear in these business documents nor their meanings. One possible solution is use a Model-driven engineering approach to *constrain* the natural language. In such an approach, a model is used to contain all of the key terms that are expected to occur in a document. A high-level functional block diagram is shown in Fig. 15.3. Note that the translations are done on a *per-pairwise-continuum* basis.

In this approach, the Model-based Data Dictionary forms the front half of a reasoning system, in which the models define "facts," and the ontologies augment these facts with formal semantics that are defined using a logic system (e.g., a type of description logic). This enables the facts to be related to other objects in the system; more importantly, it enables the system to reason about the facts using formal logic, make hypotheses, and, most importantly, mathematically prove the validity of the hypothesis. (Note that the UML standard does not provide formal semantics for its models; that is why this approach combines models and ontologies to do so.) As

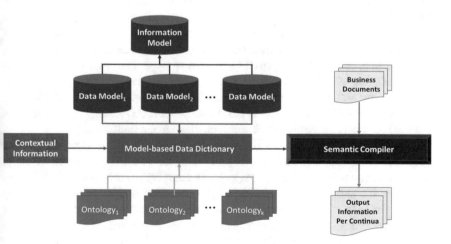

Fig. 15.3 Functional block diagram of a semantic per-continuum-level translator

will be seen, this is a key component of the *knowledge representation* used in the overall cognitive architecture (see Sect. 15.4.3 for more information).

The Model-based Data Dictionary feeds a semantic compiler, which is shown in Fig. 15.3. While compilers use semantic analysis to gather information that cannot be easily checked by parsing (e.g., type checking, ensuring that a variable is declared before use, and misusing reserved words), most compilers do not compile based on the *meaning* of recognized words and phrases. This is a key component of the above approach, since both keywords and contractual obligations must be recognized. This is also why ontologies are used in addition to models.

The use of ontologies enables facts defined in the models to be augmented by additional meaning. Conceptually, both models and ontologies may be viewed as graphs. Therefore, what is needed is to construct semantic relationships that connect one or more ontological concepts to one or more elements in the models. This forms a multigraph. Business logic in the Model-based Data Dictionary is responsible for constructing, storing, reusing, and augmenting these multigraphs. For example, a model can define a customer, which has a number of predefined attributes and relationships. The object customer, along with each of its attributes, contains a potential source of additional meaning that can be searched on by the set of ontologies to find other concepts in the ontologies that are related to the modeled object. Note that such information can appear and disappear dynamically, since the ontology will query, using a formal logic, whether that combination of information from the models and the ontologies currently exist. For example, context can be used to enable or disable all or part of a set of modeled objects to be present via, for example, the decorator pattern [9].

The semantic compiler parses a given set of business agreements (e.g., an SLA or a business rule) and automatically determine the customers, services, and, hence, the set of SLAs and SLOs that this set of business agreements refers to. This is done using named entity recognition and other semantic functions (e.g., relationship analysis and rewriting). Named entity recognition is a subtask of natural language processing. It takes a sentence or a chunk of texts and parses it to identify entities that belong to predefined categories. Categories can range from simple parts of speech (e.g., proper nouns) to important concepts that contain multiple items from the parse tree, such as an SLV, an SLO, or even an SLA. This information is sent to the ontologies, which can use these relationships to search on similar meanings. Named entity recognition works in conjunction with contextual information from the Model-based Data Dictionary to build a set of alternate meanings (e.g., based on context) for each multigraph (or objects contained in the multigraph).

Alternate meanings correspond to alternate labels, or tags. Labeling data can be very expensive. One approach to mitigating this cost is the use of active learning, which reduces the labeling cost by selectively querying the most valuable information from the annotator. This approach automatically provides the most valuable information via the multigraph. While existing multi-label active learning approaches focus on selecting object instances to be queried, this approach uses elements of the multigraph to determine what information should be queried for.

For example, a customer may have different SLAs depending on context (e.g., working versus non-working hours, or the type of access protocol used). This type of information is an example of contextual information and serves as constraints on the SLA(s) and their associated SLOs. These constraints then select what information should be used for the present analysis.

The set of multigraphs produced are then analyzed to translate important information, such as SLAs, SLOs, and SLVs. Each of these known terms has its own definition and relationships. Each business document is then recursively analyzed to discover relationships that each of these terms has. This also enables new terms to be discovered by forming hypotheses based on the meaning of each term. This is where formal logic is essential. For example, if an unknown word is encountered, then a hypothesis can be formed as to the meaning of the word. The use of ontologies to select the best meaning for the newly identified word from among a set of alternative meanings increases the accuracy and efficiency of this approach. Each business document is thus transformed into a document that contains consensually defined terms, which in turn eases its translation to other forms of the continua.

This can be made even more extensible by using roles [9] instead of the actual user, device, or application object. That way, operations that are applicable to all objects having a particular role, such as authentication, can be applied in a uniform and consistent manner using policies.

15.3.4   The Need to Incorporate Dynamicity

Model-driven engineering (MDE) [10] is a model-centric software engineering approach for building software systems that can be dynamically modified at runtime. It treats models as first-class artifacts and does design and analysis of models in place of code. Our particular variant emphasizes the use of models even more by maximizing the use of software design patterns [9]. While MDE has been used in many different domains (e.g., automotive and manufacturing) for over two decades, its use in telecommunications has been limited.

FOCALE [11] is an autonomic networking architecture, first proposed in 2006, and refined over the years. FOCALE stands for Foundation, Observe, Compare, Act, Learn, rEason, which describes its novel control loops. Model-driven means that it can dynamically generate code to reconfigure managed entities from its models, using MDE. It is shown in Fig. 15.4.

Each of the functional blocks in Fig. 15.4 is connected using semantic bus [12] that supports simple as well as semantic queries. A semantic bus is an event-driven distributed content-based message and retrieval broker; the difference between it and standard enterprise service buses (ESBs) is that it can be used to orchestrate content (and hence, route on the meaning of a message), whereas standard ESBs are limited to orchestrating messages. Its semantics and filtering capabilities are centered around the use of ontologies derived from the DEN-ng information model [13]. Messages are structured using OWL, which facilitates ontology-based

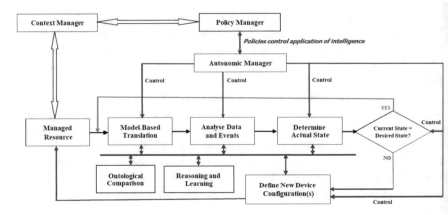

Fig. 15.4 Simplified block diagram of the FOCALE architecture

consistency checking and semantic filtering. It also enables subscriptions to be placed against any part of the message structure.

The FOCALE Autonomic Manager uses the semantic bus to orchestrate behavior. It can support different types of knowledge acquisition and distribution (e.g., push, pull, and scheduled) and perform common processing (e.g., semantic annotation, filtering and storage) before content is delivered to components. This enables components to register interest in in a more precise fashion, thus reducing messaging overhead.

The FOCALE control loops operate as follows. Data is retrieved from the managed resource (e.g., a router) and fed to a model-based translation process, which translates vendor- and device-specific data into a normalized form using the DEN-ng information model and ontologies as reference data. This is then analyzed to determine the current state of the managed entity. The current state is compared to the desired state from the appropriate finite-state machines (FSMs). If no problems are detected, the system continues using the maintenance loop; otherwise, the configuration loop is used so that the services and resources provided can adapt to these new needs.

Nodes in a FOCALE FSM represent a configuration state; each state has an associated set of one or more configuration actions that define the configuration of an entity. Edges represent state transitions and connote permission to change the configuration of a managed resource. Static behavior is thus "programmed" into FOCALE by designing a set of FSMs; dynamic behavior is defined by altering one or more FSMs. Context-aware policy management [11] governs both autonomic control loops. This enables context to select the set of policies that are applicable; policies are used to then define the functionality allowed. As context changes, policies change, and system functionality is adjusted accordingly.

The autonomic manager uses the current set of context-aware policies to govern each of the architectural components of the control loop, enabling each of the different control loop components to change how it operates as a function of context.

FOCALE develops and uses libraries of model and ontology fragments and coded behaviors, much as a library of string processing functions is used by a programming language. This library is made reusable by realizing it in the form of objects, supported by both models and ontologies. Library behaviors are associated with the application of policy actions, which in turn are selected by a particular context as previously described.

The key to FOCALE is to answer the question: "Is the current state equal to the desired state?" Here, "equal to" matches any state in the state space that is classified as meeting the system goals. This is similar to hill climbing, where instead of trying to match the highest point of the hill, a horizontal swath is cut across the hill, and any state falling within that swath is a match. Note that this does not prevent the algorithm from optimizing to a better state, which occurs in later versions of FOCALE.

The reconfiguration process uses dynamic code generation. Information and data models are used to populate the state machines that in turn specify the operation of each entity that the autonomic system is governing. The management information that the autonomic system is monitoring consists of captured sensor data. This is analyzed to derive the current state of the managed resource, as well as to alert the autonomic manager of any context changes in or involving the managed resource. The autonomic manager then compares the current state of the entities being managed to their desired state; if the states are equal, then monitoring continues. However, if the states are not equal, the autonomic manager will compute the optimal set of state transitions required to change the states of the entities being managed to their corresponding desired states. During this process, the system could encounter an unplanned change in context (e.g., if a policy rule that has executed produced undesirable side effects). Therefore, the system checks, as part of both the monitoring and configuration control loops, whether or not context has changed. If context has not changed, the process continues. However, if context has changed, then the system first adjusts the set of policies that are being used to govern the system according to the nature of the context changes, which in turn supplies new information to the state machines. The goal of the reconfiguration process is specified by state machines; hence, new configuration commands are dynamically constructed from these state machines.

15.3.5   Reacting to Context

Networks can contain hundreds of thousands of policy rules of varying types (e.g., high-level business policy rules for determining the services and resources that are offered to a user, to low-level policy rules for controlling how the configuration of a device is changed). One of the purposes of making these policy rules context-aware is to use context to select only those policy rules that are applicable to the current management task being performed.

The context of an entity is a collection of measured and inferred knowledge that describe the state and environment in which an entity exists or has existed [14]. In particular, this definition emphasizes two types of knowledge—facts (which can be measured) and inferred data, which results from machine learning and reasoning processes applied to past and current context. It also includes context history, so that current decisions based on context may benefit from past decisions, as well as observation of how the environment has changed.

Context-awareness enables a system to gather information about itself and its environment [11, 14, 15]. This enables the system to provide personalized and customized services and resources corresponding to that context. More importantly, it enables the system to adapt its behavior according to changes in context.

Context-awareness enables diverse data and information to be more easily correlated, and hence, integrated, since context acts as a unifying filter. As such, identifying contextual information is critical for understanding both ingested data and information as well as how data and information, as well as existing knowledge and wisdom, can be affected.

The contextual history of a user, a user application, or a device, as well as its prior interactions with the management system (including, e.g., session state), may be useful for driving policy decisions regarding the current and future interaction between the management system and that entity, including decisions made by the management system that affect that entity. For example, past behavior can be used to more quickly arrive at a decision. Alternatively, historical information can be used to flag anomalies that need further action to resolve.

Figure 15.5 shows the main operations required for context-based reasoning.

Context may be modeled as Big Data, since the critical factor is extracting value from Big Data. The three operations above Big Data describe a set of increasingly specific operations that can be used to semantically annotate information.

A "feature" is defined as an important characteristic or behavior that helps describe and aid in the understanding of an entity. For example, edges and corners are points of interest in an image. Reducing the number of features is important in

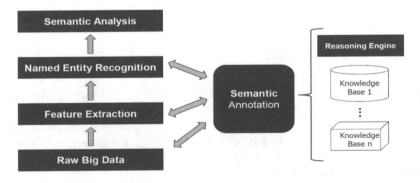

Fig. 15.5 Context-based reasoning

analysis and reasoning, since if there are too many features, it could cause overfitting in training.

Semantic analysis analyzes the information for specific semantic concepts, searching on those concepts, and then adding additional semantic relationships to enrich the information and provide more specific meaning. From a linguistic perspective, this analyzes text and finds sets of syntactic structures that are related to each other. This may be represented as a graph, or network, of related words, phrases, and other elements of a sentence. From a machine learning perspective, this computes metrics such as semantic similarity (i.e., the meaning of an object compared to the meaning of other objects, where the comparison is done using synonymy, antonymy, hyponymy, hypernymy, and other types of relationships). This is a practical and more computationally tractable approach than "absolute understanding," since the latter requires a rigorous world model, which is np-complete.

15.3.6   Incorporating Situational Awareness

As previously explained, context captures external influences of data and information being ingested by the management system. Situational awareness uses contextual information to determine how context is affecting the goals of the system.

The definition of situational awareness is, for this paper:

> The perception of data and behavior that pertain to the relevant circumstances and/or conditions of a system or process ("the situation"), the comprehension of the meaning and significance of these data and behaviors, and how processes, actions, and new situations inferred from these data and processes are likely to evolve in the near future to enable more accurate and fruitful decision-making.

Situation awareness enables the system to understand what has just happened, what is likely to happen, and how both may affect the goals that the system is trying to achieve. This implies the ability to understand how and why the current situation evolves. Briefly, situational analysis includes the comprehension of the meaning and significance of observed data and behaviors, and how processes, actions, and new situations inferred from these data and processes are likely to evolve in the near future with respect to current system goals. This is the essence of cognition and is the functional block that governs the operation of the cognitive architecture. This will be explained in Sect. 15.4.

15.3.7   Summary of Recommended Problem Solutions

Table 15.2 provides a summary of the above problems and their recommended solutions.

Table 15.2 Problem solution summary

Problem	Solution
Addressing legacy systemic problems	Use functional block design and reference points to encourage low coupling and high cohesion. Derive multiple domain-specific data models from a single information model. Add semantics, in the form of logic and/or ontologies, to modeled data
Understanding needs of different constituencies	Develop a consensual data dictionary from a set of data models and associated ontologies to represent concepts used by different constituencies. Use the Policy Continuum to share policies authored by one constituency with other constituencies to enable business needs to be transformed into a form understandable by all other constituencies
Translating business needs	Build parsers or compilers that can translate business documents, such as SLAs and SLOs, into a form that non-business people can understand. Associate those forms with policies. Use roles instead of individual objects to make more extensible
Incorporation of dynamicity	Adopt MDE principles to enable objects to be dynamically changed at runtime using a set of standard software patterns. Use one or more finite-state machines, populated by MDE, to ensure that the current state of the entity being managed is acceptable (or else, plan a set of state changes to bring that state into an acceptable state)
Incorporating contextual awareness	Use context-awareness to formalize the state and the environment in which an object exists. This is crucial for enabling the management of that entity, along with services that it received and/or provides, to be adjusted due to changes in user needs, business rules, and environmental conditions. Context should be incorporated into the knowledge representation used by the system
Incorporating situational awareness	Use situational awareness to evaluate how likely system goals can be achieved, now and in the future. This is done by analyzing contextual data, along with applicable other information (e.g., business rules, historical data, system goals, and current hypotheses), to project the current and future states of the system being managed onto a state space that classifies states as meeting, exceeding, or violating system goals. It is responsible for predicting threats to achieving its goals as well as describing how the system itself, as well as threats, will likely evolve. Situational awareness should be incorporated into the knowledge representation used by the system

15.4 Cognition Principles

This chapter describes a set of key principles for designing and understanding cognitive architectures.

15.4.1 Cognition

Cognition is the process of acquiring new data and information, analyzing those data and information to comprehend their meaning and significance, and producing new data, information, and knowledge that adds to the understanding of the operation of the system and its environment.

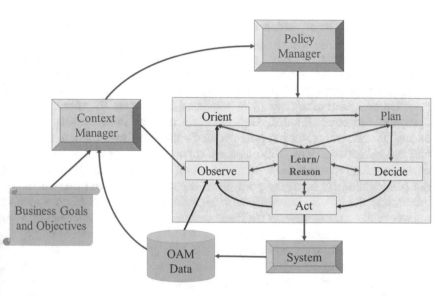

Fig. 15.6 Simplified version of the FOCALE control loop

Machine cognition is a set of processes that mimic how the human brain acquires and understands data and information and produces new data, information, and knowledge. Machines implement this process using various types of closed control loops. A *cognitive* closed control loop is one that selects data and behaviors to monitor that can help assess the status of achieving a set of goals and produce new data, information, and knowledge to facilitate the attainment of those goals. The Observe-Orient-Decide-Act (OODA) control loop [16], augmented with learning and reasoning, has been recommended as the basis for building cognitive control loops [17]. Reference [18] defines a cognitive control loop that builds on this and adds policy management is shown in Fig. 15.6. In this approach, two new functions, called "plan" and "learn," are inserted into the basic OODA framework. While the loop appears to be sequential, this is merely for convenience of representation. Observation, orientation, decision, and action occur continuously. The orientation step is critical, as it determines how observations, decisions, and actions are performed. As Boyd observed, people act according to how they perceive the world, as opposed to how the world really is. This also applies to machines.

The Observe functional block accepts input from the system being managed; this is represented by operational, administrative, and management data. These data also are sent to the Context Manager, which interprets these data according to current business goals. The result goes to the Policy Manager, which then issues policies that govern the operation of all six functional blocks of the FOCALE control loop (shown in the yellow rectangle).

The Orient functional block takes the ingested input data and normalizes them using a set of models and ontologies. Conceptually, the models supply facts, and the ontologies add meaning to those facts. This function is critical, as it enables data and information fusion from multiple different sources to get a more complete picture of the situation.

The Plan portion provides reactive, deliberative, and reflective processing and emulates how the human brain processes information to perceive, comprehend, and project how the Service behavior will evolve compared to its SLO(s). Reactive processes take immediate responses based upon the reception of an appropriate external stimulus. Such processes have no sense for what external events "mean"; rather, they simply respond with some combination of instinctual and learned reactions. This enables a cognitive system to recognize a previously encountered situation. When this is done, the system can bypass many of the computationally intensive portions of the control loop and instead follow "shortcuts" through the control loop straight to the function that issues actions to change the current state to the desired state. Deliberative processes receive data from and can send "commands" to the reactive processes; however, they do not interact directly with the external world. This process addresses more complex goals by using memory in order to create and carry out more elaborate plans. This knowledge is accumulated and generalized. Finally, reflective processes supervise the interaction between the deliberative and reactive processes. These processes reformulate and reframe the interpretation of the situation in a way that may lead to more creative and effective strategies. It considers what predictions turned out wrong, along with what obstacles and constraints were encountered, in order to prevent suboptimal performance from occurring again. It also includes self-reflection, which analyzes how well the actions that were taken solved the problem at hand.

A cognition model is created based on these three types of processing, producing one or more paths. Some of these paths may be shortcuts, which bypass one or more functions when the input is recognized and the output is either known or has a sufficiently high probability of occurrence. In FOCALE, this is realized by a set of one or more state machines, where each state corresponds to a particular service and/or network configuration. Metadata is added to appropriate states to include key situational information.

The Act portion decides on the path best suited to protect business goals given the current situation. It then uses MDE mechanisms to translate the selected set of nodes in the state machine to a series of commands to reconfigure affected resources and services, as well as monitor appropriate resource and service information.

The Learn/Reason portion uses contextual and situational awareness to understand new data and behavior, compare those new inputs to its current goals, and then formulate actions to protect and achieve those goals, while learning from the consequences of its actions. It examines the success or failure of the configuration of resources and services so that it can associate the effectiveness of each state with the actual system being managed.

15.4.2   An Adaptive and Cognitive Control Loop

Cognition Management relates each of the other functional blocks of the system to the set of end-to-end goals that the system is using at a particular time. It tries to maintain a set of end-to-end goals (such as routing optimizations, connectivity, efficiencies, security, and trust management) by modifying the directives of the other functional blocks. This modifies the FOCALE architecture, shown in Fig. 15.4, as follows.

The control loop shown in Fig. 15.7 is called an *adaptive* closed control loop, since its controlling function adapts to the object or process being controlled using parameter that are either unknown and/or vary over time. These parameters reflect both changing context and situation. A preferred implementation defines the parameters using a model that defines the desired closed loop performance; this could be augmented with statistical analysis to build a mathematical model from measured data. The control loop shown in Fig. 15.7 is also called a *cognitive* closed control loop, since it can select data and behaviors to monitor that can help assess the status of achieving a set of goals and produce new data, information, and knowledge to facilitate the attainment of those goals.

15.4.3   Knowledge Representation

There are many examples of knowledge representation formalisms, ranging in complexity from models and ontologies to semantic nets and automated reasoning subsystems. Fundamentally, knowledge representation is the expression that enables the beliefs, intentions, and judgments of a software entity to be expressed suitably for automated reasoning. This also includes modeling intelligent behavior for a software entity. Put another way, knowledge representation describes how knowledge is

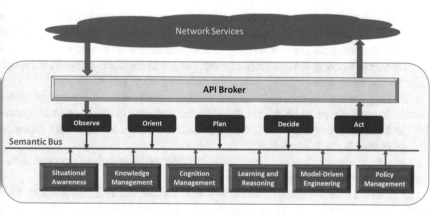

Fig. 15.7 An adaptive and cognitive set of control loops

defined and manipulated in artificial intelligence. Most importantly, knowledge representation does *not* assume that data is static! Rather, data can always be modified or augmented if sufficient evidence is present to do so.

There are a number of different types of knowledge. Procedural knowledge describes how to perform a task or activity and includes rules, strategies, and procedures. Declarative knowledge includes concepts, facts, and objects and is expressed in one or more declarative sentences. This is similar to logical knowledge, which expresses concepts, facts, and objects in a formal logic. Structural knowledge describes the composition of, and relationship between, concepts and objects. In general, all types of knowledge can be used to form a knowledge representation for a system.

A logical representation is a formal language that can define axioms, theories, hypotheses, and propositions without any ambiguity in their representation. It uses precisely defined syntax and semantics that supports different types of inferences and reasoning. Its main advantage is that it facilitates mathematically proving hypotheses and can use inferencing to define new objects from its existing objects. Its main disadvantage is that many users are not well versed using formal logic.

FOCALE also used a semantic network, which is a type of knowledge graph. Nodes represent objects and concepts, and edges describe the relationship between those objects. Later versions of FOCALE used linguistic relationships (e.g., synonymy, antonymy, meronymy, etc.) in addition to the typical IS-A and HAS-A relationships. Its main advantage is that it is a natural representation of knowledge that is easy to understand. Its main disadvantage is that it may be difficult to represent different types of relationships.

A more complete and formal definition of knowledge representation is defined in http://groups.csail.mit.edu/medg/ftp/psz/k-rep.html.

15.4.4   Memory

Cognitive systems have different types of memories and use them in a similar way as humans do [19]. The main types of memories used in digital systems can be classified as short-term memory, working memory, and long-term memory.

Short-term memory is the ability to store, but not manipulate, a small amount of information in an active, readily available store for a short period of time.

Working memory is the retention and manipulation of a small amount of information in a readily accessible form. It facilitates planning, comprehension, reasoning, and problem-solving. Hence, the information is stored in short-term memory and processed in working memory.

Long-term memory is a store that holds, but does not manipulate, data and information for as long as needed.

Cognitive systems typically use *active repositories*. An active repository is a storage mechanism that is capable of pre- and/or post-processing information that is stored or retrieved to better fit the needs of the requestor.

15.5 A Cognitive Architecture

A cognitive architecture is a system that learns, reasons, and makes decisions in a manner resembling that of a human mind. Specifically, the learning, reasoning, and decision-making is performed using software that makes hypotheses and proves or disproves them using non-imperative mechanisms that typically involve constructing new knowledge dynamically during the decision-making process.

This chapter builds on the previous chapter to create a cognitive architecture for network and service management. Its functional architecture and attendant benefits are described.

15.5.1 Overview

A cognitive system is one that can reason about what actions to take, even if a situation that it encounters has not been anticipated. It can learn from its experience to improve its performance. It can also examine its own capabilities and prioritize the use of its services and resources and, if necessary, explain what it did and accept external commands to perform necessary actions. Fundamental to cognition is the ability to understand the relevance of observed data. This is typically done by classifying data into predefined representations that are understood and relevant to the current situation. Memory is used to increase comprehension of the situation. Finally, actions are judged by how effectively they perform to support the situation.

Table 15.2 lists key functionality that should be incorporated for addressing current problems in network and service management. This leads to the simplified functional block diagram below.

The cognitive architecture shown in Fig. 15.8 is divided into three parts: the API Broker, the Input and output processing Sections, and the Cognitive Processing Function Section. There are two different closed control loops. The outer loop takes data from the system being governed, analyzes it, and changes the behavior of the system being governed as necessary to maintain system goals. The inner loop optimizes the state of the system being governed, and hence, the services being offered, at any given time.

15.5.2 The API Broker

The motivation for using an API Broker is threefold:

- The use of an API Broker enables the continuing development of the cognitive architecture to proceed independently of any specific requirements of interacting with external entities.

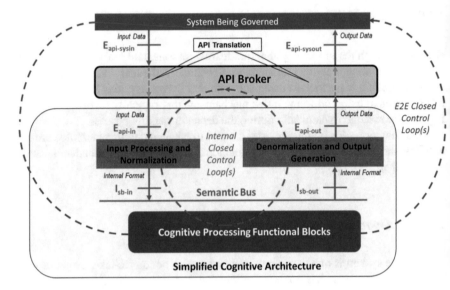

Fig. 15.8 Simplified functional block diagram of a cognitive architecture

- The use of an API Broker provides a more scalable and extensible solution, as it facilitates the use of generic (e.g., RESTful) technologies as well as custom plug-ins to meet the needs of communication with different external entities.
- The use of an API Broker enables advanced solutions, such as API composition, to be used.

The API Broker has two main functions. The first is to serve as an API gateway (i.e., an entity that can translate between different APIs). The second is to provide API management. Management of APIs includes authentication, authorization, accounting, auditing, and related functionality.

The functions of the API Gateway include the following:

- Accept incoming APIs transmitted through an appropriate external reference point and route them to the appropriate functional block(s) of the cognitive architecture
- Accept outgoing APIs transmitted through an appropriate external reference point and route them to the appropriate external entity
- Convert protocols used by external entities to protocols used by the cognitive architecture, and vice-versa
- Manage different versions of the same API

The $E_{api\text{-}sysin}$ external reference point accepts API requests from external entities and executes them, after any necessary translation by the API Broker, on the cognitive architecture.

The $E_{api\text{-}sysout}$ external reference point accepts API requests from the cognitive architecture and sends them, after any necessary translation by the API Broker, to designated external entities.

15.5.3   Input and Output Processing

The cognitive architecture must be prepared to accept a wide variety of input data using different languages. This necessitates the transformation of these input data into a single common form for more efficient and uniform processing. Otherwise, each functional block of the cognitive architecture would have to understand each type of input—its syntax and semantics. Similarly, the single (internal) format of the cognitive architecture must be subsequently transformed into a form that external entities can consume.

During these two processes, a set of common tasks are performed on all ingested data before those data reach the cognitive processing function. Similarly, a set of common tasks are performed when output commands and information are sent from the Cognitive Processing Function to any external entity. This is the motivation for having the input processing and output processing functions.

In general, input processing may include learning and inferencing from the available raw data of one or more domains; once these data are analyzed, the processing shall then decide on what knowledge is forwarded to other functional blocks. In certain cases, the processing may save the raw form of the ingested data for further use. For example, many types of trend processing require access to raw data. In most cases, the processing function may save the processed form of the data; this is both faster and more efficient. The choice of whether to save the raw or processed form of the ingested data is dependent on the current context and/or the current and anticipated situations.

The processing may include aggregation and correlation functions (e.g., to reduce dimensionality) as well as machine learning (e.g., this may yield faster results by dealing with significantly smaller data sets and enable what-if analysis and other game-theoretic algorithms to be used). In such a case, the resulting normalized data may also contain knowledge of a specific domain, or multiple domains. Input processing may include the following:

- Data filtering is the removal of unnecessary or unwanted information. This is done to simplify and possibly increase the speed of the analysis being performed and is similar to removing noise in a signal. Filtering requires the specification of rules and/or business logic to identify the data that shall be included in the analysis. Examples include outlier removal, time-series filtering, aggregation (e.g., constructing one data stream from pieces of other data streams, such as merging name, IP address, and application data), validation (i.e., data is rejected because it does not meet value restrictions), and deduplication.
- Data correlation expresses one set of data in terms of its relationship with other sets of data. For example, the number of upsells to a higher class of service may increase due to targeted advertising, which may increase even more when offering free time-limited trials. These data are usually collected using different mechanisms and, hence, are fragmented among different collection points. Data correlation can use rules and/or business logic to collect the scattered data and combine it to improve analysis. Data correlation is the first step in gaining

increased understanding of relationships between data and their underlying objects.

- Data cleansing is a set of processes that detect and then correct or remove corrupt, incomplete, inaccurate, and/or irrelevant data. Data cleansing solutions may also enhance the data, either by making it more complete by adding related information or by adding metadata. Finally, data cleansing may also involve harmonization and standardization of data. For example, abbreviations may be replaced by what they stand for, and data such as phone numbers may be reformatted to a standard format.
- Data anonymization is the process of either removing or encrypting information that can be used to identify named entities from a data set. In this document, the anonymization process is defined as irreversibly severing data that can be used to identify a named entity from the data set. Any future reidentification is no longer possible.
- Data pseudonymization is the process of replacing information that can be used to identify a named entity with one or more artificial identifiers (i.e., pseudonyms). Note that the pseudonymization process is reversible by certain trusted entities, since the identifying data was not removed, but rather substituted with other data.

The denormalization process is the opposite of the normalization process; it arranges and formats the data and information to be output so that it may be more easily and efficiently translated into a form that is understandable by the set of external entities that will consume it. This is made possible by metadata attached to ingested data that describes how the data is being used in the cognitive architecture, and any insights or hypotheses that the cognitive architecture has defined that the ingested data is a part of.

Once the data, information, or commands are denormalized, outputs may be generated. MDE techniques may be used to facilitate this transformation, as the model provides appropriate meanings of the data, information, and commands to be translated.

15.5.4 Cognitive Processing Function

Figure 15.9 shows a functional block diagram of the various functional blocks that make up the cognitive processing portion of the cognitive architecture.

There are six new functions that are required.

The Situational Awareness functional block is shown in Fig. 15.10.

The Situational Awareness functional block takes normalized input and relates that input to the current situation. Comprehension is based on fusing information from the different elements found in the situation. In particular, the fusing is done with respect to both the current situation and the system goals that apply to that situation. The projection of future status is based on the knowledge of the

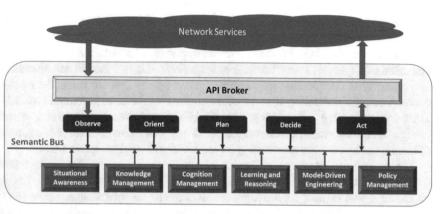

Fig. 15.9 High-level functional architecture of a cognitive network

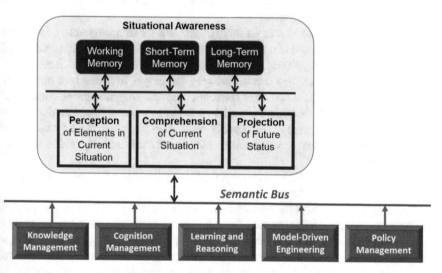

Fig. 15.10 Situational awareness functional block

characteristics and behavior of the elements in the current situation. The Cognition Management functional block then directs the interaction of the Situational Awareness functional block with the Learning and Reasoning functional block to determine how the current input has affected the most recent update of the situation. It may also review historical snapshots of the situation to learn how the situation has evolved.

The Knowledge Management functional block transforms data and information into a consistent knowledge representation that all other functional blocks can utilize. Knowledge Management contains various repositories for storing and processing knowledge. These include repositories for models, ontologies, data, and calculations, the latter of which may, for example, take the form of a blackboard. A blackboard system uses a shared workspace that a set of independent agents contribute to, which contains input data along with partial, alternative, and completed solutions. Both the blackboard and the contributing agents are under the control of a dedicated management entity. Each agent is specialized in its function and operation and typically is completely independent of other agents that are using the blackboard. A controller monitors the state of the contents of the blackboard and synchronizes the agents that are working with the blackboard. Knowledge management creates, revises, sustains, and enhances the storage, assessment, use, sharing, and refinement of knowledge assets using a consensual knowledge representation.

The Cognition Management functional block serves as the "brains" of the Cognition Architecture. It is responsible for implementing a cognition model (i.e., a computer model of how cognitive processes, such as comprehension, action, and prediction, are performed and influence decisions) that serves to guide the actions of the other functional blocks. Cognition focuses on a unified and normalized representation of knowledge. Its cognition model is constantly updated by the Learning and Reasoning functional block.

Knowledge is developed using a multigraph (or a set of multigraphs), as shown in Fig. 15.11. In this approach, the models and the ontologies are both represented as graphs; semantic edges (i.e., relationships of a semantic nature, such as synonymy and meronymy) are then created between the graphs to define how one set of concepts is related to the other set of concepts. The resulting multigraph consists of semantic relationships that join the model on the left to the set of ontologies on the right; these are represented by the double-headed arrow in Fig. 15.12 connecting them for simplicity. This semantic representation is built iteratively and is summarized below.

The process in Fig. 15.11 can be reversed, but typically, a fact has more meanings than a meaning has facts.

The information model (or a set of data models, but using the information model is more general) as well as the set of ontologies are each represented as a directed acyclic graph. A lexicon is a collection of all words, phrases, and symbols used in a language that is organized in a manner that enables each word, phrase, or symbol to have a set of meanings. This enables the most appropriate meaning of each word, phrase, or symbol to be chosen given the correct context. The lexicon serves as a mapping between the model graph(s) and the ontology graph(s) and is necessary since the nature of the knowledge in each graph is significantly different. In essence, the lexicon serves as a semantic disambiguation mechanism that enables the best meaning from the set of ontology concepts to be associated with the given set of

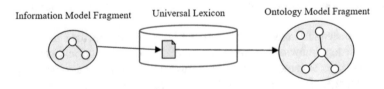

Information Model Fragment   Universal Lexicon   Ontology Model Fragment

Fig. 15.11 Knowledge processing: finding ontological matches for a model element

model elements. This is then used to search for semantically equivalent concepts in a set of ontologies.

Figure 15.11 begins with identifying one or more model elements in the information model. Then, one or more of a number of different tools, including computational linguistics, semantic equivalence, and pattern and structural matching, are used to relate the set of model elements to a set of terms in the lexicon. In general, a set of model elements can be related to a term in the lexicon, which then is related to multiple ontological concepts. Each relationship is typically either linguistic or logical, but could include other relationships as well (in this case, they would need to be weighted into a semantically equivalent form). For simplicity, the remainder of this discussion will assume linguistic relationships (e.g., hypernyms [i.e., an object whose meaning includes the meaning of other objects] and hyponyms [i.e., an object whose meaning is contained in an object], holonyms [i.e., an object that contains other objects], meronyms [i.e., an object that is contained by another object]) and custom relationships (e.g., "is similar to," which is an attributed relationship whose value is the semantic relatedness of the two objects).

Figure 15.11 shows that the search has related the set of model elements to an isolated concept plus a hierarchy consisting of four concepts, for a total of five concepts, in the ontology. This leads to the building of a new multigraph, which contains the original subgraphs from the model connected to the set of concepts in the ontologies using the set of semantic relationships discovered in the above processes. Essentially, the semantic resolution process compares the meaning (i.e., not just the definition, but also the structural relationships, attributes, etc.) of each element in the first subgraph with all elements in the second subgraph, trying to find the closest language element or elements that match the semantics of the element(s) in the first subgraph. Often, an exact match is not possible; hence, the semantic resolution process provides a ratioed result, enabling each match to be ranked in order of best approximating the collective meaning of the first subgraph. The next step is illustrated in Fig. 15.12.

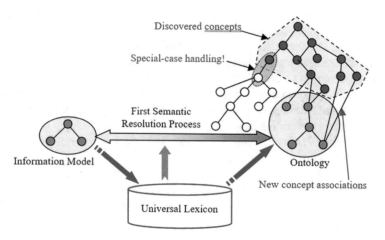

Fig. 15.12 Knowledge processing: finding new ontological concepts

In this step, each ontology concept that was identified in the semantic matching process is now examined to see if it is related to other concepts in this or other ontologies. As each new concept is found, it is marked for possible addition to the existing concepts that were already matched from the Universal Lexicon. The newly added concept is then checked to see if it is related to any of the terms identified in the Universal Lexicon. If it is, the new concept is added; this is shown in the dashed polygon in Fig. 15.12. If it is not directly related to a term in the Universal Lexicon (as shown in the dotted ellipse in Fig. 15.12), then more complex processing is required which is beyond the scope of this chapter. The addition of these new concepts serves two purposes: (1) to provide a better set of meanings of the group of model elements and (2) to verify that each new concept reinforces or adds additional support for the concept that was already selected. Hence, this process can be thought of as strengthening the semantics of the match.

A multigraph is formed by defining semantic relationships between model objects and concepts in the ontologies.

These new semantic associations, along with the new concepts discovered in the ontology, can now be used to find new model elements. This step is similar to the above, except that it is reversed. That is, each new ontology concept is first mapped to one or more terms in the Universal Lexicon, and then each of those terms is mapped to model elements. As before, the algorithm attempts to match groups of related concepts to groups of related model elements. This has the effect of increasing the semantic similarity between two concepts; as larger groups of concepts are matched to larger groups of model elements, a stronger correlation between the meaning of the grouped concept and the group of facts is established. This is, in effect, a self-check of the correctness of the mapping, and is used to eliminate concepts and model elements that match each other, but are not related to the managed entity that is being modeled.

The Learning and Reasoning functional block provides different types of learning to enable different learning algorithms to be used that are customized to particular tasks:

- Experiential learning, which is the set of processes that enable knowledge to be created through experience.
- Supervised learning, which defines a function that maps an input to an output based on example pairs of labeled inputs and outputs.
- Active learning is an iterative supervised learning algorithm where the algorithm can actively query an oracle (e.g., a human annotator) to obtain the correct label.
- Unsupervised learning, which defines a function that maps an input to an output without the benefit of the data being classified or labeled.
- Reinforcement learning uses software agents to take actions in an environment in order to maximize a cumulative reward.
- Feature learning analyses raw input data to learn the most important characteristics and behavior representations of those data that make it easier to discover information from raw data when building different types of predictors (e.g., classifiers).

- Semantic learning, which is the ability to learn by understanding the meaning of data.
- Various types of algorithms for natural language processing.

Supervised learning should be used when one or more datasets that have labeled input and output values. Supervised learning algorithms are ideal for classification and regression tasks. Classification algorithms are used to predict the category that a new datum belongs to based on one or more independent variables. In contrast, regression algorithms predict an associated numerical value for the input datum based on previously observed data.

Unsupervised learning should be used when there is a large amount of data that do not have labels, and the task is to determine the structure of the data. Clustering is a multivariate statistical procedure that collects data containing information about a sample of objects and then arranges the objects into groups, where objects in the same group are more similar to each other than to objects in other groups. Clustering identifies commonalities in the objects in each group, which can also be used to detect anomalous data that do not fit into any group.

Active learning is an iterative supervised learning algorithm where the algorithm can actively query an oracle (e.g., a human annotator) to obtain the correct label. This approach enables the learning algorithm to interactively choose the data it will learn from. Active learning iteratively selects the most informative examples to acquire their labels and trains a classifier from the updated training set, which is augmented with the newly selected examples. Unlike conventional supervised learning, it permits a learning model to evolve and adapt to new data. Active learning is concerned with learning accurate classifiers by choosing which examples will be labeled, reducing the labeling effort and the cost of training an accurate model. Active learning is appropriate for machine learning applications where labeled data is costly to obtain but unlabeled data is abundant. Active learning is especially important where objects can have multiple labels that belong to various categories (e.g., a network device has multiple roles, or an image can be labeled as containing both mountains, beach, and ocean). The main challenge is determining which set of labels is appropriate for a given context or situation.

Reinforcement learning should be used when there is no data, or the dataset is inadequate, and the task is to learn what action to take in a particular situation when interacting with a new entity. This type of learning should also be used when the only way to collect information about the entity is to interact with it. More specifically, reinforcement learning interacts with the entity to, first, negotiate its capabilities and then discover how to exchange data and commands through learning how to react with the entity.

Semantic learning uses formal logic and/or ontologies to learn based upon the meaning of the ingested data compared with the current situation. For example, semantic learning could be used to determine that the overall trend of performance data is decreasing, which may indicate that an SLA violation could occur in the future.

A cognitive architecture will include one or more algorithms for processing natural language. Some contextual and situational data may be ingested as natural language, depending on the input source. Business rules, policy regulations, and system goals are also likely to be expressed in natural language. Text Embeddings are real-valued vector representations of strings, where a dense vector is built for each word, chosen so that it is similar to vectors of words that appear in similar contexts. This enables deep learning to be effective on smaller datasets, as they are often the first inputs to a deep learning architecture and the most popular way of transfer learning in NLP. Long short-term memory (LSTM) networks introduce gates and an explicitly defined memory cell. Each neuron has a memory cell and three gates: input, output, and forget. The function of these gates is to safeguard the information by stopping or allowing the flow of it. The input gate determines how much of the information from the previous layer gets stored in the cell, while the output layer determines how much of the next layer gets to know about the state of this cell. The forget gate determines which characters are forgotten for the next layer of processing. LSTMs are currently the default model for most sequence labeling tasks. A Transformer is a deep learning model that utilizes attention, weighing the influence of different parts of the input data. The Transformer is the first transduction (i.e., convert input sequences into output sequences) model relying entirely on self-attention to compute representations of its input and output without using sequence-aligned recursive neural networks or convolution. Transformers are designed to handle sequential input data, such as natural language, but do not require that the sequential data be processed in order. Rather, the attention operation identifies context for any position in the input sequence. This enables its implementation to be inherently parallel.

The Policy Management functional block is a set of rules that is used to manage and control the changing and/or maintaining of the state of one or more managed objects. It provides a consistent and normalized mechanism for communicating data and commands within a system and between systems. Reference [7] defines a novel UML object-oriented information model for representing different types of policies. A class diagram is shown in Fig. 15.13.

The top class, MPMPolicyObject, is inherited from the MEF Core Model (MCM) (https://wiki.mef.net/pages/viewpage.action?pageId=118001933&preview=%2F118001933%2F118001934%2FMEF_78.1.pdf), which defines an overall object-oriented information model consisting of a single root class with three subclasses. These subclasses form modular hierarchies for representing managed and unmanaged entities and metadata (among other concepts).

The MEF Policy Model (MPM) [7] is made up of four types of objects. Two of them, MPMPolicyStructure and MPMPolicyComponentStructure, define hierarchies for representing policies and components of a policy, respectively. MPMPolicySource represents a set of objects that authored the policy, and MPMPolicyTarget represents a set of objects that may be affected by a policy.

There are three main types of policy paradigms that are used in the MPM: imperative, declarative, and intent policies. Additional policy paradigms (e.g., utility functions) are currently being designed.

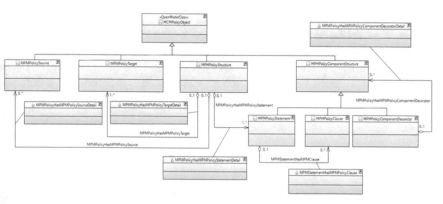

Fig. 15.13 Simplified view of the MEF policy model

Imperative Policies explicitly control the transitioning of one state to another state. In this approach, only one target state is allowed to be chosen. An example of an imperative policy is the ECA (Event-Condition-Action) policy. In this paradigm, a policy is made up of three Boolean clauses (events, conditions, and actions). The semantics of this policy are:

```
IF the event clause is TRUE
    THEN IF the condition clause is TRUE
        Execution of actions in the action clause may occur
    ENDIF
ENDIF
```

An Imperative Policy may include metadata that controls how actions are executed (e.g., execute the first action, execute the last action, execute all actions) and what happens if an error occurs (e.g., stop execution, stop execution and rollback that action, stop execution and rollback all actions).

A Declarative Policy describes the set of computations that need to be done without describing how to execute those computations. In particular, the control flow of the program is not specified. Hence, a key characteristic of declarative programming is that the order of statement execution is not defined. In the MPM, a Declarative Policy is written in a formal logic language, such as First Order Logic, and is a program that executes according to a theory defined in a formal logic. Hence, a Declarative Policy may choose any state that satisfies the theory.

An Intent Policy is a type of declarative policy that uses statements to express the goals of the policy, but not how to accomplish those goals. Each statement in an Intent Policy may require the translation of one or more of its terms to a form that another managed functional entity can understand. In particular, an Intent Policy is a policy that does not execute as a theory of a formal logic. Intent Policies are expressed in a restricted natural language and require a mapping to a form understandable by other managed functional entities. The advantage of an Intent Policy is its ability to express policies using concepts and terminology that are familiar to a particular constituency (e.g., as defined in the Policy Continuum [2–5]). Conceptually, the set of models and ontologies are used to define elements of a grammar that an Intent Policy is written in, which enables different Intent Policies written by different constituencies to be translated to a common form.

15.5.5 Achieving Goals in a Cognitive Architecture

As shown in Figs. 15.9 and 15.10, cognition is rooted in perception, comprehension, and taking action to attain or preserve a set of system goals. This implies that knowledge about situations is different than knowledge about activities that change a situation and facilitates experiential learning from how different changes in the system and/or environment affect the goals of the system. The three different types of memory structures play a critical role in reinforcing this. Specifically, logical predicates are used to relate short-term memory elements as instances of long-term memory elements. This enables each short-term element to be grounded in a fundamental goal or belief. This is facilitated by enabling more complex long-term elements to be composed from a set of simpler long-term elements, providing an inherently extensible knowledge base. This is fundamental for producing a set of ordered subgoals to achieve a particular higher-level goal. This requires minor extensions to the state machines used, where a given state may need to reflect the relationship between short- and long-term memory instances as well as the composite nature of a given long-term memory instance.

15.6 Future Work

The above framework was designed to be extensible. Some of the extensions currently under discussion deal with goal-based reasoning, spatiotemporal reasoning, learning from failure, and more complex learning from goals.

Currently, the cognitive architecture assumes that the set of system goals are ordered from most to least important. An extension to deal with system goals that are either non-prioritized or are possibly conflicting in nature is to first perform goal reasoning. This requires an extension of the long-term memory to include rules to generate and order goals and order them in terms of importance. A multi-attribute utility theory algorithm is one alternative; it provides a methodology to let the system choose the best course of action when two or more conflicting goals (e.g., cost vs. performance) are trying to be simultaneously satisfied. Alternatively, if the set of goals is not known, then a multi-objective optimization algorithm can be used to discover the most feasible solution. Simplistically, this enables the Learning and Reasoning functional block to examine the set of goals at any given time and reorder their priorities. This also requires minor extensions to the state machines used. Specifically, selected states either need to be represented using a more complex attributed structure or as an embedded state machine.

Spatiotemporal changes occur routinely in situations that change. This extension will enable the Learning and Reasoning functional block to vary the importance of concepts, beliefs, and goals over space and time. This can be accommodated by including appropriate attributes and/or metadata into objects that rank their relevance to a given situation.

Currently, the cognitive architecture is focused on learning from success. A future topic is to explore learning from failure. An analogy from linguistics is the use of antonymy in place of synonymy to understand the meaning of a new word or phrase. The idea is to infer new knowledge when a goal is not achieved. This causes three additions to the existing architecture. First, this idea needs a new knowledge representation, such as goal, status, reason, importance, and related, where:

- Goal is the name of the goal.
- Status is one of satisfied, partially satisfied, not satisfied.
- Reason is the set of facts and inferences that provided the status.
- Importance is the relative importance of the goal.
- Related is the set of other goals that depend on or interact with the current goal

Second, the above needs to be related in the representation of states in the state machine.

Finally, the Learning and Reasoning functional block needs to include the ability to check the status of each goal, and if the status was not satisfied, determine why it was not satisfied, and associate those causes with the above knowledge representation. This knowledge is then entered into long-term memory and enables such conditions to be recognized in future situations. This ability may be able to avoid situations in which similar goals were not satisfied. This could also prevent satisfying a lower-priority goal at the cost of causing a higher-priority goal to fail.

This leads to extending the concept of goal interaction. Currently, the learning and reasoning functional block evaluates goals in various ways and is focused on the successful completion of individual goals. However, sometimes goals interact. This can be accommodated by extensions to the learning and reasoning algorithms; this also requires the associated state machines to be extended to define a new type of state transition that takes into account satisfying a set of goals. This better matches the use of multi-attribute value theory or multi-objective optimization algorithms and enables the learning and reasoning functional block to evaluate different plans to solve interacting goals based on different ways the current state partially satisfies the specified goals. This makes the search space much larger. Currently, different heuristics are being explored to mitigate this problem; one promising approach is to prefer long-term memory instances that are involved with achieving more goals.

Another important extension is to incorporate the use of intent policies. This is implied by the Policy Continuum, as different constituencies, such as business users and application developers, may not be skilled in the use of different algorithms required by this cognitive architecture, how to implement commands that manage and configure networks and their services, or how to relate business needs to services offered. A promising approach is to use domain-specific languages to provide a simple and efficient means for each constituency to express their goals using a domain-specific programming environment that is tailored to their needs.

15.7 Conclusion

This chapter has reported on a new cognitive architecture that is currently being defined and prototyped. Current problems that have plagued network and service management were described, along with solutions developed from experience with the FOCALE autonomic architecture [11]. It is a model-driven architecture that incorporates different types of artificial intelligence in its closed control loops to manage the behavior of the system being managed. This results in a more robust cognition model. The underlying model-driven approach uses a combination of models and ontologies to enable semantic relationships to be defined between facts and meaning, and it was developed as an extension of the FOCALE autonomic architecture [11] and provides more sophisticated models (including a novel policy information model), code generation, and cognition model.

References

1. J. Strassner, How policy empowers business-driven device management. Third International Workshop on Policies for Distributed Systems and Networks (2002)
2. S. van der Meer, A. Davy, S. Davy, R. Carroll, B. Jennings, J. Strassner, *Autonomic Networking: Prototype Implementation of the Policy Continuum* (Broadband Convergence Networks, Vancouver, 2006)
3. J. Strassner, Management of autonomic systems—theory and practice. Network Operations and Management Symposium (NOMS) 2010 Tutorial, Osaka, Japan (2010)
4. J. Strassner, *Policy-Based Network Management* (Morgan-Kaufman, Burlington, 2003)
5. S. Davy, B. Jennings, J. Strassner, The policy continuum—policy authoring and conflict analysis. Comput. Commun. J. **31**(13), 2981–2995 (2008)
6. Please see https://portal.etsi.org/tb.aspx?tbid=857&SubTB=857#/
7. Please   see   https://wiki.mef.net/display/LSO/Policy+Driven+Orchestration+%28 PDO%29+-+Project+Home+Page
8. Please see https://wiki.mef.net/pages/viewpage.action?pageId=118001933
9. E. Gamma, R. Helm, R. Johnson, J. Vlissides, *Design Patterns: Elements of Reusable Object-Oriented Software* (Addison-Wesley, Boston, 1994)
10. A. Khalil, J. Dingel, Optimizing the symbolic execution of evolving rhapsody statecharts. Adv. Comput. **108**, 145–281 (2018)
11. Strassner, J., Agoulmine, N., Lehtihet, E.: "FOCALE - A novel autonomic networking architecture", ITSSA J. 3(1), pgs 64-79, 2007.
12. J. Famaey, S. Latré, J. Strassner, F. De Turck, An ontology-driven semantic bus for autonomic communication elements, in *IEEE International Workshop on Modelling Autonomic Communications Environments*, (IEEE, Piscataway, 2010), pp. 37–50

13. Strassner, J., Souza, J.N., van der Meer, S., Davy, S., Barrett, K., Raymer, D., and Samudrala, S.: "The design of a new policy model to support ontology-driven reasoning for autonomic networking", J. Netw. Syst. Manage. 17(1), pgs 5-32, 2009.
14. J. Strassner, J.N. de Souza, D. Raymer, S. Samudrala, S. Davy, K. Barrett, The design of a novel context-aware policy model to support machine-based learning and reasoning. J. Cluster Comput. 12(1), 17–43 (2009)
15. J. Strassner, S. van der Meer, D. O'Sullivan, S. Dobson, The use of context-aware policies and ontologies to facilitate business-aware network management. J. Netw. Syst. Manage. 17(3), 255–284 (2009)
16. J.R. Boyd, The Essence of Winning and Losing (1995)
17. R.W. Thomas, L.A. DaSilva, A.B. MacKenzie, Cognitive networks. In: Proceedings of the First IEEE International Symposium on New Frontiers in Dynamic Spectrum Access Networks, Baltimore, MD, USA, November 8–11, 2005
18. J. Strassner, The role of autonomic networking in cognitive networks, in *Cognitive Networks: Towards Self-Aware Networks*, ed. by Q. H. Mahmoud, (John Wiley and Sons, London, 2007)
19. R.C. Atkinson, R.M. Shiffrin, Human memory: a proposed system and its control processes, in *The Psychology of Learning and Motivation*, ed. by K. W. Spence, J. T. Spence, vol. 2, (Academic Press, New York, 1968), pp. 89–195

Chapter 16
Quantum Computing and Its Impact

Matthew W. Turlington, Lee E. Sattler, Dante J. Pacella, Jerry Gamble,
and Mehmet Toy

16.1 Introduction

Quantum computing is the use of quantum-mechanical phenomena such as superposition and entanglement to perform computation. Computers that perform quantum computation are known as quantum computers [1]. Quantum computers are
believed to be able to solve certain computational problems, such as integer factorization, substantially faster than classical computers.

According to [2, 3], the Google quantum computer performed in 3 min 20 s a
mathematical calculation that supercomputers could not complete in under
10,000 years.

In classical computers, the information is represented in bits (i.e., 1s and 0s). For
example, hard drives store documents by locking magnets in either the up or down
position. In quantum computers, the information is represented in quantum bits or
qubit. Qubits represent the information based on the behavior of atoms, electrons,
and other particles, objects governed by the rules of quantum mechanics. A hard
drive magnet must always point up or down, for instance, but an electron's direction
is unknowable until measured: the electron behaves in such a way that describing its
orientation requires a more complex concept—known as superposition—that goes
beyond the straightforward labels of "up" or "down."

Quantum particles can also be yoked together in a relationship called entanglement, such as when two photons (light particles) shine from the same source. Pairs
of entangled particles share an intimate bond akin to the relationship between the
two faces of a coin—when one face shows heads the other displays tails. Unlike a

M. W. Turlington (✉) · L. E. Sattler · D. J. Pacella · J. Gamble · M. Toy
Verizon Communications, Inc., Basking Ridge, NJ, USA
e-mail: matt.turlington@verizon.com; lee.e.sattler@verizon.com; dante.j.pacella@verizon.
com; Jerry.Gamble@verizon.com; mehmet.toy@verizon.com

M. Toy (ed.), *Future Networks, Services and Management*,
https://doi.org/10.1007/978-3-030-81961-3_16

435

coin, however, entangled particles can travel far from each other and maintain their connection.

It remains to be seen whether commercial computers and transmission devices based on quantum computing will be available in 2030 or not. However it is clear that quantum computing will revolutionize networking along with artificial intelligence and machine learning techniques.

For example, we expect networks to be much more decentralized. Cross-switch delays of network elements may be reduced from milliseconds to nanoseconds. With large computing power, we will be able to represent objects much more precisely. The communications will become richer.

With quantum computing, we expect Network 2030 to become fully automated and self-managed by being able to store and process large amounts of connectivity, application, and management information of a domain in a computer, instead of a number of networked computers in one or more data centers.

16.2 Technology

The use of quantum technologies allows us to enter into the world of quantum physics and behaviors. In classical physics we see causality. If we have knowledge of the past, we can perform computations on future problems. The quantum world is different. Quantum objects cannot be described in discrete terms (i.e., waves or particles). They can best be described as probabilistic functions. So, in contrast with classical physics, with knowledge of the past, we are required to make probabilistic predictions of future behavior. As an example, two identical tennis balls would bounce to the same exact height when dropped (assuming all other variables are identical). However, two discrete radioactive isotopes will decay at different rates due to the stochastic nature of the decay. Radioactive decay is a random process and cannot be predicted absolutely for a single atom. This decay becomes a probabilistic function. Physicists can approximate the decay of a large number of similar ions but cannot predict the decay of a single atom.

This section will discuss some of the underlying quantum behaviors that can be utilized in quantum computing and quantum networking.

16.2.1   Classical Data Technology

Classical data technology is built on the concept of using digital bits to store and transmit information. A classical bit is a piece of information that is represented by either a "1" or a "0." In computers, these bits are embodied as voltages in transistors. In communications, these bits are represented by pulses of electrical voltages or light. However, these bits are deterministic and can only hold one value at a time (a 1 or a 0). Adding more bits to a system allows more data to be transmitted or represented but it does not change the fact that these bits can only hold one value or state at a time.

16.2.2 Quantum Data Technology

Quantum data technology uses characteristics of subatomic particles such as photons or electrons to represent information. Instead of using bits to represent information, quantum technology uses qubits [2]. As mentioned above a classical bit is either a 1 or a 0, usually based on electrical voltages. In contrast a qubit is data that can be thought of as being both a 0 and 1 at the same time. In other words, a qubit becomes a probabilistic function. Rather than electrical voltages, a qubit might be represented by the spin of an electron or the polarization of a photon. A qubit is the fundamental building block of a quantum system. For example, a quantum computer performs operations on qubits to compute solutions and a quantum network provides the ability to transfer qubits between quantum computers. Quantum technology enables new methods of transmitting information between quantum endpoints (i.e., quantum computers or quantum sensors). The network that carries the qubits is the Quantum Internet.

In order to move towards a Quantum information world, we need to develop technologies that exploit some quantum properties of subatomic particles. These quantum properties appear strange when compared with our traditional view of the physical world. The following sections will describe some of these quantum properties that will be used in quantum communications: superposition, entanglement, teleportation, and super dense coding.

16.2.3 Quantum Superposition

A valid question is what makes a qubit so different from a classical bit. As stated previously, a classical bit can either be a "1" or a "0" at any one time. Today, information can be represented by encoding the information into classical bits. In contrast, a qubit can be thought of as a probabilistic representation of a state rather than a deterministic representation. Qubits use properties of subatomic particles such as electron spin to represent state. The spin of an electron can be represented as a probability function of either being "up" or "down" (1 or 0). In other words an electron has some finite chance of being either spin up or spin down until measured. Therefore, qubits aren't limited to two discrete states. They exist in a state called superposition which can be thought of as existing in two classical states at once. The qubits exist in superposition until measured and then they fall back to classical bits. A quantum computer can perform operations on these qubits and use superposition to its benefit. Because of the fact that qubits exist in superposition this allows for quantum computers to perform operations in a probabilistic way rather than a deterministic way. In addition, as more qubits are added to a quantum computer, the computing power expands exponentially in proportion to the number of qubits. A useful example would be determining a path through a maze. A classical computer would analyze all branch points one by one to see if they are viable. Based on normal algorithms, the number of steps to determine the one correct path increases exponentially with the number of branch points. Using superposition, a quantum computer (with the appropriate number of qubits) can analyze the outcomes of all branch points simultaneously. A quantum computer would solve the maze in N steps where N is the number of decision points. If $N = 100$, the difference in computation time would be enormous.

Qubits by their nature are very fragile and will break down when subjected to outside interference like temperature changes or vibrations. In addition, once a qubit is measured or read, the superposition breaks down and the qubit reverts to a standard classical bit value of "1" or "0." This is a very important attribute and means that qubits cannot be cloned or copied which is useful in security applications. A hacker cannot read a qubit without destroying the qubit.

16.2.4   Quantum Entanglement

Another important attribute of quantum technology is the topic of entanglement [4]. Quantum entanglement occurs when two small particles interact and influence each other. This occurs naturally for example in an atom where the electrons are entangled with the nucleus of an atom. However, it is possible to create entanglement between two subatomic particles such as two photons. Any action on one photon of the entangled pair will create a similar outcome on the other photon, even if the photons are separated by a large distance. Because qubits are based on states of subatomic particles, entangled qubits can be created and used in quantum computers and networks. Entangled qubits can affect each other instantly when manipulated by a quantum computer. The quantum computer can use superposition and entanglement to provide exponential increases in computational speed. In addition, because entangled qubits can be separated by distance, two devices could generate cryptographic keys by using entangled qubits that could not be snooped.

16.2.5   Quantum Teleportation

Quantum teleportation [5] is the method of transferring the information encoded into qubits between two quantum endpoints. This process requires two endpoints each with one of a pair of entangled qubits as well as a classical communication link between them. Quantum entanglement ensures that operations on one qubit will affect the qubit on the other side. The sending side performs quantum measurements on the data qubit in conjunction with its own entangled qubit. Quantum entanglement prepares the far end for receiving the qubit. The last step is to send the results of the measurements over the classical network. The total process in effect makes a copy of the qubit to be sent. Note that the qubit on the sender side is destroyed when measured. Also, a large amount of data can be transferred this way with very few classical bits being sent over a traditional network. This could revolutionize deployment of data networks where the bulk of data is sent with quantum methods using entanglement and control traffic is sent over lower bandwidth classical networks.

16.2.6 Superdense Coding

Superdense coding [6] can be thought of as the opposite of quantum teleportation. Superdense coding allows for the transfer of two classical bits by transmitting only one qubit. This provides a 100% increase in data transmission efficiency when compared with traditional transmission techniques. Recent research has achieved compressions results of 10 to 1 in experimental scenarios [1]. Superdense coding, just like quantum teleportation, uses quantum entanglement as part of the process. The process starts with an entangled pair of qubits at each side of a link. One side needs to send two classical bits (00, 01, 10, or 11) and performs some logic operations on its entangled qubit which encodes the classical bits into its qubit. The qubit is then transmitted over a standard optical link to the far side. The far side then performs some logic operations on the pair of entangled qubits to "de-entangle" the qubits. This process results in the knowledge of the two classical bits that were encoded on the far end. This technology will dramatically increase data network efficiency.

16.2.7 Quantum Repeaters

Transmission of single photons (qubits) across a fiber is limited in distance due to optical attenuation. In a classical network a repeater can be placed inline to regenerate the signal; however the anti-cloning theorem of quantum mechanics prevents the use of classical repeaters. Quantum repeaters are being developed for use in the Quantum Internet. Their function is to teleport the entangled state between two end nodes somewhat analogously to classical repeaters but using a very different technology. Quantum repeaters allow long-distance transmission of qubits between two endpoints and will be a key component of the Quantum Internet.

Entangled photons were sent over fiber-optic cables connecting Brookhaven National Laboratory in New York with Stony Brook University, a distance of about 11 miles [7]. The wireless transmission of entangled photons over a similar distance through the air is also tested.

16.2.8   Quantum Key Distribution

The advent of quantum computing will open new frontiers in computer modeling, artificial intelligence, and many new applications not yet thought of. The power of quantum computing will also create new cyber security issues. Algorithms which can decrypt encrypted messages have already been developed and will certainly be applied when quantum computers with enough compute power are available [8].

Quantum key distribution (QKD) [9] is a quantum safe technology which is available today. Its security is provided through the properties of quantum mechanics. Encryption keys are derived by QKD devices at two locations over a quantum channel. This technology is provably secure and has been demonstrated by Verizon at a trial in Washington, DC [10].

16.2.9   Quantum Sensors

Recently, IOT has been a significant aspect of new applications and services, and impacting networks from connectivity to management and analytics. Quantum Sensing should be viewed through the same lens. Classical networking can connect to Quantum Sensors today, but increasingly, Quantum Communications will be required to increase security for such devices and eventually Quantum Internet will carry Quantum workloads.

Quantum Sensing has natural affinity between optical and telecom networks and is expected to grow to $286M by 2024 [11]. As shown in Fig. 16.1 highlighting 2019 revenue breakdown, magnetic sensors and atomic clocks are the leading sensor categories. Quantum sensors can detect energy fields without absorbing energy from the field they're measuring. These are more precise and much faster than current systems, can be realized in photonic or solid state systems, and, in some instances, can detect otherwise unobservable phenomena.

Quantum Sensors can be used as an advanced radar, can look around corners, into rooms, detect diseases, and even detect magma flows deep underground.

One area of focus is with gravity sensors: being able to detect earth tremors, volcanic activity, cave systems, and natural resource deposits [12].

Quantum Sensors are used today to detect photosynthetically active radiation (PAR) in plant growth outdoors, in greenhouses, growth chambers (over and under leaf canopies), and in aquariums, especially in measuring light levels for healthy coral growth [13].

Microwaves are proposed for use in quantum illumination wherein low-reflective objects could be detected using a signal beam and an idler beam, such that when the signal beam is interfered with the idler beam, enough quantum correlation can be obtained to separate the reflected signal photons from the background noise.

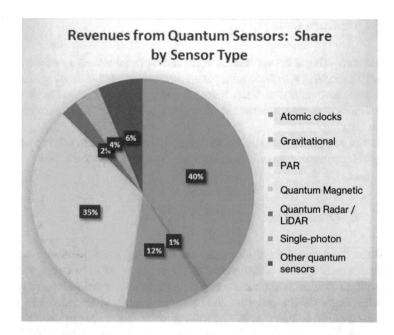

Fig. 16.1 Quantum sensor usage in 2019 [11]

16.3   Quantum Internet

All of these attributes (superposition, entanglement, teleportation, super dense coding) of quantum technology are used in conjunction to create what is called the Quantum Internet. The function of the Quantum Internet is to provide a networking technology which allows the transfer of qubits (or information) between quantum capable endpoints.

Much like classical bits, qubits can be transferred directly over optical fiber (or through air) as single photons. Each photon and its attributes represent one qubit. Using super dense coding two classical bits worth of data can be carried in one qubit. In addition quantum entanglement can be used to transfer information between endpoints using very little classical bandwidth. What this means is that a quantum-based network capable of transmitting large amounts of data between endpoints can be built while using fewer communication resources (classical bits) than are used today. The data efficiency of networks will increase as quantum networking is introduced. In the future, we expect to see communications services and products that can only be realized with quantum technologies.

16.4 Quantum Computing

Quantum computing is the application of quantum phenomena as a method of computation. The term can include all areas of quantum related technology including security and communication, but is more commonly used to refer to the application of quantum or quantum accelerated algorithms in areas such as optimization or machine learning.

Quantum computing represents a wide range of approaches to solving computational problems, and no specific paradigm has risen to dominate the field.

16.4.1 Current Status

Due to the limitations of existing quantum hardware, most quantum algorithms can't be currently applied to real-world problems. In an effort to leverage existing quantum technologies for practical solutions, researchers have developed a series of approaches that combine quantum and classical computing referred to as Hybrid Quantum-Classical computing. In a hybrid approach, one or more quantum components are combined with classical computer systems to achieve some of the benefits of quantum techniques while remaining feasible for current quantum hardware.

Hybrid quantum-classical computing includes approaches such as quantum annealing and hybrid quantum machine learning. Quantum annealing represents optimizations problems as a minimization of energy states represented as Hamiltonians. This approach can provide linear to exponential increases in performance versus purely classical approach to these optimizations. It is important to note that only specific optimization problems have been translated into representations that can be executed on quantum annealers and that the approach is not capable of executing most quantum algorithms. While substantial performance gains are possible, the scope of applications of this technology is limited. Quantum supremacy for quantum annealing has also not been established, and there are claims that purely classical approaches such as Memory-Driven Computing might achieve similar results.

Hybrid Quantum-Classical Machine Learning uses traditional machine learning topologies in conjunction with quantum representations to achieve increases in performance and capabilities. Two common approaches are replacing one or more components of a classical machine learning topology or pipeline with quantum implementations and using machine learning models to generate approximations of quantum states. In the first case, research has demonstrated that replacing classical components in a machine learning model with quantum equivalents allows models to identify patterns that purely classical models could not [14].

16.4.2   Future Expectations

Our current state of quantum technology that requires hybrid approaches to compensate for limitations in quantum hardware is referred to as Noisy Intermediate-Scale Quantum (NISQ) Computing. As quantum systems become more capable, it is expected that many hybrid techniques will be replaced with more flexible approaches. One of the leading candidates for near-term quantum computing applications is quantum circuit development. Quantum circuits are composed of quantum gates—reversible operations on registers composed of qubits. Quantum circuits and gates are analogous to classical logic gates and digital circuits.

Quantum circuits are not limited to specific types of problems as is the case with quantum annealing. They are a viable model for general purposes quantum computers, sometimes referred to as a "Universal Quantum Computer." The term "Universal Quantum Computer" lacks a precise definition, but is generally used to refer to a quantum analog to classical computers—systems that can compute solutions to a wide range of algorithms with capabilities such as processing, memory, and storage.

Due to limits on coherence time of qubits, there are limits to the complexity of quantum circuits that can be implemented. This restricts the size of problems that can be solved and most quantum algorithms cannot be applied to practical problems on current hardware. These limitations are expected to significantly decrease in the next few years as manufacturers improve hardware by better isolating superposition of particles from their environment and developing better approaches to quantum error correction. Even with current hardware limitations, limited practical applications exist in fields such as material science and machine learning [15, 16].

16.5   Quantum Networking Landscape

16.5.1   QKD Networks

The number of publicly disclosed terrestrial networks grew substantially in 2020, increasing from six to eleven last year alone. Most of these networks consisted of a combination of telecom companies and equipment vendors.

As previously discussed, the distance limitation of a QKD system poses a challenge to global networks. Satellite QKD (S-QKD) offers a mechanism that can be used to augment terrestrial QKD for global environments. With S-QKD, free space optics enables the delivery of a photonic channel from which the symmetric key can be derived. Factors that will impact the use of S-QKD include weather and other environmental factors that impact the free space optic channel. Also the key rate of the S-QKD system must be sufficient to support the key material required for the network (Table 16.1).

Table 16.1 Terrestrial QKD networks [10, 17]

Name of network	Year	Participants
DARPA	2003	BBN Technologies, Harvard and Boston Universities
SECOQC	2008	41 research and industry partners in Europe
Tokyo UQCC	2010	Toshiba, IDQ, NTT, NEC, NICT, Mitsubishi, All Vienna
Battelle Ohio	2013	Battelle, IDQ, City of Dublin
Beijing-Shanghai	2017	UBeijing, UShanghai, China Telecom, ZTE, Fiber Home
Madrid	2018	Telefonica, Huawei
NY-NJ	2019	Quantum Xchange, IDQ, Toshiba
Cambridge-Ipswich	2019	BT, IDQ, ADVA, UCambridge, UYork
Netherlands	2020	QuTech, TUDelft
Chicago	2020	DOE Argonne, UChicago, Qubitekk, Fermilab
Seoul-Daejon	2020	SK Telecom, IDQ
Seoul-Gyeonggi	2020	KT
DC Region	2020	Verizon, Toshiba & Quantum Xchange

16.5.2   Risks

Though the breakthrough and discovery pace has quickened over the past decade, several crucial problems still need to be solved. Carrier class quantum equipment beyond the readily available QKD equipment must become available. In particular quantum repeaters are a significant element that is required by the Quantum Internet to get around the distance limitations imposed by quantum technology. The challenge with quantum repeaters is that it requires physicists and network hardware manufacturers to collaborate with each other in a way that is different from the traditional hardware development. This is due to the quantum stability issues associated with the repeater.

16.5.3   United States Department of Energy Research Priorities [18]

The US government views the Quantum Internet as critical. It has published a strategic vision and budgeted $237M for 2021 to begin building the Quantum Internet. The Department of Energy (DOE) held a workshop in February 2020 which was attended by various government agencies, academia, and industry. The result was a document defining priority research directions and a potential roadmap toward building the first nationwide Quantum Internet. The milestones for a Quantum Internet are discussed later in this document. Four priority research priorities were delineated as shown in Table 16.2.

Table 16.2 Department of energy priority research directives [18]

Priority research directive	Objectives
Provide the foundational building blocks for a Quantum Internet	Usher in new devices for quantum networking which meet reliability and management standards of today's classic networking
Integrate multiple quantum networking devices	Ensure interoperability and standardization
Create repeating, switching, and routing for quantum entanglement	Increase distance limitations and implement networking functions
Enable error correction of quantum networking functions	Provide fault tolerant high fidelity networks

A fifth milestone calls out building a multi-institutional ecosystem and names DOE, NSF, NIST, NASA, DoD, and NSA as the key players.

The quantum internet will coexist with classical transport networks and in some cases over the same fibers utilizing WDM systems.

16.5.4 Internet Engineering Task Force: Quantum Internet Research Group

In 2018, the Internet Engineering Task Force (IETF) formed the Quantum Internet Research Group (QIRG) to address the question of how to design and build quantum networks. Their charter is to address problems which are very similar to those of the early Internet, such as routing, interoperability, security, and applications. QIRG is near completion of their first two documents on architectural principles and applications use cases. It should be expected that interest in Quantum Internet will grow after these are finalized.

16.6 The Road to the Quantum Internet

When the Internet was first designed and built, the full potential for its use and applications were not known. There are some ideas on what can be done with the Quantum Internet but likely these are the tip of the iceberg. Generally, potential applications fall into three buckets; cryptography, sensing, and computing. The properties of quantum mechanics provide inherent security into communications which has been demonstrated with Quantum Key Distribution. Quantum sensing utilizes single-photon detectors to capture ultra-accurate measurements. It has the potential to be utilized in a wide range of solutions including fiber sensing and light detection and ranging (LIDAR) in automobiles. Quantum computing continues to progress as computers with more qubit processors are announced. These quantum computers will require and become even more powerful utilizing the Quantum

Internet to connect them. The evolution of the Quantum Internet will occur in phases as key technological advances are made. The potential phases or milestones have been mapped in scientific literature and the Department of Energy (DOE) blueprint [18].

One vision of the Quantum Internet maps out six distinct stages based upon technology functions which increase in difficulty at each stage (Fig. 16.2). The first two stages, trusted repeater and prepare and measure, have been realized. QKD networks are an application example of the ability to prepare a qubit, transmit it over a distance, and measure at the far end. The third stage, entanglement generation, has been demonstrated over distances of a few kilometers and work still remains to extend the distances and to develop underlying protocols which will herald successful entanglement over a classical network. Quantum repeaters with quantum memory will be required to truly achieve long-distance quantum communications. Experimental work in this fourth stage is ongoing and has been demonstrated in principle but practical implementation has not yet been demonstrated. The fifth stage improves upon the previous stages providing a truly fault tolerant network through improved protocols and underlying hardware. This will open the potential of the final stage which is distributed quantum computing.

The DOE blueprint for the Quantum Internet has similarities to the six stages proposed by Wehner et al. [19] as shown in Table 16.3. The only exception is the last milestone which calls for a collaborative effort to build the Quantum Internet across academia, government, and industry. These two documents along with published

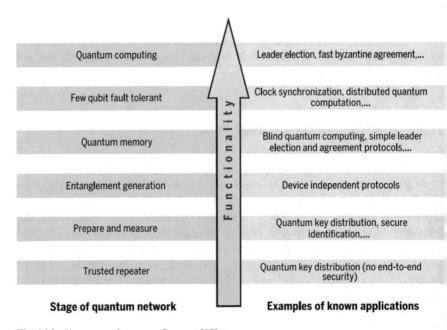

Fig. 16.2 Six stages of quantum Internet [19]

Table 16.3 Department of energy milestones [18]

Milestone	Description
Verification of secure quantum protocols over fiber networks	Prepare and measure quantum networks
Inter-campus and intra-city entanglement distribution	Entanglement distribution networks
Intercity quantum communication using entanglement swapping	Quantum memory networks
Interstate quantum entanglement distribution using quantum repeaters	Classical and quantum networking technologies have been integrated
Build a multi-institutional ecosystem between laboratories, academia, and industry to transition from demonstration to operational infrastructure	Cross-cutting collaboration

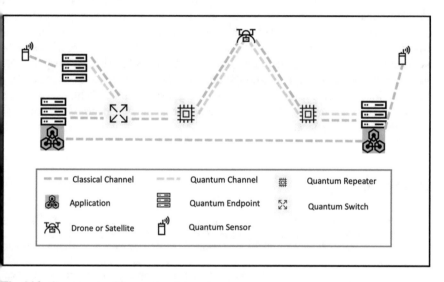

Fig. 16.3 Components of the quantum internet

research and standards work provide a template for what the road to Quantum Internet will look like.

The Quantum Internet will be built with quantum repeaters at its core which will create link local entanglement with its neighbors and through entanglement swapping extend entanglement between end nodes as depicted in Fig. 16.3. The control plane of the Quantum Internet will remain on the classical Internet which means these quantum repeaters will have both quantum and classical network links. The control plane will provide routing between the entanglement endpoints and also provide status information. There could be quantum repeaters which do not participate in the control plane and simply extend entanglement over long distances or difficult terrain. These devices may not even be connected to fiber and could be drones or satellites using free space optics. At the edges will be end nodes which receive and handle entangled pairs but would not require quantum memory or entanglement swapping capabilities. These end nodes could be quantum computers or other quantum devices and could act as gateways to classical networks. Attached to end nodes through a classical link would be non-quantum devices which may need services from the Quantum Internet.

Several quantum communication computer models have been developed to aid in design of the control protocols, potential network configurations, and applications for the Quantum Internet.

16.7 Conclusion

The next several decades will see massive growth in the quantum technology sector. Problems that were deemed intractable will be "solved" by quantum computers. The development of quantum computers will drive the need for a communications network that can provide the interconnects between quantum computers. In order to achieve networking at large distances, many research institutions, industry consortia, and commercial enterprises are working to develop cutting edge quantum technologies that will help realize the advent of a global Quantum Internet. Quantum computing and networking will change the landscape of all aspects of communication over the coming decades.

References

1. P. Benioff, The computer as a physical system: a microscopic quantum mechanical Hamiltonian model of computers as represented by Turing machines. J. Stat. Phys. **22**(5), 563–591
2. C. Metz, *Google Claims a Quantum Breakthrough That Could Change Computing* (Times, New York, 2019)
3. F. Arute et al., Quantum supremacy using a programmable superconducting processor. Nature **574**, 505 (2019)
4. The Quantum Internet Is Emerging, One experiment at a time, https://www.scientificamerican.com/article/the-quantum-internet-is-emerging-one-experiment-at-a-time/
5. Fermilab and partners achieve sustained, high-fidelity quantum teleportation, https://news.fnal.gov/2020/12/fermilab-and-partners-achieve-sustained-high-fidelity-quantum-teleportation/
6. Demystifying superdense coding, https://medium.com/qiskit/demystifying-superdense-coding-41d46401910e
7. C. Wood, Trump betting millions to lay the groundwork for quantum internet in the US, CNBC.com (2020)
8. S. Jurvetson, How to factor 2048 Bit RSA integers in 8 hours using 20 million noisy qubits. MIT Technology (2019), https://www.technologyreview.com/2019/05/30/65724/how-a-quantum-computer-could-break-2048-bit-rsa-encryption-in-8-hours/
9. S. Hardy, Quantum encryption combats threat posed by quantum computing hacks, Lightwave Online (2018), https://www.lightwaveonline.com/network-design/high-speed-networks/article/16676081/quantum-encryption-combats-threat-posed-by-quantum-computing-hacks

10. Verizon, Verizon achieves milestone in future-proofing data from hackers (2020). https://www. verizon.com/about/news/verizon-achieves-milestone-future-proofing-data-hackers
11. Inside Quantum Technology, Revenues from quantum sensors by sensor type, in *Quantum Sensors Markets, 2019 and Beyond* (2020), https://www.insidequantumtechnology.com/ wp-content/uploads/2018/09/infographic-revenues-from-quantum-sensors-by-sensor-type-e1537279635472.jpg
12. N. Metje, M. Holynski, *How Can Quantum Technology Make the Underground Visible?* (University of Birmingham, Birmingham, 2016)
13. T. Akitsu, et al., Quantum sensors for accurate and stable long-term photosynthetically active radiation observations, Science Direct (2017), https://www.sciencedirect.com/science/article/ abs/pii/S0168192317300114
14. Quantum Convolutional Neural Networks, arXiv:1810.03787v2 [quant-ph], https://arxiv.org/ abs/1810.03787
15. F. Arute, Hartree-Fock on a superconducting qubit quantum computer. Science **369**(6507), 1084–1089 (2020). https://doi.org/10.1126/science.abb9811
16. Option pricing using quantum computers, arXiv:1905.02666 [quant-ph], https://arxiv.org/ abs/1905.02666
17. M. Mehic, M. Niemiec, S. Rass, J. Ma, M. Peev, A. Aguado, V. Martin, S. Schauer, A. Poppe, M. Voznak, Quantum key distribution: A networking perspective. ACM Comput. Surv. **53**(5), 96 (2020). https://doi.org/10.1145/3402192
18. United States Department of Energy, Report of the DOE quantum internet blueprint workshop, February 5-6, 2020, SUNY Global Center, New York, NY (2020). https://www.energy.gov/ sites/prod/files/2020/07/f76/QuantumWkshpRpt20FINAL_Nav_0.pdf
19. S. Wehner, D. Elkouss, R. Hanson, Quantum internet: A vision for the road ahead. Science **362**, 6412 (2018). https://doi.org/10.1126/science.aam9288

Index

Printed in the United States
by Baker & Taylor Publisher Services